Also by Dick Teresi

Popular Mechanics Book of Bikes and Bicycling

Omni's Continuum: Dramatic Phenomena from the New Frontiers of Science

Laser: Light of a Million Uses (with Jeff Hecht)

The God Particle: If the Universe Is the Answer, What Is the Question? (with Leon Lederman)

LOST DISCOVERIES

The Ancient Roots of Modern Science—from the Babylonians to the Maya

DICK TERESI

SIMON & SCHUSTER
New York London Toronto Sydney Singapore

SIMON & SCHUSTER
Rockefeller Center
1230 Avenue of the Americas
New York, NY 10020

For information regarding special discounts for bulk purchases,
please contact Simon & Schuster Special Sales:
1-800-456-6798 or business@simonandschuster.com

Designed by Rhea Braunstein

Manufactured in the United States of America

10 9 8 7 6 5 4 3 2 1

Library of Congress Cataloging-in-Publication Data
Teresi, Dick.
Lost discoveries : the ancient roots of modern science—from the
Babylonians to the Maya / Dick Teresi.
p. cm.
Includes bibliographical references and index.
1. Science, Ancient. 2. Science—History. I. Title.
Q124.95 .T47 2002
509.3—dc21 2002075457

ISBN 0-684-83718-8

Board of Advisers

Anthony Aveni
Alfred W. Crosby
Harold Goldwhite
George Gheverghese Joseph
Robert Kaplan
David Park
George Saliba
Sheila Seaman
Barbara C. Sproul

The above scientists, mathematicians, and scholars reviewed the manuscript for scientific, mathematical, and historical accuracy. Some were chosen for a non-Western, others for a Western bias. While I deferred to these advisers on factual matters, they did not always agree with my interpretation of those facts. My point of view was greatly affected by the views expressed by my advisers, but ultimately it is my own. Where practical, I have stated differing views by the board in the endnotes.

Anthony Aveni is the Russell B. Colgate Professor of Astronomy and Anthropology at Colgate University. He is the author of *Conversing with the Planets: How Science and Myth Invented the Cosmos* and other works of archaeoastronomy.

Alfred W. Crosby is professor emeritus of history at the University of Texas. He is the author of *Ecological Imperialism: The Biological Expansion of Europe, 900–1900,* among other works.

Harold Goldwhite is professor of chemistry at California State University, Los Angeles, and is the coauthor, with Cathy Cobb, of *Creations of Fire: Chemistry's Lively History from Alchemy to the Atomic Age.*

George Gheverghese Joseph is professor of mathematics at the University of Manchester (UK) and the author of *The Crest of the Peacock: Non-European Roots of Mathematics.*

Robert Kaplan has taught mathematics at a number of institutions, most recently Harvard University. He is the author of *The Nothing That Is: A Natural History of Zero.*

David Park is emeritus professor of physics at Williams College. He is the author of *The Fire Within the Eye: A Historical Essay on the Nature and Meaning of Light.*

George Saliba is professor of Arabic and Islamic science at the department of Middle East and Asian languages and cultures, Columbia University. He is the author of *A History of Arabic Astronomy: Planetary Theories During the Golden Age of Islam,* among other works.

Sheila J. Seaman is associate professor of geology at the University of Massachusetts at Amherst.

Barbara C. Sproul is director of the program in religion, Hunter College, City University of New York. She is the author of *Primal Myths: Creation Myths Around the World* and was one of the founders of the American section of Amnesty International, which won the Nobel Peace Prize in 1977.

Contents

Lost Discoveries

1

A HISTORY OF SCIENCE
Rediscovered

THE most important scientific achievement in Western history is commonly ascribed to Nicolaus Copernicus, who on his deathbed published *Concerning the Revolutions of the Heavenly Spheres.* Science historian Thomas Kuhn called the Polish-born astronomer's accomplishment the "Copernican Revolution." It represented a final break with the Middle Ages, a movement from religion to science, from dogma to enlightened secularism. What had Copernicus done to become the most important scientist of all time?

In school we learned that in the sixteenth century, Copernicus reformed the solar system, placing the sun, rather than the earth, at its center, correcting the work of the second-century Greek astronomer Ptolemy. By constructing his heliocentric system, Copernicus put up a fire wall between the West and East, between a scientific culture and those of magic and superstition.

Copernicus did more than switch the center of the solar system from the earth to the sun. The switch itself is important, but mathematically trivial. Other cultures had suggested it. Two hundred years before Pythagoras, philosophers in northern India had understood that gravitation held the solar system together, and that therefore the sun, the most massive object, had to be at its center. The ancient Greek astronomer Aristarchus of Samos had put forth a heliocentric system in the third century B.C.[1] The Maya had posited a heliocentric solar system by A.D. 1000. Copernicus's task was greater. He had to repair the flawed mathematics of the Ptolemaic system.

Ptolemy had problems far beyond the fact that he chose the wrong body as the pivot point. On that, he was adhering to Aristotelian beliefs. A workable theory of universal gravitation had yet to be discovered. Thus hampered, Ptolemy attempted to explain mathematically what he saw from his vantage in Alexandria: various heavenly bodies moving around the earth. This presented problems.

Mars, for instance, while traveling across our sky, has the habit, like other planets, of sometimes reversing its direction. What's happening is simple: the earth outspeeds Mars as both planets orbit the sun, like one automobile passing another. How does one explain this in a geocentric universe? Ptolemy came up with the concept of epicycles, circles on top of circles. Visualize a Ferris wheel revolving around a hub. The passenger-carrying cars are also free to rotate around axles connected to the outer perimeter of the wheel. Imagine the cars constantly rotating 360 degrees as the Ferris wheel also revolves. Viewed from the hub, a point on the car would appear to move backward on occasion while also moving forward with the motion of the wheel.

Ptolemy set the upper planets in a series of spheres, the most important of which was the "deferent" sphere, which carried the epicycle. This sphere was not concentric with the center of the earth. It moved at a uniform speed, but that speed was not measured around its own center, nor around the center of the earth, but around a point that Ptolemy called the "center of the equalizer of motion," later to be called the "equant."[2] This point was the same distance from the center of the deferent as the distance of the deferent's center from the earth, but in the opposite direction. The result was a sphere that moved uniformly around an axis that passed not through its own center but, rather, through the equant.

The theory is confusing. No number of readings or constructions will help, because Ptolemy's scheme is physically impossible. The flaw is called the equant problem, and it apparently eluded the Greeks. The equant problem didn't fool the Arabs, and during the late Middle Ages

Islamic astronomers created a number of theorems that corrected Ptolemy's flaws.

Copernicus confronted the same equant problem. The birth of Isaac Newton was a century away, so Copernicus, like Ptolemy and the Arabs before him, had no gravitation to help him make sense of the situation. Thus, he did not immediately switch the solar system from geocentricity to heliocentricity. Instead, he first improved the Ptolemaic system, putting the view of the heavens from earth on a more solid mathematical basis. Only then did Copernicus transport the entire system from its earth-centered base to the sun. This was a simple operation, requiring Copernicus only to reverse the direction of the last vector connecting the earth to the sun. The rest of the math remained the same.

It was assumed that Copernicus was able to put together this new planetary system using available math, that the Copernican Revolution depended on a creative new application of classical Greek works such as Euclid's *Elements* and Ptolemy's *Almagest.* This belief began breaking down in the late 1950s when several scholars, including Otto Neugebauer, of Brown University; Edward Kennedy, of the American University of Beirut; Noel Swerdlow, of the University of Chicago; and George Saliba, of Columbia University, reexamined Copernicus's mathematics.

They found that to revolutionize astronomy Copernicus needed two theorems not developed by the ancient Greeks. Neugebauer pondered this problem: did Copernicus construct these theorems himself or did he borrow them from some non-Greek culture? Meanwhile, Kennedy, working in Beirut, discovered astronomical papers written in Arabic and dated before A.D. 1350. The documents contained unfamiliar geometry. While visiting the United States, he showed them to Neugebauer.

Neugebauer recognized the documents' significance immediately. They contained geometry identical to Copernicus's model for lunar motion. Kennedy's text was written by the Damascene astronomer Ibn al-Shatir, who died in 1375. His work contained, among other things, a theorem employed by Copernicus that was originally devised by another Islamic astronomer, Nasir al-Din al-Tusi, who lived some three hundred years before Copernicus.

The Tusi couple, as the theorem is now called, solves a centuries-old problem that plagued Ptolemy and the other ancient Greek astronomers: how circular motion can generate linear motion. Picture a large sphere with a sphere half its size inside it, the smaller sphere contacting the larger at just one point. If the large sphere rotates and the small sphere revolves in the opposite direction at twice that speed, the Tusi couple dictates that the original point of tangency will oscillate back and forth along the diameter of the larger sphere. By setting the celestial spheres properly, this theorem explained how the epicycle could move uniformly around the equant of the deferent, and still oscillate back and forth toward the center of the deferent. All this could now be done by positing spheres moving uniformly around axes that passed through their centers, thus avoiding the pitfalls of Ptolemy's configurations. A rough analogy is a steam-engine piston, which moves back and forth as the wheel is turning.

A second theorem found in the Copernican system is the Urdi lemma, after the scientist Mu'ayyad al-Din al-'Urdi, who proposed it sometime before 1250. It simply states that if two lines of equal length emerge from a straight line at the same angles, either internally or externally, and are connected up top with another straight line, the two horizontal lines will be parallel. When the equal angles are external, all four lines form a parallelogram. Copernicus did not include a proof of the Urdi lemma in his work, most likely because the proof had already been published by Mu'ayyad al-Din al-'Urdi. Columbia's George Saliba speculates that Copernicus didn't credit him because Muslims were not popular in sixteenth-century Europe.

Both the Urdi lemma and the Tusi couple are, in the words of Saliba, "organically embedded within [Copernican] astronomy, so much so that it would be inconceivable to extract them and still leave the mathematical edifice of Copernican astronomy intact."

Saliba emphasizes that plagiarism is not the issue here. Those who have been involved in a plagiarism case are probably familiar with the standard defense: *independent execution.*[3] This is an especially powerful defense in the sciences, in which there are "right" and "wrong" solutions. If Copernicus's theorem looks like al-Tusi's, perhaps that's because it's the one correct answer to the problem.

Map publishers sometimes insert fictitious islands or other features into their maps to trap plagiarists. Did Copernicus borrow al-Tusi's theorem without credit? There's no smoking gun, but it is suspicious that Copernicus's math contains arbitrary details that are identical to al-Tusi's. Any geometric theorem has the various points labeled with letters or numbers, at the discretion of the originator. The order and choice of symbols is arbitrary. The German science historian Willy Hartner noted that the geometric points used by Copernicus were identical to al-Tusi's original notation. That is, the point labeled with the symbol for *alif* by al-Tusi was marked *A* by Copernicus. The Arabic *ba* was marked *B,* and so on, each Copernican label the phonetic equivalent of the Arabic. Not just some of the labels were the same—*almost all* were identical.

There was one exception. The point designating the center of the smaller circle was marked as *f* by Copernicus. It was a *z* in Tusi's diagram. In Arabic script, however, a *z* in that hand could be easily mistaken for an *f.*

Johannes Kepler, who stretched Copernicus's circular planetary orbits into ellipses later in the century, wondered why Copernicus had not included a proof for his second "new" theorem, which was in fact the Urdi lemma. The obvious answer has eluded most historians because it is too damaging to our Western pride to accept: the new math in the Copernican Revolution arose first in Islamic, not European minds. From a scientific point of view, it's not important whether Copernicus was a plagiarist. The evidence is circumstantial, and certainly he could have invented the theorems on his own. There is no doubt, however, that two Arab astronomers beat him to the punch.

Western science is our finest accomplishment. Does any other culture, past or present, boast a scientific edifice equal to that built by

Galileo, Newton, Leibniz, Lavoisier, Dalton, Faraday, Planck, Rutherford, Einstein, Heisenberg, Pauli, Watson, and Crick? Is there anything in the non-Western past to compare to present-day molecular biology, particle physics, chemistry, geology, or technology? There's little debate. The only question is where this science came from. Who contributed to it? The consensus is that science is almost entirely Western in origin. By *Western* we mean ancient and Hellenistic Greece, and Europe from the Renaissance to the present. Greece is traditionally considered European, as opposed to being part of Mediterranean culture, which would include its neighbors in Africa. For the purposes of this book, *Western* means Europe, Greece, and post-Columbian North America. *Non-Western* means, generally, everywhere else, including the Americas of the Amerindians before Columbus. *Non-Western* thus takes in considerable area, and the prevailing opinion is that modern science owes little to the peoples of these lands.

The short form of the hypothesis is this: science was born in ancient Greece around 600 B.C. and flourished for a few hundred years, until about 146 B.C., when the Greeks gave way to the Romans. At this time science stopped dead in its tracks, and it remained dormant until resurrected during the Renaissance in Europe around 1500. This is what's known as the "Greek miracle." The hypothesis assumes that the people who occupied India, Egypt, Mesopotamia, sub-Saharan Africa, China, the Americas, and elsewhere prior to 600 B.C. conducted no science. They discovered fire, then called it quits, waiting for Thales of Miletus, Pythagoras, Democritus, and Aristotle to invent science in the Aegean.

As amazing as the Greek miracle is the notion that for over fifteen hundred years, from the end of the Greek period to the time of Copernicus, no science was conducted. The same people who stood idly by while the Greeks invented science supposedly demonstrated no interest or skill in continuing the work of Archimedes, Euclid, or Apollonius.

The hypothesis that science sprang ab ovo on Greek soil, then disappeared until the Renaissance seems ridiculous when written out succinctly. It's a relatively new theory, first fashioned in Germany about 150 years ago, and has become subtly embedded in our educational consciousness. The only concession made to non-European cultures is to Islam. The story goes that the Arabs kept Greek culture, and its science, alive through the Middle Ages. They acted as scribes, translators, and caretakers, with, apparently, no thought of creating their own science.

In fact, Islamic scholars admired and preserved Greek math and science, and served as the conduit for the science of many non-Western cultures, in addition to constructing their own impressive edifice. Western science is what it is because it successfully built upon the best ideas, data, and even equipment from other cultures. The Babylonians, for example, developed the Pythagorean theorem (the sum of the squares of the two perpendicular sides of a right triangle is equal to the square of the hypotenuse) at least fifteen hundred years before Pythagoras was born. The Chinese mathematician Liu Hui calculated a value for pi (3.1416) in 200 A.D. that remained the most accurate estimation for a thousand years. Our numerals 0 through 9 were invented in ancient India, the Gwalior numerals of A.D. 500 being almost indistinguishable from modern Western numerals. *Algebra* is an Arab word, meaning "compulsion," as in compelling the unknown *x* to assume a numerical value. (One traditional translation, that algebra means "bone setting," is colorful but incorrect.)[4]

The Chinese were observing, reporting, and dating eclipses between 1400 and 1200 B.C. The Venus Tablets of Ammizaduga record the positions of Venus in 1800 B.C. during the reign of the Babylonian king. Al-Mamum, an Arabian caliph, built an observatory so his astronomers could double-check most of the Greek astronomical parameters, thus giving us more accurate values for precession, inclination of the ecliptic, and the like. In 829 his quadrants and sextants were larger than those built by Tycho Brahe in Europe more than seven centuries later.

Twenty-four centuries before Isaac Newton, the Hindu Rig-Veda asserted that gravitation held the universe together, though the Hindu hypothesis was far less rigorous than Newton's. The Sanskrit-speaking Aryans subscribed to the idea of a spherical earth in an era when the Greeks believed in a flat one. The Indians of the fifth century A.D.

somehow calculated the age of the earth as 4.3 billion years; scientists in nineteenth-century England were convinced it was 100 million years. (The modern estimate is 4.6 billion years.) Chinese scholars in the fourth century A.D.—like Arabs in the thirteenth century and the Papuans of New Guinea later on—routinely used fossils to study the history of the planet; yet at Oxford University in the seventeenth century some faculty members continued to teach that fossils were "false clues sown by the devil" to deceive man. Quantitative chemical analyses set down in the *K'ao kung chi,* an eleventh-century B.C. Chinese text, are never more than 5 percent off when compared to modern figures.

Mohist (Chinese) physicists in the third century B.C. stated, "The cessation of motion is due to the opposing force. . . . If there is no opposing force . . . the motion will never stop. This is as true as that an ox is not a horse." It would be two thousand years before Newton would set down *his* first law of motion in more prosaic terms. The *Shu-Ching* (circa 2200 B.C.) stated that matter was composed of distinct separate elements seventeen centuries before Empedocles made the same observation, and hypothesized that sunbeams were made of particles long before Albert Einstein and Max Planck posited the ideas of photons and quanta. Big bang? The creation myths of Egypt, India, Mesopotamia, China, and Central America all begin with a "great cosmic copulation"—not quite the same as a big bang, but more poetic.

As for practical matters, Francis Bacon said that three inventions—gunpowder, the magnetic compass, and paper and printing—marked the beginning of the modern world. All three inventions came from China. The Incas of the Andes were the first to vulcanize rubber, and they discovered that quinine was an antidote for the malaria that spread among them. The Chinese made antibiotics from soybean curd twenty-five hundred years ago.

THE TEACHING OF multicultural science in the 1980s had hardly begun when it was met by a powerful backlash, much of it justified. I was part of the backlash, having accepted in the early 1990s an assignment to write an article about faulty multicultural science being taught in schools. While there was plenty to expose, the most egregious program was called the Portland African-American Baseline Essays, developed by the Multnomah County, Oregon, school board.

The scientific portion of the curriculum was a disaster. It cited "evidence of the use of gliders in ancient Egypt from 2500 B.C. to 1500 B.C.," adding that the Egyptians used their early planes for "travel, expeditions, and recreation." The Portland essays speculated that these gliders were made from papyrus and glue. The evidence cited for this ancient Egyptian air force was the discovery in 1898 of a birdlike object made of sycamore wood. It sat in a box of other birdlike objects in the Cairo Museum's basement until 1969, when an archaeologist and his flight-engineer brother concluded that the object was a model glider with a distinctive resemblance to an American Hercules transport aircraft because of its "reverse dihedral wing." The Portland essays insisted that this fourteen-centimeter-long object was a scale model of full-sized gliders that once filled the skies over the Great Pyramids, which, one can therefore assume, served as platforms for ancient air-traffic controllers.[5]

The Portland essays also claimed that the ancient Egyptians and Mesopotamians knew how to make batteries. Clay pots found in 1962 in Baghdad contained five-inch-long cylindrical sheet-copper cores with a lead-tin alloy at the bottom. Inside the copper tube was an iron or bronze rod thought to have been surrounded by a solution of sulfate, vinegar, acetic acid, or citric acid. A General Electric laboratory demonstrated that ten such batteries connected in series could produce up to two volts. Were these really batteries? It's possible, though the Portland essays do not explain how it was known that acid was used in the pots. Nor do we know to what use the batteries were put.

The Portland essays also touted the Egyptians as masters of psi: precognition, psychokinesis, and remote viewing. The essays make a distinction between magic, which they disregard, and psi, or psychoenergetics, which they describe as being "science." We will not take time here to discuss the Egyptians' alleged accomplishments in psychoenergetics.[6] One can only wonder why this ancient civilization, with airplanes and telekinesis at its disposal, bothered with swords and spears to fight its battles.

Some multiculturists claimed that eleventh-century Chinese warriors were armed with machine guns, and that the Incas frolicked above the Nasca plains in hot-air balloons. Certain Afrocentric scholars have made some dubious claims: that the Greek mathematician Euclid was black, for example, and that the Olmec heads, huge sculpted heads with Negroid features found in Mexico, are proof that Nubians visited the Americas.

In its issue of April 18, 1999, the *New York Times Magazine* chose the best inventions, stories, and ideas of the previous one thousand years. Richard Powers wrote that the most important scientific event of the last millennium occurred at its very beginning, around A.D. 1000, when the Arab scientist Alhazen solved a centuries-old problem: how does vision work? Alhazen, who was born as Abu Ali al-Hasan ibn al-Haytham in Basra, in what is now Iraq, dispatched the "ray theory," which had been around since ancient Greece. This theory, espoused by Euclid, Ptolemy, and others, held that the eye sent out a ray to the object in order to "see" it. The ray theory seems ridiculous today because we know the speed of light and how far away the stars are. If our eyes had to send out rays, we'd be waiting years before we could see even the nearest stars.

In 1000, the ray theory seemed reasonable. Alhazen conducted a simple experiment: he and others looked into the sun; it hurt. Clearly, if there were rays, they were coming into the eye, not going out of it. He developed a comprehensive theory of vision that dominated optics in Europe until 1610, when Kepler improved upon it. Alhazen may not have been smarter than Euclid and Ptolemy, but he worked quite differently. The latter two followed a classic Greek method of announcing a set of axioms, then reasoning from them. Alhazen began with his observations of and experiments with light, then reasoned toward a theory.[7] Ptolemy and Euclid also collected measurements and made observations, but the Greek ideal made the data subservient to the precept.

Powers was reaching, perhaps, when he stated that Alhazen's challenge of the old optical theory "has led to the certainties of electron microscopy, retinal surgery, and robotic vision," but he was correct in stating that the "vesting of authority in experiment" and the "skeptical rejection of concept in favor of evidence" began not in Europe but in the Islamic world.[8]

For some, the failure to acknowledge the successes of non-Western cultures derives not just from ignorance but from a conspiracy. Martin Bernal, a professor of government studies at Cornell University, is the author of *Black Athena,* a series of books that challenges our Greek-rooted view of history. Bernal believes that the roots of Greek civilization are to be found in Egypt and, to a lesser extent, in the Levant—the Near East of the Phoenicians and the Canaanites. Using linguistic analysis, he determined that 20 to 25 percent of the Greek vocabulary derived from the Egyptian. The roots of European civilization are Afro-Asiatic. The Greeks knew this and wrote about it, telling of Egyptian colonies in Greece during the Bronze and even the Iron Ages. The great Greek wise men, including Pythagoras, Democritus, and even Plato, traveled to Egypt and brought back Egyptian ideas and knowledge. (We have Democritus's own writings to acknowledge that his math skills were honed in the shadow of the pyramids.) The Greeks acknowledged their debt to Egypt. This "ancient model" held that the Greek culture had arisen as the result of colonization, in around 1500 B.C., by Egyptians and Phoenicians, and that the Greeks continued to borrow heavily from Near Eastern cultures. It was the conventional wisdom among Greeks in the classical and Hellenistic ages. This ancient model, writes Bernal, was also embraced by Europeans from the Renaissance through the nineteenth century. The Europeans, says Bernal, were enamored of Egypt.

For several centuries, Europe believed that Egypt was the cradle of civilization. This began to change in the eighteenth century when Christian apologists worried about Egyptian pantheism, and ideas of racial purity began taking hold among Locke, Hume, and other English thinkers. This led to the "Aryan model" in the first half of the nineteenth century. This view denied the existence of Egyptian settlements. Later, as anti-Semitism grew during the late nineteenth century, proponents of the Aryan model also denied Phoenician cultural influences.

The Aryan model was refined throughout the years to establish ancient Greece as distinctly European. Accordingly, there had been an invasion from the north—unreported in ancient tradition—that had overwhelmed the local Aegean or pre-Hellenic culture. Thus, Greek civilization was now seen as the result of the mixture of the Indo-European-speaking Hellenes and their indigenous subjects. It is this Aryan model that most of us were taught during the twentieth century. Bernal advocates a return to a modified ancient model, which is supported by the historian Herodotus and other ancient Greeks.

In its January 14, 2000, issue, on the occasion of the beginning of the third millennium, *Science* magazine, in conjunction with the American Association for the Advancement of Science (AAAS), published a time line, called "Pathways of Discovery," that detailed ninety-six of the most important scientific achievements in recorded history. The *Science* time line included some sophisticated choices that many educators would have missed: William Ferrel's 1856 work on ocean winds and currents, the 1838–39 cell theory of Matthias Schleiden and Theodor Schwann, and William Gilbert's 1600 theory that the earth behaves like a huge magnet.

Of those ninety-six achievements, only two were attributed to non-white, non-Western scientists: the invention of zero in India in the early centuries of the common era and the astronomical observations of Maya and Hindus in A.D. 1000. Even these two accomplishments were muted by the editors of *Science.* The Indians were given credit only for creating the "symbol for zero," rather than the concept itself. The Mayan and Hindu "skywatchers" (the word *astronomer* was not used) made their observations, according to the journal, for "agricultural and religious purposes" only.

Most interesting is the first entry in the time line: "Prior to 600 B.C.,

Prescientific Era." *Science* proclaimed that during this time, before the sixth-century B.C. pre-Socratic philosophers, "Phenomena [were] explained within contexts of magic, religion, and experience." *Science* thus ignored more than two millennia of history, during which time the Babylonians invented the abacus and algebra, the Sumerians recorded the phases of Venus, the Indians proposed an atomic theory, the Chinese invented quantitative chemical analysis, and the Egyptians built pyramids. In addition, *Science* gave Johannes Gutenberg credit for the printing press in 1454, though it was invented at least two centuries earlier by the Chinese and Koreans. An essential precursor to the printing press is paper, which was invented in China and did not reach Europe until the 1300s.[9] *Science* cited Francis Bacon's work as one of its ninety-six achievements, yet ignored his opinion that inventions from China created the modern world.

Pre-Columbian achievements in the New World have long eluded traditionalists. The Maya invented zero about the same time as the Indians, and practiced a math and astronomy far beyond that of medieval Europe. Native Americans built pyramids and other structures in the American Midwest larger than anything then in Europe.

MANY TRADITIONAL Western historians believe that little original science was conducted after the collapse of the Greek civilization; that the Arabs copied the work of Euclid, Ptolemy, Apollonius, et al.; and that eventually Europe recouped its scientific heritage from the Islamic world. During the Middle Ages, Arab scholars sought out Greek manuscripts and set up centers of learning and translation at Jund-i-Shapur in Persia and Baghdad in Iraq. Western historians don't often like to admit that these same scholars also sought manuscripts from China and India, and created their own science.

Scholarship moved to Cairo and then to Córdoba and Toledo in Spain as the Muslim empire expanded into Europe. When the Christians recaptured Toledo in the twelfth century, European scholars descended upon the documents.[10] They were interested in all Arabic documents—

translations of Greek works but also original Arabic writings and Arabic translations of other cultures' manuscripts. Much of the scientific knowledge of the ancient world—Greece as well as Babylonia, Egypt, India, and China—was funneled to the West through Spain. George Saliba has found that there was an intense traffic in Arabic manuscripts between Damascus and Padua during the early 1500s, and more and more scientific documents, written in Arabic, are being rediscovered in European libraries. Saliba has documented that many European scholars in the Renaissance were literate in Arabic. They read the Islamic papers and shared the information with their less literate colleagues.[11]

One example is Copernicus, who studied at Padua. Saliba points out that if Copernicus did borrow from Islamic astronomers—and the jury is still out—he had good reason not to acknowledge his intellectual debt. It would have been impolitic, says Saliba, to mention Islamic science when the Ottoman Empire was at the door of Europe. Another European scholar who studied at Padua was William Harvey, who established the geometry of the human circulatory system in 1629, another landmark in science according to the AAAS's *Science* time line. A 1241 Arab document, notes Saliba, lays out the same geometry, including the crucial assertion that the blood must first travel through the lungs before passing through the heart, contrary to the opinion of the ancient Greek physician Galen and past medical scholarship.[12]

Historian Glen Bowersock of the Institute for Advanced Study writes that "the classical antecedents of western civilization have long served to justify the study of ancient Greece and Rome," but he admits that "the porousness of Greek culture and the parallels to its achievements in other cultures have never been a secret. . . . The Greeks did not emerge, like Athena from the head of Zeus, fully equipped with their arsenal of culture. . . . An expression like 'the Greek miracle' was a catchy phrase for great drama, heroic statues and the Parthenon, but all this had its historical context. For the Greeks themselves, the context was Phoenicia and Egypt."[13]

The AAAS and *Science* magazine, in their "Pathways of Discovery" time line, acknowledge that from the ninth to the fifteenth centuries,

"The flow of science and technology is mostly *into* Europe *from* Islam and China" (italics theirs). Yet *Science* reports that the contributions of Islam and China are among those events that "represent the countless twists, turns, ironies, contradictions, tragedies, and other unkempt historical details that have synthesized into the far more complex and multitextured reality of the scientific adventure." Other such events they list are Isaac Newton's practice of alchemy, the false discovery of "N-rays," and the failure of geologists to accept the theory of continental drift.

This shall be a book of "unkempt historical details"—a tale of the non-Western roots of science. I began to write with the purpose of showing that the pursuit of evidence of nonwhite science is a fruitless endeavor. I felt that it was only responsible, however, to attempt to find what meager legitimate non-European science might exist. Six years later, I was still finding examples of ancient and medieval non-Western science that equaled and often surpassed ancient Greek learning.

My embarrassment at having undertaken an assignment with the assumption that non-Europeans contributed little to science has been overtaken by the pleasure of discovering mountains of unappreciated human industry, four thousand years of scientific discoveries by peoples I had been taught to disregard.

There is no good definition of science. The AAAS, for example, does not have one. After many trials, the American Physical Society (for physicists) finally decided upon a definition. The APS found that if the definition was too broad, pseudosciences like astrology could sneak in; too tight, and things such as string theory, evolutionary biology, and even astronomy could be excluded.

For the purposes of this book, science is a logical and systematic study of nature and the physical world. It *usually* involves both experiment and theory. Those theories normally arise from or are verified by experiment. That's a bit squishy, but most definitions of science are. I put *"usually"* in italics because if we absolutely require experiment, we might have to exclude astronomy, the oldest science, since one cannot re-create new stars or galaxies in the laboratory or reenact the formation of the solar system. Yet the observations in astronomy are often as good as experiment. Halley's comet returns with stunning regularity; the sun comes up each morning.

The philosopher Karl Popper introduced the requisite of "falsification." Science is falsifiable; religion is not. A scientific theory or law can never be proved absolutely, but it should be able to be falsified. For example, Newton said that force equals mass times acceleration (F = ma). We cannot prove that every object in every galaxy obeys this law or that all objects will *always* obey this law. We can prove it wrong, however, in an experiment. (And some of Newton's concepts *have* been proved wrong, by Albert Einstein and by quantum physicists.) So scientists must come up only with theories that can be falsified, as Popper put it. They must be testable. There is no such requirement for religion.

All this said, there remain problems with such a definition. Astrology, for instance, is falsifiable. If your astrologer says you will meet a handsome stranger on Tuesday, you can test this. On the other hand, superstring theory, posited by some physicists as "the theory of everything," would require a particle accelerator ten light-years in diameter to falsify it. Most of evolutionary biology cannot be verified experimentally either. One cannot reenact the evolution of a new species or re-create the dinosaurs beginning with a one-celled animal. If we follow the falsification rule too closely, we have to include astrology and exclude evolutionary biology, string theory, and maybe even astronomy.

So let's not take falsification too seriously. Otherwise we might have to exclude all science practiced by the ancient Greeks. The Greeks not only avoided experiments, they abhored them, trusting reason over empirical evidence.

We will confine ourselves to the hardest sciences here: physics, astronomy, cosmology, geology, chemistry, and technology. We shall include math also, as it is indispensable to science and inextricably entwined with it. We will leave the softer disciplines—anthropology, agronomy, psychology, medicine, and the like—for another time.

One thing we won't consider is the pragmatism of the science or the motivation of the scientist. These have often been used to discredit non-Western science: *yes, it's good work, but it wasn't "pure";* or, conversely, *it wasn't practical.* As for motivation, many scientific discoveries were driven by religion: Arab mathematicians improved algebra, in part, to help facilitate Islamic inheritance laws, and Vedic Indians solved square roots to build sacrificial altars of the proper size. This was science in the service of religion, but science nonetheless.

Stigler's law of eponymy, formulated by statistician Stephen Stigler, states that no scientific discovery is named after its original discoverer. Journalist Jim Holt points out that Stigler's law itself is self-confirming, given that Stigler admits that it was discovered by someone else: Robert K. Merton, a sociologist of science.[14]

The most famous Stiglerism is the Pythagorean theorem, which holds that the sum of the squares of the perpendicular sides of a right triangle equals the square of the hypotenuse. Or, in math parlance, $a^2 + b^2 = c^2$, where a and b are the sides and c is the hypotenuse. Jacob Bronowski writes:

> To this day, the theorem of Pythagoras remains the most important single theorem in the whole of mathematics. That seems a bold and extraordinary thing to say, yet it is not extravagant; because what Pythagoras established is a fundamental characterisation of the space in which we move, and it is the first time that it is translated into numbers. And the exact fit of the numbers describes the exact laws that bind the universe. In fact, the numbers that compose right-angled triangles have been proposed as messages which we might send out to planets in other star systems as a test for the existence of rational life there.

The only problem is that Pythagoras is not the first mathematician to come up with the theorem. By Bronowski's own admission, the Indians, Egyptians, and Babylonians used "Pythagorean triplets" in order to determine right angles when constructing buildings. A Pythagorean triplet is a set of three numbers that describes the sides of a right triangle. The most common triplet is 3 : 4 : 5 ($3^2 + 4^2 = 5^2$, or $9 + 16 = 25$). Others you probably learned in high school include 5 : 12 : 13, 12 : 16 : 20, and 8 : 15 : 17. Pythagoras invented his theorem around 550 B.C. The Babylonians, Bronowski concedes, had cataloged perhaps hundreds of triplets by 2000 B.C., long before Pythagoras. One of the triplets the Babylonians found is the enormous 3,367 : 3,456 : 4,825.

Nevertheless, Bronowski dismisses Babylonian triplets (as well as Egyptian and Indian triplets) as being merely "empirical." That is, he believes that they somehow arrived at triplets (or triples) such as 3,367 : 3,456 : 4,825 by trial and error. Yet there is considerable evidence that the Babylonians used various algebraic techniques derived from $a^2 + b^2 = c^2$ to generate Pythagorean triplets. "There's no way even God could come up with all Pythagorean triples by trial and error," says mathematician Robert Kaplan.

What Pythagoras arguably did that impressed Bronowski and others—and justifiably so—was to construct a geometric *proof* of the theorem. The concept of the proof as more important than the theorem itself was promulgated two centuries later by Euclid. Thus, non-Western mathematics has been viewed as second-rate because it is empirically based rather than proof based. Both methods are useful. The Euclidian geometry most of us learned is axiomatic. It begins with an axiom, a law assumed to be true, and deduces theorems by reasoning downward. It is deductive and assumptive. Centuries later, Alhazen in the East and, notably, Galileo in the West helped popularize an inductive, empirical method for science, much as the Babylonians, Egyptians, and Indians had used. One begins not with assumptions but with data and measurements, and then reasons upward to overarching truths.[15] Most of what we call science today is empirical. When Isaac Newton collected data on the passage of comets, on the moons of Jupiter and Saturn, and on the tides in the estuary of the Thames River to construct his great syntheses in *Principia,* he was being empirical and inductive.

Math is slightly different, but many mathematicians see a need to include both proof based and empirically based work. A case in point in the present century is the great Indian mathematician Srinivasa Ramanujan, whose notebooks contain the germs of superstring theory and whose work has been used to evaluate pi to millions of digits past the decimal point. According to his wife, Ramanujan did his calculations on a handheld slate, then transferred the final results to his notebooks, erasing the slate; thus, we have few clues as to how he arrived at these equations, yet no one doubts that they are true.[16]

According to one historical account, Pythagoras brought back his eponymous theorem from his travels to the East and founded the tradition of proof because his less numerate countrymen refused to accept the theorem. Consider, too, the naming of Fermat's last theorem, the work of the Frenchman Pierre de Fermat in the seventeenth century. The last theorem is a remote derivation from the Pythagorean theorem, but Fermat neglected to leave us a proof—at least not one we could find. Yet for more than three hundred years, Fermat's last theorem has worked. A few years ago, Andrew Wile, of Princeton University, finally devised a proof. Still, we have yet to hear an outcry to change the name of Fermat's last theorem to Wile's first theorem. (It is an in-joke among mathematicians that the correct name is Fermat's last conjecture, a conjecture being an unproved theorem.)

In 1915, about the time, according to Otto Neugebauer, that the Germans were rewriting their encyclopedias to edit out the Phoenicians from Greek history, the English science historian G. R. Kaye admonished "western investigators in the history of knowledge" to look for "traces of Greek influence" because the "achievements of the Greeks" form "the most wonderful chapters in the history of civilisation."[17] Our pop science historians—Bronowski, Daniel Boorstin, Carl Sagan, et al.—have certainly been faithful to that directive. Western historians have also criticized past non-Western scientists, such as the Maya and Egyptians, for their strange religious beliefs, implying that acute religiosity disqualifies the work of a scientist. Then again, when Pythagoras finally proved "his" theorem, he offered a hundred oxen to the Muses in thanks.[18]

Science is science. It can be practical or impractical. The Danish physicist Niels Bohr owned a cabin retreat, to which he invited his scientist friends for long intense discussions about the meaning of quantum physics. Over the door of the cabin was hung a horseshoe on a nail.

His guests often viewed this with a roll of the eyes. Finally, one screwed up the courage to say, "Come on now, Niels. You don't believe in this nonsense, do you?"

According to legend, Bohr replied, "That's the beauty of it. It works whether I believe in it or not." For our purposes, science embraces those facts about the physical world that work . . . whether we believe in them or not.

2

MATHEMATICS
The Language of Science

THE Mark's Meadow School is a public elementary school in Amherst, Massachusetts, in the western region of the state. Located across North Pleasant Street from the University of Massachusetts (UMass), it has served as a laboratory for the university's school of education. Education majors can sit in a darkened, elevated corridor and secretly observe the students through two-way mirrors in the ceiling while eavesdropping through the use of a hidden sound system. In the future, they may want to listen more carefully during math lessons.

Recently I took a group of Mark's Meadow fourth graders to the local mall, where we stopped to eat at a Taco Bell. The kids read the menu and started laughing. The joke was this: there were three sizes of drinks—small, medium, and large; twelve ounces, sixteen ounces, and twenty ounces—and three prices, \$1.19, \$1.49, and \$1.79. The kids were laughing at the sign beneath the prices: UNLIMITED REFILLS!

Then a group of college students wearing UMass sweatshirts joined the line. They studied the sign. "Hey, let's get the large drinks," said one.

"Yeah," said another. "Then we'll really clean up on the unlimited refills."

What the fourth graders understood that the college kids did not is the concept known as "infinite sets." In the above case, one infinite set is equal to another. Take a ruler and cut it into infinitesimally small segments from the 1-inch line to the 2-inch line. There would be an infinite number of slices. Do the same with the ruler from the 2-inch line to the 12-inch line. Will there be ten times as many slices? No. If we're dealing with rational numbers, infinity is infinity. The same principle operates with our soft drinks: twelve, sixteen, and twenty times infinity all equal infinity. (In other cases, however, infinite sets are *not* equal.)[1]

The concept of infinite sets of rational numbers was grasped by Jaina (Indian) thinkers in the sixth century B.C. and by Alhazen in the tenth century A.D. It entered Europe nearly a thousand years later, when the nineteenth-century German mathematician Georg Cantor refined and categorized infinite sets. Here in the twenty-first century, the idea has crossed the Atlantic to the Mark's Meadow School. It has yet to make the giant leap across North Pleasant Street to the University of Massachusetts.

UMass students shouldn't feel too bad; Galileo was stumped by the problem in the 1600s. He envisioned a row of all integers, starting at 1 and going off to infinity. Then he envisioned the squares of those same integers, starting at 1^2 and going off to infinity. He realized that if he placed the squares side by side with the set of all integers (1^2 next to 1, 2^2 next to 2, 3^2 next to 3, and so on), he would have enough squares to pair with all the numbers in the integers column. How is this possible? Galileo decided to put the problem aside and return to something easier—astronomy.

IMAGINE YOURSELF as a German merchant living in the fifteenth century. You want your son to learn enough math for a career in commerce. A professor you know suggests a good German university where your son will be taught addition and subtraction. But, you ask, what about multiplication and division? The professor explains that study of such "advanced" mathematics is not available locally; your son must travel to Italy, the only European country in which such operations can be learned.

"Reckoning schools," in which arithmetical operations using Indo-Arabic numerals were taught, had begun to spring up in Italy.[2] However, what your son would most likely find at his Italian university would be a sort of multiplication that hardly resembles what we call multiplication today. In medieval Europe, multiplication was simply a succession of doublings. For example, we multiply 9 times 11 in a simple operation, like so:

$$\begin{array}{r} 11 \\ \times 9 \\ \hline 99 \end{array}$$

But in Italy in the Middle Ages, a mathematician would commonly envision the multiplication of a number by 9 as eight doublings and a single. Let's multiply 11 by 9, medieval style:

11 times 1 equals **11**	**(1×)**	➤
11 doubled equals 22	(2×)	
22 doubled equals 44	(4×)	
44 doubled equals **88**	**(8×)**	➤

The medieval scholar looks at the numbers in the right-hand column to find a combination of multiples that add up to the desired multiplier, in this case 1× and 8× to equal 9. Then the mathematician adds the two products, 11 and 88, to get the answer, 99.

Now try something a bit more complicated, 46 × 13. Today we do it like this:

$$\begin{array}{r} 46 \\ \times 13 \\ \hline 138 \\ 46 \\ \hline 598 \end{array}$$

The medieval European mathematician might do it like this:

46 times 1 equals **46**	**(1×)**	◄
46 doubled equals 92	(2×)	◄
92 doubled equals **184**	**(4×)**	
184 doubled equals **368**	**(8×)**	◄

Again, he finds the combination of doublings that add up to 13, the ones indicated above, 1×, 4×, and 8×. Then he adds up the three resultant sums to solve the problem: $46 + 184 + 368 = 598$. Remember, all this must be done with roman numerals. (Keep the above technique in mind. We will encounter it again.) Similarly, division was a tedious process of "halving" the divisor until one arrived at the divider, or close to it.[3]

Meanwhile, in India about a thousand years earlier, mathematicians were doing multiplication and division the "modern" way, as well as algebra and even a crude form of calculus.

Now, imagine yourself again in fifteenth-century Italy. You are, let's say, a bookseller. You need to keep track of sales and inventory. You need to pay your suppliers, total your sales, calculate your overhead, determine your profit or loss. How would you do this? Certainly not with roman numerals; even the simplest arithmetic using roman (or Greek) numerals was beyond all but advanced scholars. Furthermore, there is no roman numeral for zero; in fact, there is no concept of zero, of nothingness, in European math of this era. How do you get your accounts to balance?

Like other merchants, you keep a secret set of books, in the gobar, or Gwalior, numerals, the so-called Hindu-Arabic numerals, which date from approximately first- to eighth-century A.D. India. They look something like this: 0, 1, 2, 3, 4, 5, 6, 7, 8, 9. You would keep these books secret because in 1348 the ecclesiastical authorities of the University of Padua prohibited the use of "ciphers" in the price lists of books, ruling that prices must be stated in "plain" letters. A century earlier, a Florentine edict had forbidden bankers to use the "infidel" symbols.[4]

Numbers were dangerous; at least these Indian numbers were. They were contraband. The zero was the most unholy: a symbol for nothingness, a Hindu concept, influenced by Buddhism and transplanted to Christian Europe. It became a secret sign, a signal between fellow travelers. *Sunyata* was a well-established Buddhist practice of emptying the mind of all impressions, dating as far back as about 300 B.C.[5] The Sanskrit term for zero was *sunya,* meaning "empty" or "blank." Flashing a zero to another merchant let him know that you were a user of Hindu-Arabic numerals. In many principalities, Arabic numerals were banned from official documents; in others, the numbers were prohibited altogether. Math was sometimes exported to the West by "bootleggers" in Hindu-Arabic numerals. There is plentiful evidence of such illicit number use in thirteenth-century archives in Italy, where merchants used Gwalior numbers as a secret code.[6]

Imagine yourself an out-of-work mathematician in Italy in the late Middle Ages. How could you support yourself? Assuming you could do multiplication and long division, there was an obvious answer: you could become an itinerant math performer. Traveling from town to town, you would set up in the village square and perform "magic" tricks for the public. Multiplying 27 by 14 was considered as entertaining in that era as sword swallowing or juggling, and fewer people could do it. The public would toss coins in your cup. You would count your take at the end of each performance—secretly using Hindu-Arabic numerals, of course.[7] Or you could find employment in one of Italy's new reckoning schools.

THERE IS NO good definition of mathematics. We will not improve upon that situation here. As laymen, we know that math involves numbers, symbols, and logic and includes things such as arithmetic, algebra, geometry, trigonometry, and calculus. Professionals don't do much better in defining it.

University of Manchester (England) mathematician George Gheverghese Joseph calls math "a worldwide language with a particular kind of logical structure." He goes on to say that it "contains a body of knowledge relating to number and space, and prescribes a set of methods for reaching conclusions about the physical world."[8] Physicists might take issue with the last statement, arguing that journals of theoretical physics are filled with lovely math that says little about the physical world.

Harvard University's Barry Mazur, despite being a mathematician for more than forty years, declined to give a definition. Mathematician Robert Kaplan calls math "an activity about activity." Pressed further, Kaplan came up with a tantalizing morsel. "Math," he said, "describes what generalizes." For example, the commandment "Thou shalt not kill" is not a generalization. It allows killing in self-defense or during wartime. Math deals with generalizations, universal truths such as the Pythagorean theorem.

People sometimes try to fool us. Economists and psychologists, for example, fill their papers with curves, numbers, and equations. It looks like math, but it usually isn't. For example, economists use an equation called "the utility function" to explain why people buy home insurance even though the insurance companies are virtually guaranteed to profit. The function is expressed in curves and numbers. It doesn't always work; it predicts, for example, that people won't gamble or play lotteries.[9] Economists blame the breakdown of the utility function on people's foolishness. Math describes what can be generalized, not human behavior.

The ancient Egyptians had no word for mathematics. Our primary source for Egyptian math is a school textbook called the Ahmes (or Rhind) Papyrus. (Non-Western scholars prefer to call the papyrus Ahmes, after the scribe who composed it; Western scholars prefer Rhind, after the British collector who acquired it.)[10] Its title is *The Right Method for Entering into Things, for Knowing Everything That Is, Every Obscurity . . . Every Secret.* It's as good a definition as any.

Math describes the fall of rocks and the orbits of planets, and it is a common conceit among scientists to say that "math is the language of nature." This is unlikely. "I don't think nature speaks," says Mazur. "*We* speak."

The greatest of all the English experimenters, the nineteenth-century physicist Michael Faraday, spoke no math. Faraday learned science as a bookbinder by reading the books he bound, and despite being a high school dropout, he went on to invent the dynamo (electrical generator), which led to the electrification of the industrial world. Faraday wrote out all the results of his experiments in plain English. Yet he never claimed that "nature speaks English."

Though math may not be the language of nature, it is certainly the language of science. It's what most scientists speak. James Clerk Maxwell, of Maxwell's equations fame, made his mark by translating Faraday's work into math, a far more useful language for physicists. This is why we begin with math.

There are modern sciences that don't lend themselves to math. Biology, for example, deals with large systems of interactive cells that defy a numerical approach. Evolutionary biologist Paul Ewald, of Amherst College, says that numbers, while useful, ultimately cannot be used to explain evolution. There are no biological equivalents of Maxwell's equations that explain the platypus or the giraffe.

Let us accept for our purposes that mathematics is an essential foundation for science. We are forced to accept this idea because it is a long-cherished Western notion. If we are to say that non-European cultures had science long before the Europeans exported it to them, we must prove they had math. Even in America, sciences whose principles cannot be reduced to mathematical formulas have often been dismissed as "soft sciences." These include anthropology, medical science, certainly psychology, and, until this century, biology and chemistry. Chemistry first made the "hard" club in the 1920s, when the useful but mysterious order of the periodic chart of the elements was finally explained by quantum physics and the Pauli exclusion principle; biology became rigorous (or "hard") with the deciphering of the DNA molecule and the advent of molecular biology and its rigorous mathematical codes.

We might expect to find non-Western cultures to be mathematically weak throughout history. Yet nowhere is non-Western science stronger than in math. The mathematical foundation of Western science is an intellectual gift from the Indians, Egyptians, Chinese, Arabs, Babylonians, and others. The Maya, too, developed powerful mathematics, their priests judged as much for their ability to calculate as to pray. In their civilization, numeracy was next to godliness.

George Gheverghese Joseph, who was born in India but teaches in the United Kingdom, cites this line from the *Vedanga Jyotisa,* the oldest (500 B.C.) extant Indian astronomical text: "Like the crest of a peacock, like the gem on the head of a snake, so is mathematics at the head of all knowledge." Few modern Western scientists would disagree with that sentiment.

The traditional Western story is that math was created by the ancient Greeks around 600 B.C. and elaborated by Greco-Roman culture until A.D. 400, at which time the discipline fell dormant for a thousand years, only to be revived in post-Renaissance Europe. There is ample evidence, however, that nonwhite, non-Western cultures made significant contributions to European mathematics—or, at the very least, developed mathematical techniques that predated Western discoveries. For example:

- The Indians developed the use of zero and negative numbers perhaps a thousand years before these concepts were accepted in Europe. The Maya invented their own zero—in fact, a whole slew of them—at about the same time as the Indians.
- Clay tablets dated a thousand years[11] before the Greek civilization reveal traces of a sophisticated algebra among the Sumerians. Papyri of the eighteenth century B.C. and earlier show that the Egyptians used simple equations to deal with problems in distribution of food and other supplies.[12]
- In the third millennium B.C. the Babylonians developed a place-value system.[13] (In our base 10 system, 348, for example, stands for 8 ones, 4 tens, and 3 hundreds.) The Babylonian sexagesimal (base 60) number system may at first appear cumbersome, but Copernicus used sexagesimal fractions to construct his model of the solar system, and we still use the system for keeping time and measuring angles (60 minutes per hour, each minute divided into 60 seconds).

• The priestly scribes of Egypt knew the formula for calculating the volume of a cylinder—and thus recognized the existence of the mysterious factor π (pi) long before the Greeks—in fact, long before there were literate Greeks.[14] The Egyptians also developed the concept of the lowest common denominator, as well as a fraction table that modern scholars estimate required twenty-eight thousand tedious calculations to compile.[15]

• In 2000 B.C., the priestly astronomers of Mesopotamia, in the area now known as Iraq, kept extensive tables of squares. We know this from the clay tablets of cuneiform script found in temple libraries.[16] Remember that Europeans in the fourteenth century did not even keep times tables.

• Gottfried Leibniz, the coinventor of the calculus, claimed to have discovered the secret of deciphering the diagrams of the ancient Chinese sage Fu Hsi. Leibniz maintained that Fu Hsi's diagrams corresponded to his own modern binary mode of arithmetic.[17]

• The Indians invented a nascent form of calculus centuries before Leibniz invented calculus in Europe.[18]

• The Arabs coined the term *algebra* and invented decimal fractions: .25 for ¼, etc.[19]

• Aristotle credited the Egyptians with developing math before his countrymen, in a somewhat backhanded manner: "The mathematical sciences originated in the neighborhood of Egypt because there the priestly class was allowed leisure."[20]

Despite this, America's most prominent modern historian of mathematics, Morris Kline, wrote, "Compared with the achievements of their immediate successors, the Greeks, the mathematics of the Egyptians and Babylonians is the scrawling of children just learning how to write as opposed to great literature."[21] In his classic work *Mathematics: A Cultural Approach,* Kline acknowledges that the Babylonians and Egyptians pioneered mathematics long before the Greeks, but he dismisses them as pragmatists.[22] "The Egyptians and Babylonians did reach the stage of working with pure numbers dissociated from physical objects.

But like young children of our civilization, they hardly recognized that they were dealing with abstract entities." The Greeks, he said, were the first to recognize numbers as "ideas" and emphasized that this is how they must be regarded.[23]

The rules keep changing. When we discuss ancient Indian physics, in chapter 5, Western physicists will insist that it is meaningless because it was abstract, with no empirical backup. In the case of math, Kline seems to be saying the opposite, that the Babylonians and Egyptians were unsophisticated because they *used* their math. Because these civilizations saw math as "merely a tool in commerce, agriculture, engineering," says Kline, hardly any progress was made in the subject in a period of more than four thousand years.[24] As for the math required to build the pyramids, Kline writes, "A cabinetmaker need not be a mathematician."[25]

Another common charge is that non-Western mathematicians did not employ the ancient Greek custom of constructing proofs for their work. For example, Pythagoras gets credit for the Pythagorean theorem, say Western scholars, even though the Babylonians had the concept centuries earlier. This is because he, or his followers, constructed the first proof for this overarching principle, while the Babylonians did not. Critics find the Greek-style proof so important that its nonexistence in non-European cultures, they contest, discredits thousands of years of mathematics. The controversy over proof is a thorny one. Some mathematicians claim that non-Western peoples *did* have proofs, while others doubt that one can really "prove" any concept for eternity and throughout the entire universe. For a brief debate on the topic, see note.[26]

Skepticism is appropriate to all research, but the researcher in non-Western mathematics must often face a high hurdle. Ayele Bekerie, of Cornell University, who has studied ancient Ethiopian number systems, describes how Western scholars once refused to accept that this African civilization had developed its own numerals. Ethiopian numbers resemble, not surprisingly, the more ancient Egyptian numbers and, to a lesser extent, ancient Greek numbers—again not surprisingly, because of Ethiopia's geographical proximity to Egypt, and because Egypt influenced Greek mathematics. The controversy involves letters written by Ethiopians to Greeks. These letters contain both Ethiopian and Greek numbers. One explanation is that the letters were written in both languages so the Greeks could understand. Western skeptics maintained that Africans were not capable of such sophistication, that these letters had actually been written by Greeks, who thus introduced the Ethiopians to a crude alphabet and number system that they now claim as their own. Of course, this makes little sense, since the letters were found in Greece. If the Greeks had written to the Ethiopians, the letters should have been found in Ethiopia. The dispute, according to Bekerie, was finally solved by chemists. The ink on the pre–Christian era parchment in question was of an unusual hue. Chemical analysis showed that the ink had been made from berries indigenous to Ethiopia.[27]

Our Western mathematical heritage and pride are critically dependent on the triumphs of ancient Greece. These accomplishments have been so greatly exaggerated that it often becomes difficult to sort out how much of modern math is derived from the Greeks and how much is from the Babylonians, Egyptians, Indians, Chinese, Arabs, and so on. The math of the Greeks was wonderfully imaginative, and a great debt is owed to them. But if our math today were based entirely on Pythagoras, Euclid, Democritus, Archimedes, et al., it would be a highly deficient discipline.

BEFORE WE GET into the mathematical history of ancient non-Western peoples, let us first briefly discuss how the math we study arrived in Western classrooms of the twentieth century. The different paths described by scholars are often in violent disagreement. We shall pass no judgment here on the correct solution.

The "traditional" Western view—and I put "traditional" in quotes here because this tradition is hardly a century old—is best summed up by two respected mathematical historians, Rouse Ball and Morris Kline. In 1908 Ball wrote, "The history of mathemmatics cannot with certainty be traced back to any school or period before that of the Ionian Greeks."[28] In 1952, Kline wrote, "[Mathematics] finally secured a new grip on life in the highly congenial soil of Greece and waxed strongly for a short period. . . . With the decline of Greek civilization the plant remained dormant for a thousand years . . . [until] the plant was transported to Europe proper and once more embedded in fertile soil."[29] Fleshed out, this is often interpreted to mean that there have been three stages in the history of mathematics:

1. Circa 600 B.C. the ancient Greeks invent math, which thrives for a thousand years until approximately 400 A.D., at which time it disappears from the face of the earth.
2. A dark age of mathematics ensues, lasting over a thousand years. Some scholars concede that the Arabs kept Greek math alive during the Middle Ages.
3. Greek math is rediscovered in sixteenth-century Europe, and mathematics flowers again from then until the present.

This view is controversial. Our modern numerals—0 though 9—were developed in India during stage 2, the so-called dark age of mathematics. Mathematics existed long before the Greeks constructed their first right angle. We can perhaps excuse Rouse Ball, writing in 1908, for being unaware of the Greeks' mathematical predecessors. On the other hand, George Gheverghese Joseph points out that Ball should have been aware of the early Indian mathematics contained in the *Sulbasutras* (The Rules of the Cord). Written somewhere between 800 and 500 B.C., the *Sulbasutras* demonstrate, among other things, that the Indians of this period had their own version of the Pythagorean theorem as well as a procedure for obtaining the square root of 2 correct to five decimal places. The *Sulbasutras* reveal a rich geometric knowledge that preceded the Greeks.[30]

Kline's statement, says Joseph, is more problematic, ignoring a rich body of non-European mathematics that had been unearthed by the mid–twentieth century, including math from Mesopotamia, Egypt, China, India, the Arab world, and pre-Columbian America.[31] There is

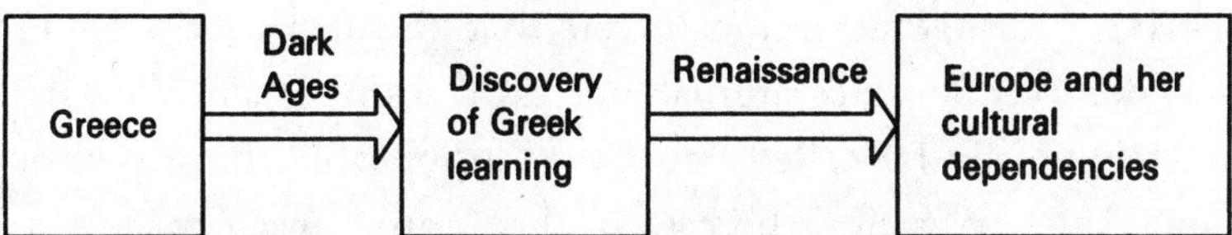

The classic math trajectory gives no credit to non-Western civilizations for the development of mathematics. (As per Joseph.)

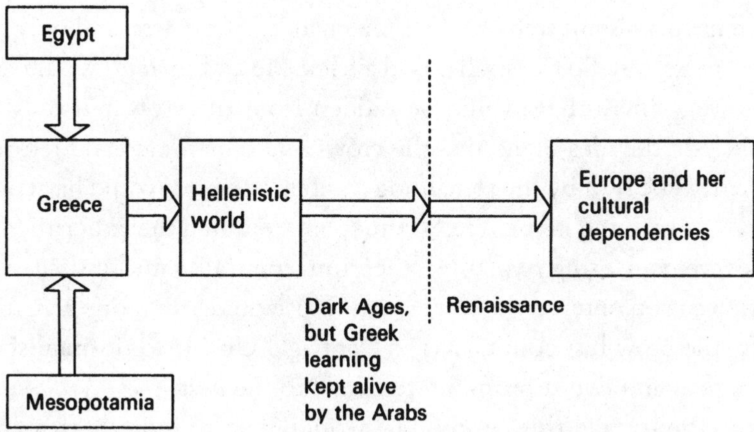

Recently, a modified Eurocentric trajectory has been taught. It acknowledges non-Western mathematics, but paints it as subservient to European math. It also is oversimplified. (As per Joseph.)

the problem, too, that the Greeks themselves—Democritus, Aristotle, Herodotus—lavished praise upon the Egyptians, crediting them as their mathematical gurus (though not in those words). The fact is that many people were counting before the Greeks.

It is impossible to think of a culture that didn't have some form of counting, that is, a method of matching a collection of objects with a set of numbers, markers, or other tallying symbols, whether written or in the form of beads, knots, or notches on wood, stone, or bone. Counting is math, and not everyone can do it, but every culture has contained at least some individuals who could.

This is a safe assumption, given that the mathematician Tobias Dantzig demonstrates that animals possess a "number sense" even if they cannot technically be called "counters" (they don't have numbers or tally sticks). Dantzig cites the case of the "counting crow" in his 1930 book *Number: The Language of Science.*

A squire, writes Dantzig, wanted to shoot a crow that had made its nest inside his watchtower. When he entered the tower and approached the crow, the bird knew what was coming and left the nest. It would watch from a distant tree until the man had left the tower, and only then return to its nest. So the squire used a trick. He and a friend would enter the tower, where they would be hidden from the crow, but only one would exit, the idea being that the crow would be fooled and return to the nest, to be shot by the remaining hunter. The bird would have nothing of it, remaining in the tree. So the squire repeated the experiment on successive days, using two, three, then four men, all without success. Finally, five men entered the tower, and four exited while one remained. "Here the crow lost count," writes Dantzig. "Unable to distinguish between four and five, it promptly returned to the nest."

On the surface, this famous tale would seem to indicate that crows can "count" only to four. We cannot question the crow to find out what prompted it to return to the nest prematurely, but it would seem obvious that a crow's number sense is inferior to that of humans.

Dantzig points out that it is very difficult to test a human's number sense because our species has relied for so long on counting that it has become "an integral part of our mental equipment."[32] Humans are always consciously or unconsciously aiding their innate number sense with artifices such as counting, mental grouping, and symmetric pattern reading. Psychologists have, with great difficulty, devised tests that eliminate such artifices. They have come to the conclusion that our direct visual number sense is . . . four. We do no better than the crow.

The magician Harry Houdini knew this about humans, at least intuitively. One of his tricks was called "Walking Through Walls." Houdini explains to the audience that he will walk through a brick wall. He says there is no trapdoor in the stage, and to demonstrate that, he unrolls a wide carpet from the rear of the stage to the front. This will block any trapdoor. To safeguard the stage further, a long, heavy steel beam is placed atop the carpet, again from the rear to the front, pointing toward the audience. Bricklayers appear, and they build a brick wall on the beam. The audience is facing the end of the wall. Houdini announces that he will walk *through* the wall, left to right. Curtained screens are placed on either side of the wall. Houdini disappears behind the left curtain and, at a signal, magically rematerializes on the right side, exiting through the right curtain to the applause of the audience. Clearly, he didn't go over or under the wall. The trick: he walked around the wall, in full view of the audience.

The trick works because the many bricklayers, ten or more, all wear identical overalls as they scurry about the stage. When Houdini disappears behind the first screen, he dons one of the outfits hidden there, making him look like a bricklayer. Houdini simply walks around the wall in plain sight and joins those bricklayers who are moving the second screen into place. He goes behind the screen, strips back to his original clothes, and walks through the curtain. No one notices the extra bricklayer.

Houdini realized the limitations of the human number sense. He could fool the audience with ten people. If an audience member had bothered to count the bricklayers, the trick would have failed. But who's counting? He needed more than double the human number of four, however, because of what Dantzig calls "symmetric pattern recognition." If there were only eight bricklayers, for example, with at some point four on a side, the asymmetry of five and four could be noticed when Houdini joined the workers.[33]

The Great Houdini inadvertently made a significant contribution to science, proving that humans don't have any greater innate sense of numbers than crows do. Numbers, tally sticks, and other artifices are required. Let me add that some interesting research conducted since Dantzig's era has shown that some animals *may* be sophisticated counters as well. We'll get to that shortly.

The German mathematician Karl Menninger had a liberal definition of counting. He points to the Wedda, a tribe living on the island of Ceylon. If a Wedda wishes to count coconuts, he assembles a pile of sticks and matches up the sticks with the coconuts. The Wedda have no words for numbers. "Does that mean he is unable to count?" asks Menninger. "Not at all. He translates the pile of coconuts he has laid out into the auxiliary quantity of sticks." He can tell if anyone has stolen one of the coconuts by arranging the nuts and sticks in a one-to-one order. But how can he describe the total number of coconuts? He points to the pile of sticks, explains Menninger, and says, "That many!"[34]

In any case, counting—whether with numbers, sticks, or some other device—extends innate number sense far beyond its modest limits, propelling humans above other species. Other scholars believe we may not be the only species that counts. The Clark's nutcracker, for instance, a large bird that lives high in the mountains where food is sparse during the winter, hides thousands of seeds in good weather, digging them up months later. In a lab at Northern Arizona University, the birds retrieved seeds from a large sandbox with an accuracy rate of 90 percent. Experts in animal intelligence believe the nutcracker uses a kind of nearest-neighbor system, choosing a focal point for the first cache, then hiding successive seeds in a geometric pattern that it somehow memorizes.[35] If a Clark's nutcracker is comparing set one (the seeds) to set two (the pattern), one could consider the bird to be counting.

THERE IS SOME debate over whether counting, or even calculating, qualifies as math, but George Joseph says that mathematics arose initially from a need to count and record numbers: "As far as we know, there has never been a society without some form of counting or tallying, i.e. matching a collection of objects with some easily handled set of markers, whether it be stones, knots, or inscriptions such as notches on wood or bone."[36]

The Ishango Bone is evidence—controversial evidence—of one of the first counting societies, about twenty thousand years ago. Ishango is an area around Lake Edward, in the mountains of central equatorial Africa on the border between Uganda and Zaire. Ishango is sparsely populated today, but twenty thousand years ago a small community fished the lake and gathered food and farmed by its shores. The Ishango society lasted only a few hundred years before being buried by a volcanic eruption.

The Ishango Bone itself is a dark brown object, like a bone tool handle. It features a sharp piece of quartz at one end, which may have been used for engraving, tattooing, or perhaps writing. More interesting are three columns of notches. They are asymmetrically grouped, which makes Joseph and others believe that they are functional rather than decorative. The groups of notches line up like this:

Row 1: 9, 19, 21, 11

Row 2: 19, 17, 13, 11

Row 3: 7, 5, 5, 10, 8, 4, 6, 3

Tally sticks predate the Ishango Bone. Notches on sticks or bones (or knots on strings or cuts on stones) have been found worldwide. These are records of counts, perhaps kills by hunters. A thirty-seven-thousand-year-old baboon fibula with twenty-nine notches was found in Swaziland. A thirty-two thousand-year-old wolf shinbone, marked with fifty-seven notches—the first fifty-five grouped in fives—was found in Czechoslovakia. These tally sticks are similar to calendar sticks still used today in Namibia to record time. The grouping—the Romans did something similar—may have been the first step toward constructing a numbering system.

J. de Heinzelin, the archaeologist who unearthed the bone, speculated that the Ishango not only had a number system but that this system, through the transmission of harpoon heads and other tools, spread north to Egypt and led to Egyptian mathematics. Joseph comments, "A single bone may well collapse under the heavy weight of conjectures piled upon it." One has to accept the skepticism of Joseph and others, given the ambiguity of the evidence, though Joseph points out that the builders of Stonehenge have been credited by scholars with monumental mathematical skills on the basis of a few large rocks.[37]

There are other examples of unwritten mathematics, such as Incan quipus, knotted strings used for recording numbers in a decimal base system, and the cowrie shells of the Yoruba of southwestern Nigeria. But let us go directly to the first culture to make strides in written mathematics.

Egypt

Like the civilization itself, the history of Egyptian math is a long one, beginning somewhere around 3200 B.C., when a system of writing was invented, and stretching to 332 B.C., when Alexander the Great conquered and Hellenized Egypt. Our sources are meager, since papyrus deteriorates under humid conditions. The only readable documents have been found in cemeteries and temples in the desert fringe along the Nile valley. Few papyri have been recovered from major towns or cities in fertile areas around the Nile or in the delta. Most date from the Middle Kingdom period, between 2000 and 1700 B.C. In total, there are but five papyri, a pair of wooden exercise tablets, and a stone flake.[38] Yet we find a rich mathematical tradition. Who knows what was being done with numbers in the major cities?

The Egyptians used three different number systems: the hieroglyphic and hieratic systems early in the civilization's history, and the demotic toward the end, during the Greek and Roman periods. The hieroglyphic numbers were, obviously, pictorial, each character readily recognizable as a common object, from ropes to man to the sun. All numbers could be stated using combinations of only eight figures, representing powers of ten from 1 to 10^7:

1 10 10^2 10^3 10^4 10^5 10^6 10^7

The first three symbols, for 1, 10, and 10^2, were variations of rope: a short length of rope, rope in a U shape, and a coil of rope, respectively. Perhaps the rope imagery was inspired by the *harpedonaptai,* the "rope stretchers," or surveyors, who regularly surveyed the lands of the Nile

River valley. One thousand (10^3) looked like a lotus; 10,000 (10^4), a crooked finger; 100,000 (10^5), a tadpole; 1,000,000 (10^6) a man with upraised arms; 10,000,000 (10^7), a sun, perhaps Ra, the sun god.

The Egyptians wrote any number they pleased by grouping the above symbols together. For example, 1,321 would be written:

$$1 + 2(10) + 3(10^2) + 1(10^3) =$$

As the Egyptians had no zero or placeholder, the hieroglyphs could be arranged in any sequence. Separate symbols for each power of ten made any system of place notation redundant. Generally, the hieroglyphs were placed left to right in ascending order of magnitude, as above—in other words, opposite to the way we write numbers today. Addition was the process of adding up the various symbols, then replacing those symbols of which there were more than ten with the next largest symbol. For example, 547 + 624 = 1,171 would be written as:

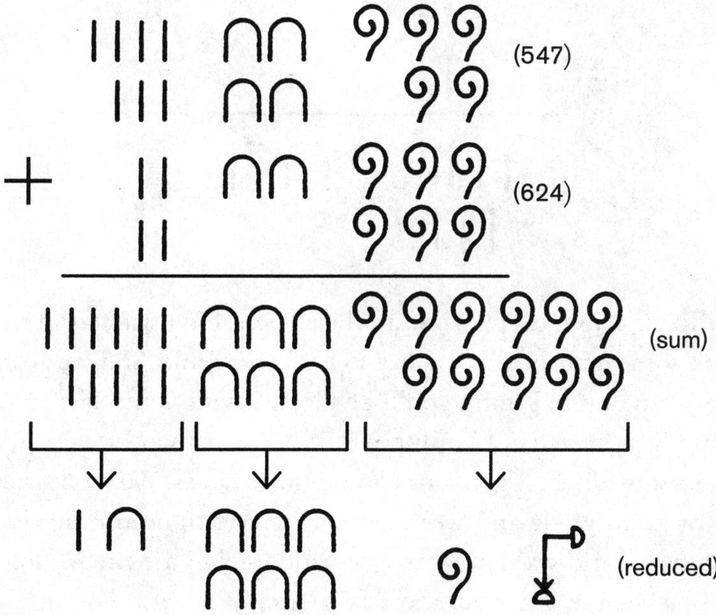

Which in turn would be reorganized to:

Subtraction is the reverse process, but in this case, a large hieroglyph must often be replaced with ten smaller ones wherever necessary. Take 32 − 5:

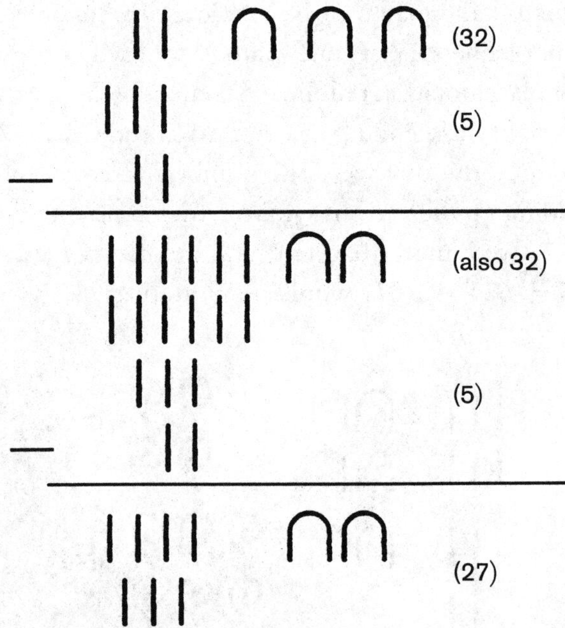

To explain multiplication in the ancient Egyptian world requires us to do some actual math. It's not as hard as it looks, and we won't spend much time on it. I think you'll be pleased with yourself if you give it a try, perhaps even feel a kinship with the ancient world.

First of all, the Egyptians had no times tables. We memorize the tables in third grade and we're set for life. Multiplication in the ancient Egyptian world was similar to a method used in the late Middle Ages in Europe, where multiplication was simply a series of doublings, as demonstrated at the beginning of the chapter. Let's take the same problem, 13 × 46, which the Egyptian mathematician would break down into a series of integral powers of 2, or doublings.

46 times 1 equals 46	**(1×)**	➤
46 doubled equals 92	(2×)	
92 doubled equals **184**	**(4×)**	➤
184 doubled equals **368**	**(8×)**	➤

Again, he finds the combination of doublings that add up to 13, the ones checked above, 1×, 4×, and 8×. Then he adds up the three resultant sums to solve the problem: 46 + 184 + 368 = 598. Of course, it's a bit more difficult here as he's working with hieroglyphics rather than our Indian numerals. The Egyptian would double like so (note: the 0 power of any number is 1, and the 1st power of any number is the number itself, so $2^0 = 1$ and $2^1 = 2$):

46 × 1 (2^0) would be written:

46 × 2 (2^1) would be:

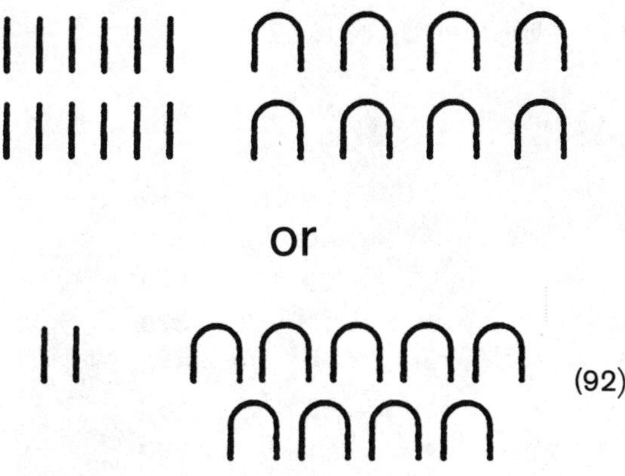

46 × 4 (2^2) would be:

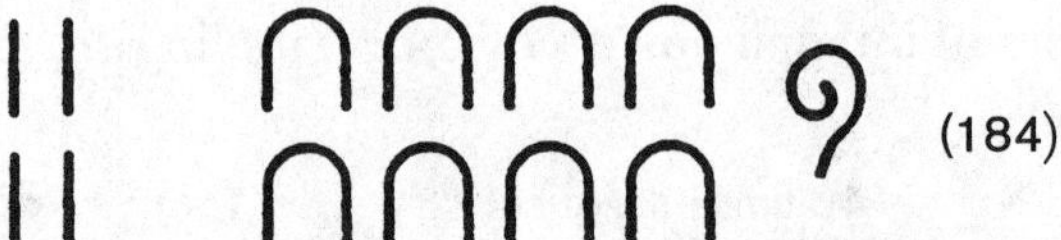

And so on. You get the idea. It was *relatively* easy to do the doublings. Each step simply required writing two sets of the hieroglyphs from the step above. Admittedly, this is a cumbersome process as the numbers get larger and one has to constantly reduce the hieroglyphic terms. The Egyptian method of multiplication works because of a basic principle of mathematics: every integer can be expressed as the sum of selected integral powers of 2. No matter what number the multiplier, one can assemble it by picking and choosing from a list of powers of 2.

Trust me on this. Or try it, but set aside plenty of time. The above rule is well known today. The question is, Were the ancient Egyptians aware of the rule? It's at the heart of their math, but Western mathematicians are skeptical that non-Europeans living five thousand years ago came to this conclusion. The integral-powers rule also lies at the heart of the multiplication method below, a relatively modern variation of the Egyptian method. Let's do an easy problem, 180 × 20 (which obviously equals 3,600), to show how it works. One puts the 180 in the left column, the 20 in the right. Then one successively doubles the right column, while halving the left. (When halving a figure into a noninteger—say 11 into 5.5—one rounds off to the lower number, in this case 5.) Like so:

	HALVE		DOUBLE
	180	×	20
	90		40
➤	45		**80**
	22		160
➤	11		**320**
➤	5		**640**
	2		1,280
➤	1		**2,560**

Today, we can do the above problem in our heads: 180 × 20 = 3,600. Ancient Egyptians and medieval Europeans couldn't, so the above represents a shortcut from the classic Egyptian method. After doubling the right column above, and halving the left, go down the left column and choose all the odd numbers, then add together the corresponding sums on the right. Thus 180 × 20 = 80 + 320 + 640 + 2,560 = 3,600. Why does this work? The odd numbers on the left correspond to those powers of 2 that the multiplicand (180) comprises. Note:

2^0		180 ×	20	
2^1		90	40	
2^2	➤	45	80	(2^2 = **4**)
2^3		22	160	
2^4	➤	11	320	(2^4= **16**
2^5	➤	5	640	(2^5 = **32**)
2^6		2	1,280	
2^7	➤	1	2,560	(2^7 = **128**)

The powers of 2 equal 180: 4 + 16 + 32 + 128 = 180. I've taken an easy problem here, 180 × 20, so that you could do it, using modern methods, in your head. The technique works for any multiplication problem. Try some.

George Gheverghese Joseph says the above modern variation of the original Egyptian method is still popular among rural communities in Russia, Ethiopia, and the Near East, where multiplication tables have yet to catch on. It is sometimes referred to as the "Russian peasant method." When I first came across the technique, it hit me the way the the madeleine hit Proust.

My father, a high school dropout, used to teach me odd methods of arithmetic, including the Russian peasant method, though he never called it that. A first-generation American, he had picked up a variety of exotic calculating techniques from his father, who had been a farmer in Sicily. My father was a fruit peddler, delivering produce to small stores and restaurants. He refused to use an adding machine. His bills contained strange numerical scribbles, not unlike the calculations above. His accountant feared the day that an IRS auditor would glower over such nonsense, despite the fact that the totals were always correct.

How did a fruit peddler working in a small Minnesota town in the twentieth century come to be using mathematical techniques pioneered by ancient Egyptians? A possibility—and I just throw this out—concerns the Greek mathematician Pythagoras, who was educated in numbers in Egypt and emigrated in the sixth century B.C. to Italy, where he founded the Pythagorean school of math. Could he have spread ancient Egyptian techniques throughout Italy and Sicily, where it was passed down by peasants for millennia? It doesn't matter, really. What's interesting is that here in the twenty-first century, people around the world still count like an Egyptian.

WE CAN'T BEGIN to enumerate all the mathematical accomplishments of the ancient Egyptians. We'll examine just one more for the time being. The Ahmes Papyrus, a leather manuscript discovered in 1927, revealed that the Egyptians were the first culture to master fractions.[39] In 1927, Egyptologists, waiting with great expectations for the first translations of the papyrus, were disappointed when they learned that the manuscript contained only twenty-six rudimentary mathematical identities, such as $\frac{1}{10} + \frac{1}{40} = \frac{1}{8}$. The first translator quipped that if the Ahmes had any value it would be in providing insight into leather-making techniques of the era. In the West in the twentieth century, fractions are taken for granted. The Ahmes Papyrus, however, reveals that the Egyptians were the only ancient culture to operate with unit fractions. The Egyptians did not use money. They bartered, and fractions helped them exchange goods, divide food and land, and calculate the percentages of foods in recipes.

The Egyptians loved tables (the lack of a times table remains curious) and kept copious numbers of them to speed up calculations. The papyrus presents several problems, one of which challenges the reader to divide 9 loaves of bread among 10 men. (Beer and bread were common standards of exchange.) Our modern approach is straightforward—each man gets $\frac{9}{10}$ of a loaf—but gives us no satisfactory method for actually dividing the physical loaves. Today we would cut a tenth off each loaf. Nine men would get a $\frac{9}{10}$th loaf. The tenth man would get nine heels. Mathematically fair, but hardly equitable unless one is partial to crust.

The ancient Egyptian would instead go to a unit fraction table and find that $\frac{9}{10} = \frac{2}{3} + \frac{1}{5} + \frac{1}{30}$. He would then cut the nine loaves into various segments representing thirds, fifths, and thirtieths.

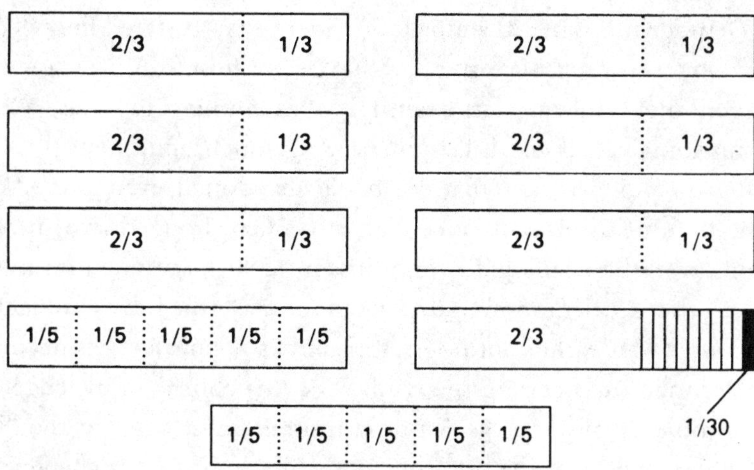

Bread cutting, Egyptian style: the Ahmes Papyrus explains how to divide 9 loaves among 10 men so that nobody gets stuck with all heels. (After Joseph)

Seven men would each receive three pieces of bread: a $\frac{2}{3}$, $\frac{1}{5}$, and $\frac{1}{30}$ segment of a loaf. The other three men would get four pieces: two $\frac{1}{5}$s, a $\frac{1}{5}$, and a $\frac{1}{30}$. The Egyptian method requires a lot of cutting but divides the bread according to form as well as substance.

Let us stop breaking bread for the moment and press on to examine the influence of the Egyptians, the first known culture to promulgate a full-blown written mathematics.

THE MAIN COMPLAINT against the Egyptians is that their contributions were trivial. What did the Egyptians do wrong? According to Morris Kline, the math popularizer Lancelot Hogben, and others, the Egyptians and Babylonians depended on empirical evidence, on experimenting with numbers and forms. The Greeks found their answers through logic.[40]

Where does one begin to dissect this argument? Kline's attack on empiricism—that is, relying on evidence rather than logic—is an ancient Greek point of view. Around 400 B.C. the great mathematician Democritus of Abdera set forth the proposition that the mind is superior to the senses. Through logic we tap into "trueborn" knowledge, said Democritus, whereas empirical evidence is "bastard" knowledge, colored by the unreliable senses.[41] What tastes sweet to A may taste sour to B. A homely child appears beautiful to his mother. How can we trust information gleaned via taste, sight, hearing, touch, and smell?

It was a point of view that was being abandoned, even in the West, by the time of the Renaissance. When Galileo dropped two unequal weights from the Leaning Tower of Pisa in 1589, he was demonstrating not only that acceleration is independent of mass (the heavier object hit the ground a bit earlier but not significantly so), but that it is necessary to experiment to ascertain the truth. The "trueborn" knowledge that heavier objects fall faster, as Aristotle insisted, must defer to the "bastard" knowledge that they do not. We shall see in a later chapter that even Democritus would succumb to bastard knowledge, his greatest achievement the result of smell, not logic.

The Greek historian Herodotus refers to geometry as the "gift of the Nile." Because the annual overflow of the Nile River wiped out the boundaries of farmers' lands, the Egyptians developed geometry to redetermine plot lines. (It is a colorful reference, but only partially valid; the Egyptians were practicing geometry long before the 1400 B.C. date cited by Herodotus.) The logic of Herodotus and Kline is somewhat flawed. Because there is a use for an invention doesn't necessarily imply a causal relationship.

Kline, denigrating the Egyptians for their pragmatism, credits them for applying math to astronomy, calendar reckoning, and navigation. Motions of heavenly bodies, he says, give us our fundamental standard of time, and their positions at given times enable ships to determine their locations and caravans to find their bearings in deserts. The Egyptians needed to predict the flooding of the Nile so that farmers could move their belongings and cattle. The Egyptian calendar was eventually adapted by the Romans, and passed on to Europe (our present calendar is essentially the Julian calendar, commissioned by Julius Caesar).

Mesopotamia

Math in Egypt and Mesopotamia spans roughly the same time period. The various Mesopotamian civilizations stretched from 3500 B.C., when the Sumerians established the first city-states, to 539 B.C., when the area was conquered by the Persians.

A string of different peoples populated the land between the Tigris and the Euphrates. The Sumerians were first, building Ur, perhaps the best-known city of antiquity, on the banks of the Euphrates. Biblical references to Sumer abound. The *Epic of Gilgamesh* was written here, and ziggurats were erected. The Sumerians gave way to the Akkadians, from the surrounding desert, who in turn were squelched by the First Babylonian Empire in around 1900 B.C., which was then overrun by the Assyrians in 885 B.C., who were conquered by the Chaldeans, thus initiating the Second Babylonian Empire in 612 B.C., which gave way to the Persian invasion in 539 B.C. Interspersed were Hittites and Hurrians and other interlopers. For convenience, when speaking of mathematics, the period is known generically as the Babylonian era. When it can be pinpointed to the earliest period, we use the term Sumerian.

Fortunately, our records of Babylonia are indelible. The Babylonians wrote on tablets formed from clay from the banks of the Tigris or the Euphrates. Scribes made wedgelike impressions with a reed. These tablets, dried in the sun or baked in kilns, are still readable, unlike many of the Egyptian papyri.[42] There are plenty of errors. The scribes had to write fast, before the clay dried. Half a million tablets have been found, but fewer than five hundred contain math. Though the cuneiform script of the Sumerians was decoded about 150 years ago, the math tablets have been studied only since the 1930s.

The tablets that have been deciphered tell an interesting tale. In one sense, the Sumerian/Babylonians developed a more sophisticated mathematics than the Egyptians. On the other hand, it was an imperfect sophistication, leading to ambiguity, as opposed to the clarity of the Egyptian system.

The triumph of the Babylonians is their place-value notation system, in which the position of a numeral determines its value (our number 111, for example, stands for 1 hundred, 1 ten, 1 one). The Babylonian notation system was not unlike our own but differed in a number of ways. First, the Sumerians had two base systems operating within each other. They counted by both 10s and 60s. This may seem odd until one recalls that we have a sexigesimal system at play also: our 60-minute hour, our 60-second minute, the 360 (6 × 60) degrees of the compass. Copernicus also made use of sexagesimal fractions in the sixteenth century.

The Sumerian system was incomplete. They used positional notation in base 60 only and did not have a positional system for all powers of 60. Here's what they did have:

1 10 60 600 3600

Around 2000 B.C., the Babylonians devised a simpler system, using a place-value system with only two symbols, a pin shape for 1 and a wing-like figure for 10. Here is how they'd write three numbers under 60:

4: 28: 59:

From 2500 B.C. on, the Babylonians figured out that they could give multiple values to their two symbols depending on their relative positions. This put them ahead of the Egyptians, who, each time they wanted to go up a power of ten, had to invent a new hieroglyph. Like us, the Babylonians wrote left to right, and wrote 95 thus:

$$95 = 60(1) + 35:$$

The first "pin" is worth 60, the 3 "wings" are worth 30, and the final 5 "pins" are worth 5, for a total of 95. They made fractions similarly, with the denominator off to the right. How could you tell fractions from larger numbers? You couldn't with any certainty. The Babylonians had no symbol for zero at this time and no decimal point to distinguish between the integer and fractional parts of a number. Joseph points out that in this system, for example, 160, 7,240, $2\frac{2}{3}$, and $\frac{2}{45}$ are all written the same way.

In the Babylonian system, a symbol written slightly larger would have a different value than its smaller brethren.[43] The Babylonians didn't have zero to designate "empty columns" in their numbers (as we differentiate 202 from 22 with the zero in the middle). So they left extra space to account for the "empty columns." These subtleties may evade the modern eye. Slight unintentional variations in scribing numbers with a reed, or later with a three-sided stylus,[44] or misinterpretations of size or spacing by the reader could lead to mistakes. At least that's our present perspective, and that perspective must be somewhat accurate because the Babylonians kept improving their system, eventually removing the ambiguity. Somewhere between 700 and 300 B.C.[45] they started using a placeholder consisting of two little triangles or wedges inserted in the empty columns:[46]

These triangles (often written in other forms) meant "nothing in this column." It was a limited kind of zero. So one could now write the number 7,240 like so:

Without the zero points, the number would be 160; that is, 2 pin shapes (2 × 60) for 120 plus 4 "wings" (4 × 10) for 40 more. But the wedges fill the 60s column, so the pin shapes get promoted a power of ten, from 60 to 60^2 (or 3,600). One has two pin shapes for 7,200 (3,600 × 2) plus 4 wings for 40 more. Result: 7,240.

The Babylonians never turned their wedges into a real zero, a real number, as they used them only in the middle of numbers, never at the end. They had a flexible but ambiguous system.

The pioneer scholar of Babylonian math in the 1930s, Otto Neugebauer, added commas, zeros, and semicolons to the Mesopotamian number system in order to study Babylonian mathematical operations. Modernized (actually, "Indianized," as our modern system comes from ancient India), Babylonian operations differ little from addition, subtraction, multiplication, and division done today.[47]

Of course, the Egyptians could do all this, and with greater clarity. What one could do with the Babylonian number system, though, is algebra. Even during the First Babylonian period, mathematicians were solving equations. The Babylonians had a kind of "x," which they called *sidi,* for side (as in the side of a square to be found), and they used *mehr* ("square") for x^2. They were able to do linear and quadratic equations. A typical problem from the First Babylonian period asks: "Multiply two-thirds of [your share of barley] by two-thirds [of mine] plus a hundred qa of barley to get my total share. What is [my] share?" The technique used to solve the problem is identical to the one we now use.[48]

Babylonians kept multiplication tables as well as tables for reciprocals, squares, cubes, and square and cube roots.[49] They even kept tables for the values of $n^3 + n^2$ for the integers from 1 to 20 and for 30, 40, and 50. These values help solve quickly a type of equation called a mixed cubic equation. Such equations can be used, for example, to calculate how long it would take for an amount of money to double, given various interest rates.[50]

Unlike the Greeks, who abhorred them, the Babylonians dealt routinely and comfortably with irrational numbers. Formally, rational numbers are those that can be expressed as the ratio of whole numbers; irrational numbers are those that cannot. Rational numbers are everyday numbers like 3 (which can be expressed as 3⁄1), .25 (25⁄100), and 1⁄3. Perhaps the best way to describe irrational numbers is to say that when expressed as decimals, they have no repeating pattern of digits past the decimal point. The rational 1⁄3, for example, comes out 1.33333333 . . . ; it goes on forever, but there's a pattern. The first discovered irrational number, the square root of 2, on the other hand, is 1.41421356237309 . . . , with no repeating pattern.[51] Another famous irrational is pi, or 3.14159265358979 . . . .[52] One Old Babylonian tablet contains the number 1.41421297, the square root of 2 accurate to five places.[53] (Not all mathematicians agree that the Babylonians truly understood irrationals. For an opposing view, see note.)[54]

The Babylonians were poor geometers, certainly compared to the Egyptians and, later, the Greeks. To calculate the area of a circle, for example, they at first used 3 as the value of pi. Not very close. Later they improved it to 3.125. What they did have was the "Pythagorean theorem"—about 1200 to 1500 years before Pythagoras was born.[55]

In the Plimpton collection at Columbia University is a Babylonian tablet dated to 1700 B.C. The tablet is filled with columns of numbers and marred by a deep chip on the right side. It appears there was more to the tablet, that it was broken when excavated. Still, there's plenty left to think about: fifteen sets of three numbers each. Some scholars insist that the numbers are Pythagorean triplets. That is, the numbers that describe the lengths of the sides and hypotenuses of various right triangles, conforming to the Pythagorean theorem that the sum of the squares of the two perpendicular sides must equal the square of the hypotenuse.

We know some of these Pythagorean triplets from high school geometry: 3 : 4 : 5; 5 : 12 : 13, 8 : 15 : 17, and so on. Critics of Babylonian math insist that the existence of fifteen Pythagorean triplets on a tablet does not indicate that the inscribers had the concept of $a^2 + b^2 = c^2$ vis-à-vis right triangles in 1700 B.C., 1200 years before the birth of Pythagoras.

Yet it's a bit of a coincidence: fifteen Pythagorean triplets unbroken by other numbers. Some point out that the triplets are not organized the way we now would organize them, numerically, beginning with the lowest numbers. Does this necessarily discredit the Babylonians' understanding of the theorem?

Other mathematicians see a more subtle pattern. The triplet numbers themselves are contained in columns 2, 3, and 5. Column 4 contains the number of the row (1 through 15). The numbers in column 1 remain mysterious. One explanation, put forward by George Joseph, is that the numbers in column 1 have something to do with the derivation of Pythagorean triplets used in the construction of right-angled triangles with rational sides. He feels that it is unlikely that the triplets were reached via trial and error because they are far too complicated (row 15, for example, is 56 : 90 : 106). Joseph has demonstrated how the numbers in column 1 can generate triplets, and he finds it similar to the method used by the Greek mathematician Diophantus in A.D. 250. He notes that Diophantus is known for introducing Babylonian algebraic techniques into Greek mathematics.[56]

There is a more intriguing possibility. Harvard's Mazur points out that the numbers in that first column may be related to trigonometric functions. The numbers in the first column are close to the squares of the secants of the triangles in question. The secant is a trigonometric function; it is the ratio of the length of the hypotenuse to the length of one of the other sides of a right triangle, and obviously is dependent on the angles involved. Thus, it appears that the fifteen triplets are arranged by the angles of incidence. The first number in the first column is close to the square of the secant of a 45-degree angle. The last number in the column is close to the square of the secant of a 31-degree angle. The thirteen numbers in between correspond to angles from 44 degrees to

32 degrees. That is, the angles descend from 45 to 31 one degree at a time.[57]

Perhaps this is coincidence. If it is a kind of "trig table," says Mazur, "then they [the Babylonians] had to have a lot of theoretical framework."

LET'S TAKE a brief look at what we have all been taught is the pinnacle of achievement in the ancient world: Greek mathematics. As we shall see, Democritus, Pythagoras, and others viewed the Egyptians as their mathematical masters. One of the problems of Greek mathematics arises from the numbers themselves. The Greeks, like the Romans after them, used cumbersome numerals such as this notation, from the fifth century B.C., for 318:

$$\overline{\tau\iota\eta}$$

Here τ, the twenty-first letter of the ancient Greek alphabet, stood for 300; the tenth letter, ι, stood for 10; and the eighth letter, η, stood for 8. A line was drawn over the whole affair to distinguish $\overline{\tau\iota\eta}$, the number 318, from the Greek word τιη, or "why."[58] Number theorist Tobias Dantzig wrote that neither the Greeks nor the Romans were capable of "creating an arithmetic which could be used by a man of average intelligence." The rules for operating within Greek and Roman systems were so complex that any man skilled in the art was regarded as "endowed with almost supernatural powers." In a sense, Greek mathematics was a mystical endeavor. The Greeks, writes Dantzig, "never completely freed themselves from this mysticism of number and form."[59]

Dantzig confirms the Greeks' contempt for applied science. Some Greeks may also have regarded math as not terribly important, leaving their slaves to teach their children the discipline.[60] Despite the accomplishments in geometry by Euclid and others, the Greeks never developed even a rudimentary algebra.[61] The Egyptians had already made breakthroughs in this area.

Mathematical geniuses reigned throughout the Greek civilization. Yet, as Dantzig points out, Greek mathematics "stopped short of an algebra in spite of Diophantus, stopped short of an analytic geometry in spite of an Apollonius, stopped short of an infinitesimal analysis in spite of an Archimedes."[62] Greek mathematics came up short in many respects, in part, says Dantzig, because of the lack of a notational symbolism. Unlike the Sumerians and, later, the Babylonians, the Greeks had no positional notation system (ones, tens, hundreds, and so on) to simplify their numbering. Neither did the Romans.

Mathematician D. H. Fowler points out that while the ancient Greeks had numbers and a labeling system for the positive number line, their math was "completely non-arithmetised."[63] That is, they really couldn't do significant operations, as they had no mathematical language, no conceptual machinery, for the mathematician to work with. In other words, the mathematician had to use natural language to confront his problem. Today we state the Pythagorean theorem as: *In a right triangle the square of the hypotenuse is equal to the sum of the squares of the other two sides.* Note, by contrast, Euclid (*Elements* 147): "In right angled triangles the square on the side subtending the right angle is equal to the squares on the sides containing the right angle." To Euclid the Pythagorean theorem of right triangles meant literally that the square can be cut into two and manipulated into two other squares. When Euclid used the expression "square of the hypotenuse," he meant a literal square. The big square made from the hypotenuse is equal to the two smaller squares made from the sides.

Today, the above theorem is usually interpreted as:

$$p^2 + q^2 = r^2 \quad \text{or} \quad a^2 + b^2 = c^2$$

We must explain what the p's, q's, and r's represent (the lengths of sides of a right triangle) and how they can be multiplied and added. This seems natural to us today. We translate triangles into numbers and then manipulate those numbers in a variety of ways. We've turned those triangles into useful abstractions. All the evidence points to the conclusion that Greek mathematicians did not contemplate such thoughts. Rather, instead of imagining p squared (p^2), they actually visualized squares.

Proof of this to Fowler is the inability to find fractions in Greek mathematics. "Note that fractions," he wrote, "being 'numerical quantities,' belong to what I have called the arithmetised style of mathematics, and the non-arithmetised approach may often be signalled by the use of the alternative terminology of ratios." The Greeks of the classical period spoke of ratios, not fractions. They couldn't manipulate sides of a right triangle with, say, fractions like $p/q \times r/s = pr/qs$, as we do today, or as the Egyptians could do. It seems odd, yet true, to say that the Greeks may have been skilled mathematicians but, according to Fowler, not much good at arithmetic.

Take exponents, which we manipulate with ease today. The Greeks understood 2^2 and 2^3, 2 squared and 2 cubed. They saw 2 squared as an actual square, a line 2 units long, squared. Two cubed was an actual cube, which could be formed in three dimensions, measuring 2 units in each direction. Such numbers, however, as 2^4, 2^5, or 2^{12} had no meaning to the Greeks, as there are no counterparts in a three-dimensional world for exponents of 4 and higher. Multiplying $2 \times 2 \times 2 \times 2$ to get 16, or 2^4, seems commonplace today, but it requires taking a concept, 2^x, that originally represented a physical shape, and abstracting it so that it can be manipulated arithmetically.

In Plato's school (he died in 347 B.C.), teachers of geometry repudiated its roots as a practical discipline for solving real-world problems, made it an end in itself, and in effect banished measurement from mathematics.[64] The obsession with purity kept the Greeks from embracing irrational numbers. They discovered that the square root of 2, for example, is neither a whole number nor a fraction. The square root of 2 is somewhere between 1.41 and 1.42 but cannot be pinpointed, no matter how many decimal places we employ. The Greeks were appalled that there is a large collection of such numbers—the square root of any number not a perfect square, the cube root of any number that is not a perfect cube, and so on. Pi, the ratio of the circumference of a circle to its diameter, is also an irrational number. The term "irrational" in mathematics today is applied to numbers that cannot be expressed as a ratio of whole numbers. In Pythagoras's times the term meant "unmentionable and unknowable." Legend has it that the Pythagoreans threw overboard the person who discovered irrational numbers to keep the loathsome discovery secret. (In fact, points out mathematician Robert Kaplan, these ancient Greeks may not have been that far off. Irrationals are apparently "unknowable" in the sense that we can't "know" them to the degree we know rational numbers.)[65]

The Greeks never developed an arithmetic of irrational numbers, nor did other Western mathematicians until the Renaissance.[66] The Egyptians and Babylonians, familiar with irrational numbers long before the Greeks, remained unruffled by them, employing approximations, such as 1.4 for the square root of 2, or 3 for pi. This allowed them to do calculations never attempted by the Greeks.[67]

The Greeks were important mathematicians. Diophantus (third century A.D.), an early number theorist, finally developed a form of algebra for the Greeks. Long before René Descartes, he was using letters in equations. Euclid's *Elements,* a work on plane and solid geometry written about 300 B.C., is a "best of" book containing the best results produced from 600 to 300 B.C. by dozens of great mathematicians—the Pythagoreans, Hippias, Hippocrates, Eudoxus, members of Plato's academy, and others—and its concepts are still taught in junior high and high schools today.

If the Greeks borrowed heavily from the Egyptians and others, then an innumerate Greek civilization would not speak well for its forebears. What can be said is that the Greek penchant for exactness in thinking was satisfied, perhaps too much so, by geometry rather than algebra. Kline praises the Greeks for converting the scattered geometric facts of the Egyptians and Babylonians into a "vast, systematic, and thoroughly deductive structure."[68] They built a system of theorems and deductive reasoning. In geometry, one can draw pictures to represent what one is thinking about. The Greeks saw matter as formless; to mold matter into the shape of a triangle gave it significance.

Kline admits that the Greeks' penchant for using geometrical methods for performing what are naturally algebraic processes was a step backward, that Greek geometry was so complete, so admirable, that later mathematicians continued to see it as the pinnacle of math, and this delayed the development of algebra. Dantzig takes a slightly different—perhaps even contrasting—view, saying that Greek thought was too concrete to be algebraic. Algebra deals with symbols, objects stripped of their physical content. Dantzig contends that the Greeks were too intensely interested in the objects themselves; hence their fixation on geometry. The Greeks didn't get on the algebraic track until Diophantus, around 250 A.D., who was the first Greek to recognize fractions as numbers, and the first to handle equations in a systematic way.[69]

In the West, geometry remained the mathematics of choice. For utilitarian reasons, our textbooks are often written as if Europeans used algebra long before they did. A prime example is that of Galileo's work in the early seventeenth century with falling and moving objects. He showed how objects accelerate by rolling a ball down an inclined plane fitted with lute strings. The ball made a click as it passed over each string, and Galileo, a musician with excellent timing, kept rearranging the strings until the clicks came at equal intervals. He sang a marching tune to help him get the timing right. Then, when all the clicks were equally spaced in time, he measured the distances between the strings, and found that acceleration increased geometrically. That is, if the first interval is 1 inch, the second will be 4 inches, then 9, then 16. In modern notation, this progression would be stated as squares: 1^2, 2^2, 3^2, 4^2, and so on.

Thus Galileo is often credited with the following equation:

$$s = At^2,$$

where s is the distance a falling body falls, t^2 is the time taken to cover the distance, squared, and A is a number that changes depending on the angle of inclination of the plane.[70] Galileo never wrote $s = At^2$. He was considered a great mathematician, but he didn't speak algebra. What he wrote was: "If a movable descends from rest in uniformly accelerated motion, the spaces run through in any times whatever are to each other as the duplicate ratio of their times; that is, they are as the squares of those times."[71]

Peter Machamer, a professor of the history of science at the University of Pittsburgh, writes, "For Galileo, mathematics meant geometry." To translate Galileo's proofs and theorems into algebra, says Machamer, destroys "the mind set, the schema" with which seventeenth-century mathematicians were working. Galileo stated in *Il Saggiatore,* "[The universe] is written in the language of mathematics, and its characters are triangles, circles, and other geometrical figures."

Galileo rarely tried to find out the real speed or weight of anything. He was concerned with ratios, with measuring one thing by showing its relationship to another.[72] Also rarely did he concern himself with absolute numbers, as the Egyptians had. He was a product of the ancient Greeks, many centuries removed.

Science would move toward the concrete, toward experiment, in the eighteenth century and onward. A modern scientist, measuring lengths in angstrom units and time in femtoseconds, might find himself more comfortable in third-millennium B.C. Egypt than in third-century B.C. Greece or even seventeenth-century A.D. Italy. The Greeks may have taken math down the wrong road with their rarified approach to geometry. There is little disputing, however (though many have tried), that the mathematics they did construct they did so with verve, beauty, and intellectual rigor.

Who were their mathematical benefactors? The logical answer has always been the Egyptians, around the bend of the Mediterranean Sea. Three of the earliest Greek mathematicians reportedly studied in Egypt and beyond. Thales of Miletus (c. 600 B.C.) was the first to suggest, according to Aristotle, that all the widely varied matter in the universe had a common constituent. He said the common denominator was water, and he was wrong, but the basic concept of reductionism still rules physics today. Thales, according to most Western scholars, developed a taste for, and his knowledge of, geometry from traveling in Egypt.[73]

Pythagoras, who founded his school in the Greek-speaking section of southern Italy in the sixth century B.C., is a more mysterious figure. He left no writings and swore all of his students to secrecy. Yet historians have concluded that, like Thales, Pythagoras traveled to Egypt, and also to Iran and the East—possibly to India. One Greek philosopher, Iamblichus, wrote in the fourth century A.D. that Pythagoras spent twenty-two years in Egypt.[74] In any event, he returned to Greece with newfound knowledge of mathematics.

Democritus (c. 430–370 B.C.) was also a prodigious traveler. According to Western historians, Democritus studied with the Chaldean Magi in Babylon, the priests of Egypt, and the Naked Sages of India, among others.[75] That the early Greek mathematicians imported much of their discipline from Egypt has rarely been disputed.

India

The earliest recorded Indian mathematics was found along the banks of the Indus. In the early 1920s, archaeologists uncovered an urban center at Harappa, in northern India, dating back to 3000 B.C. The precise mathematical expertise of the Harappan culture, which lasted from 3000 to 1500 B.C., is difficult to pinpoint, as Harappan script has never been deciphered. There are physical clues, though.

The Harappan civilization was primarily agrarian—the people grew wheat and barley and raised livestock—and measurement appears to have been important. Archaeologists have uncovered several scales, instruments, and other measuring devices. The Harappans employed a variety of plumb bobs that reveal a system of weights based on a decimal scale. For example, a basic Harappan plumb bob weighs 27.584 grams. If we assign that a value of 1, other weights scale in at .05, .1, .2, .5, 2, 5, 10, 20, 50, 100, 200, and 500. These weights have been found in sites that span a five-hundred-year period, with little change in size.

Archaeologists also found a "ruler" made of shell lines drawn 6.7 millimeters apart with a high degree of accuracy. Two of the lines are distinguished by circles and are separated by 33.5 millimeters, or 1.32 inches. This distance is the so-called Indus inch. Some speculation: a

Sumerian *shushi* equals half an Indus inch, exactly, and adds to the long-held suspicion that there was a link between the Sumerian and Indian cultures.[76]

The ruins of Harappa contributed to a hundred-mile railway line between Multan and Lahore in the nineteenth century. The British dug up Harappa for its bricks, which they used as ballast on the railbed. The bricks speak a story greater than ballast. The Harappans learned to exploit the annual flooding of their farmland to grow crops without need for plowing, fertilization, or irrigation. To do so they had to control the flooding with flood walls, so they developed kiln-fired bricks, less permeable to rain and floodwater than mud bricks. Harappan bricks contain no straw or binding material and are still in usable shape after five thousand years. Most interesting are their dimensions: while found in fifteen different sizes, their length, width, and thickness are always in the ratio of 4 : 2 : 1.[77]

Bricks and religion are at the root of the Vedic period of Indian mathematics. Vedic literature, one of the largest and oldest literary collections—from 1000 to 500 B.C.—encompasses works of hymns and prayers, songs, magic formulas and spells, and, most important to us here, sacrificial formulas. One collection of Vedic literature, called the *Brahmanas,* spells out the rules for conducting sacrifices. Another collection, known as the *Sulbasutras,* meaning "the rules of the cord," dictates the shapes and areas of altars (*vedi*) and the location of the sacred fires. Square and circular altars were okay for simple household rituals, but rectangles, triangles, and trapezoids were required for public occasions.

These altars sometimes took extravagant forms, such as the falcon altar, made from four different shapes of bricks: (a) parallelograms, (b) trapeziums, (c) rectangles, and (d) triangles.

Making a sacrifice on such an altar allowed the supplicant a swift ride to heaven on the back of a falcon. The *sulbasutras* were written between 800 and 500 B.C., making them at least as old as the earliest Greek mathematics.[78] According to George Joseph, researchers in the nineteenth century made a point of emphasizing the religious nature of the *sulbasutras*—and certainly they are religious—but ignored their mathe-

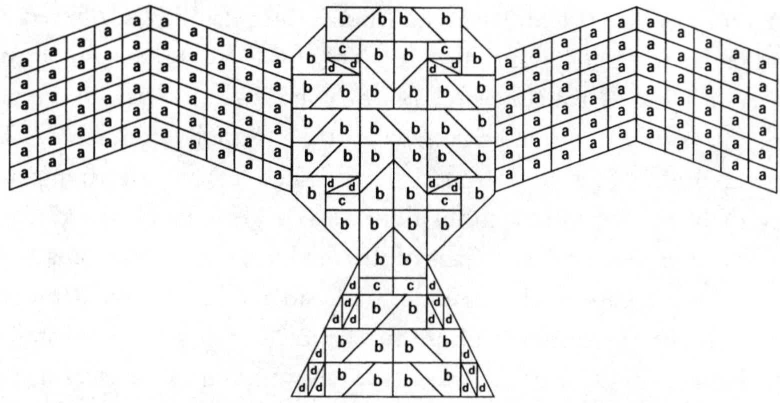

This first layer of a Vedic sacrificial altar in the shape of a falcon combines the ritual and religious features of altar construction in the *Sulbasutras* with brick technology and geometry: 196 bricks in four different shapes (a,b,c,d) are used. (After Joseph and Thibaut.)

matical content. Joseph sees in the *sulbasutras* a link between the Harappan culture and the highly literate Vedic culture, by means of the Harappan brick technology, which was put to geometrical and religious uses in Vedic sacrifices. To ignore the mathematical component of Vedic rituals is akin to characterizing the Gregorian calendar as a religious exercise rather than a mathematical and astronomical accomplishment.

The earliest *sulbasutras* were composed by the priest-craftsman Baudhayana somewhere between 800 and 600 B.C. and include a general statement of the Pythagorean theorem and a procedure for obtaining the square root of 2 to five decimal places.[79] Baudhayana's motivations were religious and practical; he needed a mathematics that would help scale altars to the proper size depending on the sacrifice. His version of the Pythagorean theorem is: "The rope that is stretched across the diagonal of a square produces an area double the size of the original square." Another *sulbasutra* states: "The rope (stretched along the length) of the diagonal of a rectangle makes an (area) that the vertical and horizontal sides make together."[80]

The *sulbasutras* contain instructions for the building of a *smasana,* a cemetery altar on which soma, an intoxicating drink, was offered as a sacrifice to the gods. (You may recall that Aldous Huxley borrowed "soma" for his dystopian novel *Brave New World*. It was the narcotic drink given to the proletariat to keep them happily distracted.) The *smasana*'s base was a complicated shape called an isosceles trapezium, which comprised, among other figures, six right triangles of different sizes. It's obvious that the Indians of this era knew the Pythagorean rule.

The most basic right triangle, with sides of 3, 4, and 5 units in length, might be stumbled upon by chance. Using a rope marked off with knots at 3, 4, and 5 units would allow builders to ascertain the squareness of corners, and the Egyptians, for example, did just that. Mathematicians have pointed out to me that ancient nonwhite people might by accident come up with a triangle with sides of 3, 4, and 5 and note that it always formed a right angle.

However, the instructions given for a *smasana* in the *Sulbasutras* dictate that six right triangles be used in the construction, consisting of sides of 5 : 12 : 13, 8 : 15 : 17, 12 : 16 : 20 (a multiple of 3 : 4 : 5), 12 : 35 : 37, 15 : 20 : 25 (another multiple of 3 : 4 : 5), and 15 : 36 : 39.[81] That's a lot of luck. In addition, the *Sulbasutras* employed right triangles with sides of fractional and even irrational lengths.

The Vedic sacrificers figured out a method of evaluating square roots. Joseph suspects the technique evolved from a need to double the size of a square altar. Say you wish to double the area of an altar with sides 1 unit long. Obviously, doubling the lengths of the sides would result in an altar four times the size. It becomes clear that one needs a square whose sides are the square root of 2, and thus one needs a technique for calculating square roots. The *Sulbasutra* square root of 2 is 1.414215 . . . ; the modern value is 1.414213 . . . . No one is certain how the Indians arrived at their method, but it probably involved positing two equal squares with 1-unit sides, then cutting the second square into various strips and adding those strips to the first square to make a square with twice the area, then converting the strips to fractions to construct a numerical formula.[82] This may have been the first recorded method of evaluating square roots.

Early Indian geometry is filled with fantastic and phantasmagorical dynamic constructions, such as the *sriyantra,* or "great object," which belongs to the tantric tradition. In it nine basic isosceles triangles form forty-three others, encircled by an eight-petaled lotus, a sixteen-petaled lotus, and three circles, which in turn are surrounded by a square with four doors. The meditator concentrates on the dot, called a *bindu,* in the center, and moves outward, mentally embracing more and more shapes, until he reaches the boundary. Or the meditation can be done in reverse.

The *sriyantra* is typical of Indian geometry, with its religious origi-

The *sriyantra.* To meditate, concentrate on dot (*bindu*) in center, then take in smallest triangle, then move outward through larger triangles and other shapes. (After Joseph and Kulaichev.)

nality, mysticism, and even playfulness, qualities we rarely see in Greek geometry, which remains "uncontaminated" by religion. Various special "numbers" are integrated into the *sriyantra,* such as pi and another irrational number, the golden ratio, or approximately 1.61803. The golden ratio is found in the pyramids at Giza and in the later construction of the Parthenon and other classical Greek buildings.[83]

Is 1.61803 a better number when found in later secular Greek architecture than in earlier Indian religious patterns? Interestingly, as Vedic sacrifices declined around 500 B.C., so, too, did the practice of mathematics among Indians.[84]

THE ANCIENT INDIANS practiced a very sophisticated form of mathematics. They had the usual arithmetic operations—addition, subtraction, multiplication, division—but also algebra, indices, logarithms, trigonometry, and a nascent form of calculus. In fact, Joseph claims that India and Japan are the only countries outside of Europe that ever developed calculus.

The most obvious contribution of Indian mathematics is the gift of our Western numbers. Indian numerals went through a lengthy evolution, from the roman-numeral-like Kharosthi numbers dating back to the fourth century B.C., to Brahmi numerals, a mixed bag of lines and squiggles circa A.D. 100, and finally to the Gwalior system. Note the similarity to our own numerals.

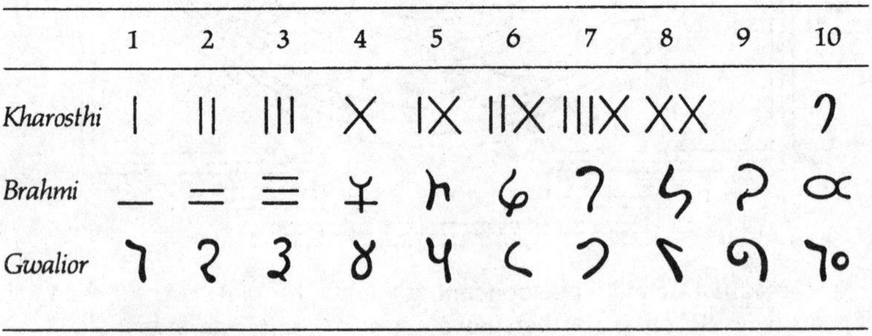

	1	2	3	4	5	6	7	8	9	10
Kharosthi	I	II	III	X	IX	IIX	IIIX	XX		𐩅
Brahmi	𑁒	𑁓	𑁔	𑁕	𑁖	𑁗	𑁘	𑁙	𑁚	𑁛
Gwalior	१	२	३	४	५	६	७	८	९	१०

The three Indian ancestors of our modern numbers, in chronological order. There is no Kharosthi numeral shown for 9. It is unknown.

Found in the town of Gwalior, India, these numerals date back to at least A.D. 876.[85] These ten numerals, including a zero, in a decimal place-value system, are capable of expressing any number, no matter how large. Our so-called Arabic numerals clearly came from India.

The richest era of Indian math is the classical period, from A.D. 500 through the Middle Ages. Doubts about Indian accomplishments during this period remain, however. The critical document is the Bakhshali manuscript.

In 1881, in the village of Bakhshali in northwest India, a farmer was digging in a ruined stone enclosure when he unearthed seventy leaves of birch bark. They made up a manuscript written in an old form of Sanskrit, the now famous Bakhshali manuscript. Much of the manuscript fell apart when examined, but enough remained to tell an amazing tale of Indian math in the early centuries A.D., a precursor to the math of the classical period.

The manuscript, whose author or authors are unknown, covers such topics as fractions, square roots, and even profit and loss and interest, as well as simultaneous equations, quadratic equations, and arithmetic and geometric progressions. A key section covers the critical area of notation. Curiously, negative numbers are noted with a plus (+), a point that becomes important later. Division is noted by putting a numerator over the denominator, as in present-day fractions, except that in the Bakhshali system, there is no horizontal line between them. Multiplication is denoted by placing numbers side by side.

The Bakhshali manuscript now sits in a library at Oxford University, England. Its importance was diminished considerably by the English scholar G. R. Kaye, who completed the first full translation in the 1930s.[86] Unfortunately, Kaye's knowledge of Sanskrit was flawed, and, as stated earlier, his agenda was to trace achievements in mathematics to the Greeks whenever possible. Kaye's biggest error was to date the Bakhshali manuscript to the twelfth century A.D. This conflicts sharply with estimates before and after his pronouncement. The first English translator, Rudoph Hoernle, judged the manuscript to be a copy. In essence, Kaye was dating the copy of a document rather than the document itself. Joseph believes the Bakhshali manuscript dates to 400 A.D.

or earlier,[87] in part because of the use of plus signs for negative numbers (abandoned after 500 A.D.), and in part because its naïveté about certain types of equations, developed later, would indicate an early origin.[88] Takao Hayashi, a respected contemporary Japanese expert, dates the manuscript to no later than the early seventh century.[89]

Most important is that the Bakhshali manuscript is the first document depicting a form of Indian mathematics devoid of religious associations.[90] We cannot leave the Bakhshali manuscript without noting a peculiar form found therein. It is a heavy black dot. It is often used in operations to denote the unknown value we are seeking, much as we use the letter *x* today. The dot was called *sunya,* the Sanskrit word for "empty" or "void." It would become, perhaps, the most important number ever invented.

China

In the movie *Infinity,* based on the life of the physicist Richard Feynman, the actor Matthew Broderick, in the lead role, belittles a Chinese merchant doing math on an abacus. "He knows everything about beads," says Broderick/Feynman, "but nothing about numbers."

Chinese math, even ancient Chinese math, is not as weak as the movie character insists. (The real-life Feynman may not have been quite so arrogant, though reporters had a habit of covering up for him, making this mythic figure seem cuddlier than perhaps he was.)

Still, it is difficult to assess Chinese mathematics. We are considering it out of order here. Chronologically, we can trace its beginnings back to the third millenium B.C.—before the Indians—and extend it to the Ming dynasty, A.D. 1260 to 1644.[91] Yet we're not sure what to make of Chinese math. It doesn't attain the sophistication of Indian or even Mayan math, in some respects.

The ancient Chinese did devise a sophisticated number system somewhere between 1500 and 1200 B.C. Some nineteenth-century farmers tilling a field found the Shang oracle bones, a collection of tortoise shells and animal bones—they were first thought to be the bones of a dragon—with numerals representing 1 through 10 (no zero) inscribed on them. Presumably, Shang nobles recorded prophesies gleaned from the spirts of their ancestors regarding the best times for traveling, harvesting, and the like. With these ten numerals and a few other symbols in a decimal system, the Chinese of that era could represent any number. It was the most advanced number system of its time save for one—that of the Babylonians.[92]

However, ancient Chinese math seems less interesting because it appears to be devoted to practical matters: teaching minor bureaucrats how to make calculations. There appears to be little interest in the underlying logic of procedures, not "pure" math as we perceive it today. The most influential text is the *Jiuzhang suanshu,* from the Han dynasty (206 B.C. to A.D. 220). It contains 246 problems and their solutions, addressing all the problems with which reckoning clerks in the bureaucracy might be confronted. The text was used for more than a thousand years and inspired great commentaries.[93] It was evidently followed blindly, with little interest in the underlying principles.[94] It fits into the Feynman "bead characterization." Then again, perhaps there is a deeper level that eludes our gaze from twenty centuries distant.

Underlying their many practical workbooks was considerable theoretical foundation. Claims that the Chinese had discovered the Pythagorean theorem in 1000 B.C., five hundred years before Pythagoras, are now fading. There is evidence, though, that the Chinese had the concept somewhere between 700 and 400 B.C., either a couple of centuries before or no later than a century after, the Greek mathematician. In the *Chou pei suan ching,* a text of the period, Duke Chou Kung and a person named Shang Kao discuss the properties of right triangles, including the Pythagorean theorem. In a dialogue, they demonstrate how it works geometrically.[95]

The Chinese are known for the abacus, though it came late to them. The Babylonians had it earlier; other cultures used the device as well. China's greater accomplishment was the counting board, of which the abacus was a kind of simpler, laptop version. The counting board looked like a smallish chessboard on which rods were placed to represent the numbers. Color was important. Red rods represented positive (*cheng*) numbers; black rods, negative (*fu*).[96] We, of course, reverse that system today in our bookkeeping notation; red ink means negative numbers, or losses.

Placing the rods in different positions—vertically, horizontally, adjacent, and so on—on the counting board, the Chinese could represent any number. They did not have an official circular zero until Chin Chiu Shao's work in 1247,[97] but for many centuries Chinese mathematicians simply left a square blank on the board instead of a zero. It may have denoted zero or, as Kaplan suspects, simply meant "no number here."[98]

Counting rods evolved beyond calculating devices, ultimately being used for algebraic operations. They could be arranged on a counting board to represent a system of equations. For example, the following board, from the first century A.D.,

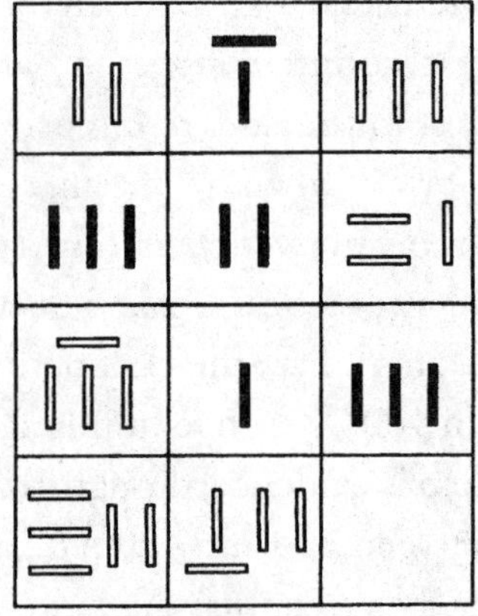

represents (reading vertically):

$$\begin{aligned} 2x - 3y + 8z &= 32 \\ -6x - 2y - z &= 62 \\ 3x + 21y - 3z &= 0 \end{aligned}$$

On the counting board, the equations read down the columns, and the coefficients of each unknown read across the first three rows, the last row showing the right side of the equations. The light rods above are red, signifying positive numbers; the dark rods are black, signifying negative numbers. Some speculate that this counting-board method was an early precursor of matrices,[99] a mathematical technique developed in the West in the nineteenth century and used most notably by Werner Heisenberg in his theory of quantum mechanics. Matrices remain one of the most widely used tools in mathematics today—in economics, geography, demography, and sociology.[100]

We have not done the ancient Chinese justice here, and their math remains mysterious and perhaps more profound than we can appreciate. Certainly that is the feeling of the great seventeenth- and eighteenth-century mathematician Gottfried Leibniz.

Leibniz, born in Leipzig in 1646, invented the calculus almost simultaneously with and independently of Isaac Newton. He was a polymath and a seer. He saw what shape mathematical logic would take, and he imagined a universal computing machine. Some biographers find in his work premonitions of quantum mechanics and DNA grammar.[101]

Leibniz was fascinated with the Chinese culture and shared the Jesuits' view that introducing European science into China would help convert the Chinese to Christianity. Leibniz was working with a binary arithmetic (the only numbers are 0 and 1), which today makes computers possible. In 1716, in a letter to Peter the Great of Russia, Leibniz claimed to have deciphered the diagrams of the ancient sage Fu Hsi and to have discovered correspondences between his binary code of arithmetic and Fu Hsi's diagrams.

The Jesuits of his day believed, incorrectly, that it was Fu Hsi who had constructed the famous "Prior to Heaven" diagram of sixty-four figures, each with six hexagrams, derived from the *I ching* (The Book of Changes), the venerable Confucius-era sixth-century B.C. work. (The diagram, according to more recent scholarship, has now been attributed to a later neo-Confucianist, not Fu Hsi.)

Leibniz came to believe that his binary system of double geometrical progression corresponded to the system of Fu Hsi, that Fu Hsi had derived a binary arithmetic from a sixth-century B.C. religious and mystical work, and that he himself had reinvented the arithmetic centuries later. Do computers owe an ancient debt to the *I ching*? Probably not. But the great mathematician liked thinking about it. Leibniz came to believe that the universe was binary, that God created the world out of units of 0 (nothing) and 1 (God). He called the process "the secret of creation." [102]

The Arab World

Religion as a driving force behind science is evident in the Arab world during the golden age of Islamic civilization. From A.D. 750 onward, after the proliferation and then consolidation of Muslim rule to cover half of the Old World, the sciences flourished in a largely peaceful Islamic culture. The Muslims dominated from North Africa north to France, spanned Persia and the central Asian plains to the borders of China, and extended their rule down to northern India.[103] In one of the great syntheses of mathematics, statistics, and linguistics, the Arabs invented cryptanalysis, the craft of unscrambling decoded messages.

The impetus was the revelations of Muhammad, who more than a century earlier had dictated messages he had received from the archangel Gabriel. These revelations were assembled into a single text, the Koran, but not necessarily in the order in which Muhammad had received them. To determine the proper chronology, theologians counted the frequencies of words in each revelation. Since some words were more recently coined than others, those passages with a higher frequency of these words must have been written later.

Al-Kindi, a ninth-century Arab philosopher and scientist, devised a similar technique, called "frequency analysis," for breaking codes. In *A Manuscript on Deciphering Cryptographic Messages,* al-Kindi said that one simply counts the occurrences of each letter in a language's alphabet in common usuage, ranking each: most common, second most common, and so on. Then one ranks the letters in the encrypted message. In English, for example, the most common letter is *e;* the second most common is *t.* So if the most common letter in the code message is *x* and the second most common is *p,* then *x* probably corresponds to *e* and *p* to *t.* Al-Kindi's is not a foolproof method, as it's based only on averages. The most common letter in an encrypted English message may not be *e* (for example, "Sink the Bismarck").[104]

British physicist and science popularizer Simon Singh, author of *The Code Book,* writes, "While al-Kindi was describing the invention of cryptanalysis, Europeans were still struggling with the basics of cryptography. The only European institutions to encourage the study of secret writing were the monasteries, where monks would study the Bible in search of hidden meanings, a fascination that has persisted to modern times."[105]

It isn't necessary to explain Arab mathematics in great detail. The math practiced by medieval Arab mathematicians differs little from what the average American encounters from grade school through high school, and perhaps a bit into college. Arithmetic, geometry, algebra, trigonometry, and some higher mathematics—math as most nonscience majors know it—was being used on a regular basis in the Arab world in the Middle Ages. The medieval Arab would breeze through most modern U.S. secondary schools, perhaps even attempting a bit of calculus. He'd use the same numerals, 0 through 9.

That Islamic mathematicians and scientists during the Middle Ages were far more advanced than their European counterparts has never been a matter for debate. Their precise role, however, has been passionately argued. The word favored by Western scholars, and reviled by non-Westerners, is "custodial." They were considered Islamic Xerox machines whose sole function in life was to preserve the Greek culture, to translate it, and to keep it well oiled and maintained, until Europe could rouse itself from its medieval slumber and reclaim its intellectual heritage.

This hypothesis, though well disseminated in American, European, and other schools, can't possibly be true, on at least three levels. First, how did the Arabs preserve Greek algebra and trigonometry when the ancient Greeks had no such disciplines? Second, the hypothesis ignores the fact that Arab math was based primarily on Indian numbers, 0 through 9, not Greek or Roman numerals.[106] Finally, more than mere scribes, the Arabs developed a mathematics of their own.

The Arabs were, in fact, wonderful preservers and, yes, custodians, of other cultures' intellectual accomplishments. The golden age of Islam provided a stable society in which scholarship thrived. Islam demands justice, and justice requires knowledge, so Muslims set about translating Babylonian, Egyptian, Indian, Greek, Chinese, Farsi, Syrian, Armenian, and Roman texts into Arabic. A series of caliphs from 762 to 833 turned Baghdad into a Muslim version of Alexandria, building an observatory and a library. In 815, the Bait al-Hikmah, or House of Wisdom, was built as a center for translation and research; it would remain the intellectual epicenter of the Arab world for two hundred years.[107]

ONE OF THE early directors of the House of Wisdom was al-Khwarizmi, who was likely from Khwarizm, east of the Caspian Sea.[108] Our word *algorithm,* for any special method of solving a problem, is derived from his name.[109] Al-Khwarizmi introduced Hindu-Arabic numerals—essentially a modernized version of the Gwalior numerals—to the Arabic world around A.D. 830. The ten numerals, including zero, were not an overnight success. The Arabs resisted the Indian numbers almost as much as the Europeans would during the Renaissance, though not for so long. Arabs continued using the sexagesimal system invented by the Sumerians for astronomical calculations. Even in algebra, Islamic mathematicians used words rather than numbers ("three squares and four is equal to seven things" rather than $3x^2 + 4 = 7x$). The Hindu numbers eventually gained acceptance when large numbers were required, the zero showing its dexterity in these matters.[110]

Among those mathematicians preferring "prose algebra" was al-Khwarizmi himself. It was he who named this branch of mathematics in his book *Hisab al-jabr w'al-muqabala* (Calculation by Restoration and Reduction).[111] *Al-jabr* is the word he used for the operation of solving equations. The traditional explanation is that *jabr* shares a common root with the Arabic word for "the setting of a broken bone." (*Algebrafista* is Spanish for bone setting.) As mentioned in chapter 1, Columbia's George Saliba says the connection to bone setting is incorrect, that algebra is an Arab word meaning "compulsion," as in compelling the unknown "x" to assume a numerical value.[112]

By "restoration" (in the title of his book) Al-Khwarizmi is referring to the transfer of negative terms from one side of an equation to the other. By "reduction" he means the reduction of many terms into a single one,[113] as is commonly done in algebra. Al-Khwarizmi introduced the 0 though 9 in another book and was careful to credit the Indians, but various translations deleted this point, which is why today these numbers are called "Arabic numerals."[114] Al-Khwarizmi also introduced to the Arab world the decimal positional system developed by the Indians.[115]

Al-Khwarizmi assigned the latitude and longitude of twelve hundred important places on the globe and corrected Ptolemy's overestimate of the length of the Mediterranean.[116] Al-Khwarizmi is still required reading in some Arab states because he applied algebra to Islamic inheritance laws, which can be arcanely complicated.[117] For example, one-fourth of a woman's estate goes to her husband, and the rest is divided among the children, except that sons must receive twice as much as daughters. If a share is left to a stranger, he may not receive more than one-third of the estate without the permission of the natural heirs, and if so, those who approved the legacy must compensate the nonapprovers for the amount exceeding one-third.[118] Muslim estate lawyers obviously had to know algebra.

Another Arab algebraist, born two centuries later, was the poet Omar Khayyam. Born in Khurasan, now part of Iran, Omar most likely came from a line of tent makers. We know him primarily as the author of *The Rubaiyat of Omar Khayyam,* but he also wrote a book on algebra in 1070, classifying equations according to their degree and providing techniques for solving quadratic equations[119] (the Babylonians were solving quadratics two thousand years earlier).[120] He did a number of other flashy things, devising, for example, a calendar in which eight out of every thirty-three years was a leap year. This provides a more accurate approximation of a solar year than does the Gregorian calendar devised centuries later.[121] (The Gregorian employs a forgettable system of leap years every four years, excluding century years, yet including century years divisible by 400. This last rule was ignored by some computer programmers in A.D. 2000, resulting in a missing February 29 in many computers for that year.)

A more important contribution of Omar Khayyam was his reinterpretation of Greek geometry. Euclid and other Greeks did not consider certain ratios capable of being expressed as arithmetic numbers: for example, the ratio of a circle's circumference to its diameter (pi) or the ratio of a diagonal of a square to its length (the square root of 2). Omar extended the concept of number to include positive irrational numbers, an idea explored later in Europe by René Descartes in the seventeenth century and Georg Cantor in the nineteenth. Omar's seminal contribution was his geometric solution to cubic equations.[122]

THE MEDIEVAL ARABS approached mathematics with the rigor of the ancient Greeks combined with the playfulness of the Indians. They were fascinated, for example, with perfect numbers. A number is perfect if it is equal to the sum of its own divisors. The first perfect number is 6 (the sum of 1, 2, and 3). In the ancient world, only four perfect numbers were known: 6, 28, 496, and 8,128. By the thirteenth century, the Arabs had added three more—not a trivial feat given that these perfect numbers contain eight, ten, and twelve digits, the third one being 137,438,691,328.[123] The practical purposes of perfect numbers? No one has thought of any yet. Arab work in this area is thus evidence of an interest in pure mathematics.

Western scholars can point to the respect the Arabs showed their Greek predecessors. One of Islam's prominent mathematicians was Abul Hassan al-Uqlidisi, in the tenth century. His name was evidence of his reverence for the Greeks. He copied the works of Euclid, hence the name al-Uqlidisi. One of his legacies is paper-and-pen mathematics. It was common in India and the Islamic world to do calculations in the sand or dust, wiping out intermediate steps as one proceeded. Al-Uqlidisi recommended paper and pen instead. His motivation was not intellectual; he simply wanted to separate himself from street astrologers and others who did their calculations in the street.[124] Written arithmetic preserves the process by which one arrives at a solution, an important issue to the Greeks.

The Arabs were also reluctant to abandon the limited Western concept of number, which to the Greeks was always a positive integer or a fraction composed of integers. The Arabs clung to this antiquated idea for centuries. Despite introducing the zero to half the civilized world, it appears that the medieval Arabs never quite accepted it as a rank-and-file number since it isn't a positive integer (and the Greeks never had it).[125] This attitude began breaking down, however, with the Arab mathematician-astronomers, who needed more numbers than the meager assortment provided by Euclid et al.

George Gheverghese Joseph points out that the Arabs broke the "straitjacket of Greek mathematical tradition."[126] They brought together the geometry of the Greeks with the algebra, trigonometry, and numeration of the rest of the world.

Mesoamerica

A painting on a classic Mayan vase depicts a pair of deities. One god has the facial features of a monkey and carries a codex. The second god rests a hand on the back of the first, with a scroll containing bar and dot numerals streaming from his armpit. The god with the codex represents writing. The god with numbers flowing from his armpit represents math. The implication is that the Maya were not mere counters and reckoners but recognized mathematics as a separate discipline, on a par with writing.[127] The significance of numbers emanating from the armpit, a common image in Mayan art, remains unclear.

We have limited knowledge of the great Mayan civilization that once occupied an area that includes present-day central and southern Mexico, Belize, Gautemala, El Salvador, and parts of Honduras. Sixteenth-century Spanish conquerors destroyed most of their writings. Fortunately, a few codices escaped, as well as cave paintings, cave writings, and hieroglyphic inscriptions over the stelae, upright stone monuments that were erected every twenty years and contained the date of their construction.[128] These meager clues are enough to tell us that the Mayan culture, whose golden age spanned A.D. 200 to 1000, embraced mathematics with a passion.

As the stelae confide, 20 was an important number to the Maya. They utilized a base 20, or vigesimal, number system with nineteen numerals and a zero. The basic numerals were composed of two symbols, a dot for 1 and a vertical bar for 5. These were used in various combinations to form the numbers 1 through 19. A zero consisting of an elliptical form resembling a snail's shell also stood in for 20. It was not a straightforward system like ours. Adding zero to a number did not necessarily multiply it by 20.

The Maya clearly loved numbers, and they had several forms of them besides the stick, ball, and snail variety. There were the "head variant" numbers, for example, in which 1 through 20 took the form of the typical Mayan "comic book" style of caricatures.

Mayan math, so far as we know, was based on integers. The Maya appear to have taken great pains to avoid using fractions.[129] In keeping track of time, Maya scribes used a positional notation that was written vertically, with the smallest units, *k'ins,* or days, at the bottom.[130] It appears that positional notation is very ancient in the New World, predating even the Maya, having been found on monuments in Mesoamerica dating back to 36 B.C.[131] It is a profound and important concept, says Marjorie Senechal, a professor of mathematics at Smith College. "Positional notation is to a mathematician," she says, "what laboratory equipment is to an experimental scientist."[132]

The Maya were obsessed with counting because they were obsessed with time—obsessed with the notion that they might run out of the lat-

The Mayan "head variant" numbers, from 1 to 19 and 0, based on deities. These numbers were often used on calendars. (After Joseph and Closs.)

ter, and the universe would end.[133] They had at least six calendars, including a 584-day Venerean calendar, based on Venus years. The Maya offset their three primary calendars—the 260-day *tzolkin,* or "sacred year," the *haab,* or "civil year," and the *tun,* or "long count"—to avoid cosmological disaster. They feared that when a calendar came to its end, so might the universe, but with calendars of different lengths running simultaneously, they felt safer.

If the *tzolkin* year was ending, well, the *tun* might still have time left. On those rare days when the end of one calendar coincided with another—as the *haab* and the *tzolkin* did every fifty-two years—the Maya offered up blood sacrifices to fend off annihilation. The *haab* calendar employed another trick. Each of its twenty-day months began with 0.[134] When one hits the zero day, is one at the beginning or the end? With death and rebirth occurring simultaneously, the end is smeared.

George Joseph comments, "Isn't it interesting that the two mathematical cultures that had the most developed positional number system with a zero were both obsessed with time, but in very different ways? The Indians saw time as never-ending, and so measured eras in vast periods of time (*mahayugas*), and the Maya were fearful of running out of time, so they had to undertake sacrifices to the gods to try to avoid this calamity."[135]

Did the Maya practice geometry? The surviving codices contain no clues. Much has been made, however, of the fact that the sites of three major temples at Tikal form an isosceles right triangle.[136] Perhaps so, but let's hope these clever people, if they did practice geometry, did so in a manner more efficient than building three temples every time they needed a triangle.

Once again, religion presents itself. A holy class of scribes was in charge of mathematics in the Mayan civilization. Not all scribes, however, could do the math. Those who could manipulate numbers enjoyed the greatest prestige.[137] In our scientific culture, we say that the physicists defer only to the mathematicians, and the mathematicians defer only to God. (Though some would argue that few mathematicians are that modest.) Among the Maya, the route to the gods was through the scribes, and the route taken by the scribes was through mathematics and astronomy.

Zero

Let us turn to a major concept that can be traced through many different peoples: zero.

It is a troublesome number even today. A few years ago, I wrote an article about the fact that the years in our Gregorian calendar are misnumbered. When the the Scythian monk Dionysius Exiguus instituted the anno Domini (A.D.) system in the sixth century, he began with A.D. 1, the year in which, according to his calculations, Jesus was born. No problem there—1 is a good number to start with. Two centuries later, however, the Venerable Bede, a Northumbrian Anglo-Saxon monk, picked up Dionysius's system and popularized it in his classic work the *Ecclesiastical History of the English People,* completed in A.D. 731.[138] To preserve history before Jesus, Bede extended the system backward, creating the B.C. years.

There was a major hitch. Bede had only Roman numerals to work with, which means that he had no zero. Therefore, he constructed a calendar that went directly from A.D. 1 to 1 B.C., with no zero in between. I wrote that the lack of a year zero causes immense problems. For example, if one reckons the number of years from A.D. 5 to 15, the answer is ten. Reckon from 5 B.C. to A.D. 5, however, and you get . . . nine! Skipping the year zero creates a messy situation for mathematicians, astronomers, calendarists, and others. In fact, because it is so confusing to date ancient eclipses and comets with the Gregorian calendar, astronomers devised their own, one with a year 0. In 1740, the French astronomer Jacques Cassini replaced B.C. and A.D. with a − and + system, in which 0 replaces 1 B.C., 2 B.C. becoming −1, 3 B.C. becoming −2, and so on. Zero is a leap year.[139]

For years after the article was published, I received passionate letters from educated people who insisted that there was no need for a year 0.

Zero equals nothing, they wrote, and one can thus skip over zero, as one is skipping over nothing. A group of teachers said I'd struck a blow against math literacy by treating nothing as something.

In fact, zero is more than nothing, but even today it remains a difficult concept for many of us. I had earlier called the math department at the Massachusetts Institute of Technology, asking if a mathematician could confirm that zero is a valid counting number. A department spokesman declined to comment, suggesting that I needed a number theorist, and that I should call Harvard University. Harvard confirmed that zero was indeed a number, and could not be skipped with impunity any more than one could skip 3 or 8 or 412 when counting. A number theorist there, however, cautioned that "zero is a very modern concept"[140] and people still have trouble with it.

To understand why zero is not synonymous with nothing, take this classic example: grade point average (GPA). In a four-point system, an A equals 4, B equals 3, and so on, down to F, which equals 0. If a student takes four courses and gets A's in two but fails the other two, he receives a GPA of 2.0, or a C average. The two zeros drag down the two A's. If zero were nothing, the student could claim that the grades for the courses he failed did not exist, and demand a 4.0 average. His dean would laugh at such logic.

The concept of zero was less alien to the ancient non-Western world than it is to us today. "In the history of culture," wrote Dantzig in 1930, "the discovery of zero will always stand out as one of the greatest single achievements of the human race." Zero, he said, marked a "turning point"[141] in math, science, and industry. He also noted that the zero was invented not in the West but by the Indians in the early centuries after Christ. Negative numbers followed soon thereafter.[142] The Maya invented zero in the New World at approximately the same time.[143] Europe, says Dantzig, did not accept zero as a number until the twelfth or thirteenth century.[144]

There are many "biographies of zero," and Dantzig's concise and spirited account of the birth of a number is adequate for most of us. He sees zero's invention appearing on an Indian's counting board in, say, the first or second century A.D. The Indian counting board had columns for the ones, tens, hundreds, thousands, and so on. To "write" 302, for instance, a mathematician would put a 2 in the first (right) column and a 3 in the third, leaving the second column empty. On one fateful day, as Dantzig sees it, an unknown Indian drew an oval in the second column. He called it *sunya,* for "empty" or "blank." *Sunyata,* an important concept in Buddhism, is often translated as "emptiness" or "void."[145]

The Arabs turned *sunya* into *sifr* ("empty" in Arabic), which became *zephirum* in Italy, and eventually zero. In Germany and elsewhere, *sifr* became *cifra,* and then, in English, *cipher.*[146] In other words, it took over a thousand years for Western civilization to accept a number for "nothing." Dantzig blames the Greeks. "The concrete mind of the ancient Greeks could not conceive the void as a number, let alone endow the void with a symbol."[147]

That's the short version, and not a bad one. You don't want to hear the long version, so let's suffice with a medium-sized tale.

Zero lay rustling in the weeds for many centuries before that Indian drew it on a counting board. It was an unnamed, unwritten force. It took many more centuries after the Indians and the Maya dared speak its name before zero was promoted to a full-fledged number.

The U.S. Library of Congress defends our calendar and its missing zero. "There has never been a system of recording reigns, dynasties, or eras," the library states, "that did not designate its first year as the year 1."[148] In fact, Pol Pot began the Khmer Rouge calendar with the year 0. The Maya had both years 0 and days 0.

The Babylonians had no zero, but they knew something was wrong. If they numbered the first year of each king's reign as year 1, then added up the number of years of each separate reign, they'd end up with too many years unless each king died just before midnight on New Year's Eve and his successor took the throne after midnight. Thus, the Babylonians called a king's first year the *accession year.* The following year was year 1.[149] The accession year was a kind of year 0. The Babylonians, so far as we know, never articulated zero, but seemed aware that there was a missing number in their system.

The contemporary mathematician who has conducted the most rigorous research on nothing is Robert Kaplan, the author of *The Nothing That Is: A Natural History of Zero.* Zero turns up throughout history in different cultures as a series of dots and circles, and Kaplan writes of following "the swarm of dots we find in writings from a host of languages, across great spans of time, and on topics mathematical and otherwise."[150]

Kaplan traces the roots of zero to Sumer and Babylonia. The Sumerians counted by tens and sixties, a system adopted by the Babylonians, who eclipsed them in Mesopotamia. The Babylonians, far ahead of the Romans and Greeks to come, imposed a positional notation on the old Sumerian sexigesimal system. Writing their numbers on clay, the Babylonians needed a symbol to put in the "empty" columns, just as we today use zero to differentiate between 302 and 32.

Somewhere between the sixth and third centuries B.C., the Babylonians began using two slanted tacklike symbols to insert in the empty columns. They borrowed the slanty tacks from their language, where they were used as periods, among other things.[151] However, the Babylonians used their "zero" only in the middle of numbers, never at the end.[152] Clearly, this was not a full-fledged zero.

Kaplan argues that when Alexander invaded the Babylonian empire in 331 B.C., he hauled off zero along with the women and the gold. Shortly thereafter we find the symbol 0 for zero in the papyri of Greek astronomers,[153] but the mathematicians never pursued the concept. Why not? Kaplan, Hogben, and others mention the old reason: the ancient Greeks had little respect for counting, leaving such micromanagement to tradesmen.[154] Kaplan also has another theory: that the Greeks kept zero to themselves, a secret they were loath to put into writing.[155] It is not such a far-fetched idea. The zero continued to have a bad reputation among Westerners. In A.D. 967, the monk Gerbert, who would become Pope Sylvester II, fashioned a zero sign for one of the counters on his counting board. For dabbling in nothingness, Gerbert was accused of having intercourse with evil spirits.[156]

The Greeks often placed stones on their counting boards and then sprinkled the boards with dust or sand, presumably to make a temporary record of the numbers when the stones were removed. Kaplan speculates that the 0 symbol was the impression left when the stone was extracted. Further, he finds zero in an A.D. 270 Indian work called *The Horoscopy of the Greeks.* The Indian author was translating into verse an earlier Indian work from A.D. 150, which in turn was probably a translation of a Greek original.

Kaplan's hypothesis is that the ancient Greeks invented zero, declined to use it themselves, and then inadvertently donated it to the Indians. He hastens to add that he is bridging a chasm "on the slenderest threads of evidence."[157] We're in a generous mood here. Why not give the Greeks credit for inventing one of the most important concepts in math . . . and then for their generosity in giving it away, virtually unused? It reminds one of the man who throws his great-aunt's old bureau out on the sidewalk; a passing shopkeeper picks it up, scrapes it down, and discovers a priceless heirloom under all the dust.

This is what the Indians found in the Greeks' discarded zero. Zero's journey from a circle in the dust to a full-fledged number was a long one, and Kaplan feels that zero never achieved full citizenship until seventeenth-century Europe.[158] During the first millennium A.D., mathematicians referred to the nine Indian numerals *and* zero, as if it were separate. Even the European mathematician Fibonacci, who grew up in North Africa and was tutored by Arabs, wrote in the thirteenth century about the "nine [Indian numerals] with the sign 0."[159]

Was this distinction between "number" and "sign" meaningful? Probably, but it didn't stop the Indians (or the Maya) from operating with zero—using it in arithmetic operations, though division was a quandary.[160] Clearly, the Indians started realizing that zero was more than a sign. In A.D. 600 Brahmagupta set down the rules for addition and subtraction with zero, including operations with negative numbers.[161] The Indians figured out that the square root of zero is zero, and that 0^2 is also 0, but they couldn't figure out how to divide with zero. (One *cannot* divide by zero.)

Hogben gives the Indians credit for inventing negative numbers, though in a backhanded way. "Perhaps because the Hindus were in debt more often than not," he writes, "it occurred to them that it would also be useful to have numbers which represent the amount of money one owes."[162] Perhaps we can excuse the ancient Greeks for not inventing zero or negative numbers because they were too fiscally responsible to conceive of being broke or in debt.

I've relied heavily on the version by Kaplan, a traditional Western scholar, for this brief history of zero in the Old World. Other scholars tell a slightly different story. For a brief alternative history, from George Joseph, a mathematician more sympathetic to Eastern philosophy and history, see note.[163]

COMBINED WITH positional notation, zero becomes quite a bit more than nothing. By adding zeros to the end of a sum, the Indians and Maya were suddenly capable of describing monstrous quantities. One cannot take this for granted.

Consider the plight of Archimedes, around 250 B.C., writing to Gelon, King of Syracuse. In his letter, Archimedes proposed to count the number of grains of sand required to fill the entire universe. Greek mathematicians had no zero with which to construct large numbers. What Archimedes had was the myriad, which is 10,000 in our system. So he thought of a myriad myriads, then created multiple orders of a myriad myriads, to get to his estimate of the number of grains of sand in the universe. His answer, in our notation, was a 1 with fifty-one zeros, or 10^{51}.[164]

He was impressed with himself for being able to think of such a large number, and today we would respond with a resounding "So what?" When we want a big number, we just pile on some zeros, or, faster still, use scientific notation (10^{23}, 10^{51}, and so on). The ancient Indians may not have understood all the mathematical ramifications of their new toy, but they knew that zero could be enlisted in the construction of big numbers. For example, the Indians quickly dated (we're not sure how) the earth at 4,300,000,000 years old, very close to current estimates. The Maya were more extreme. As part of their effort to convince themselves that there was plenty of time, the Maya reckoned the age of their universe at 2×10^{27} years.[165] (Modern cosmologists date our big bang universe at a mere 1.5×10^{10} years old. The Maya universe beats ours by seventeen zeroes.) Ancient Indians were also fascinated by huge numbers and had names for numbers with increasing powers of 10 up to 17 as early as 500 B.C.[166]

This gives one added respect for Archimedes. Thinking of big numbers was no mean feat. Without zero, Archimedes had to build his numbers with much smaller bricks. "Once zero appears," says Robert Kaplan, "the calculation gets easier, but it kills the imagination."[167] Bureaucrats and cosmologists like to toss around numbers and statistics that are too large for the mind to grasp. The difference between 10^{48} and 10^{52} is huge, but do we discern it? Chemists complain that Avogadro's number, 6.022×10^{23}, a constant used to calculate the number of molecules in a chemical, is too large. That is, if one makes a mistake and multiplies by 10^{22} or 10^{24} instead of 10^{23}, it's a huge error but one not noticeable to humans. If zero has been considered the work of the devil, perhaps the fear is justified. Our brains are no more evolved than Archimedes'. Zero is relatively new software. Perhaps our hardware isn't up to it.

To those who fear the mix of religion and science, zero holds even greater horrors. With the Maya, zero reached the proportions of a cult religion, and a bloodthirsty one. Not content with one simple symbol, like 0, for the number, the Maya needed many. There was the basic snail shape,[168] a flower,[169] a tattooed man with his head thrown back,[170] and many others. Barbara Fash, of Harvard University's Peabody Museum, says the flower's center point is a "bed of creation," the zero signifying both the beginning and the completion of a cycle. The zero was an affirmation of life.

There was a dark side. When the end point of the sacred calendar coincided with the end point of the civil calendar, the Maya felt the need to fend off the death of the universe by killing Death himself. So they played a ball game, a kind of deadly Super Bowl, in which one opponent was the Hero and the other was the God of Zero, or Death. The game was staged, but the players were real people, and the injury and death they suffered were real also. The "ball" was an important hostage, such as a captured king, who had been saved for the event. He was trussed up into a ball, and the Hero and the God of Zero kicked him back and forth, eventually killing the "ball," sometimes by rolling him down a long flight of stairs. The fix was always in, and the Hero always won. The God of Zero would then be sacrificed by having his lower jaw torn off.

FROM WHAT WE have seen, the standard trajectory of mathematics, can't possibly be true. That is, that math was invented by the Greeks, who gave it to the Arabs for more than a thousand years of custodial care, after which it was turned over to the Europeans during the Renaissance. Clearly, the Arabs served as a conduit, but the math laid on the doorstep of Renaissance Europe cannot be attributed solely to ancient Greece. It incorporates the accomplishments of ancient Sumer, Babylonia, Egypt, India, China, and the far reaches of the medieval Islamic world.

The "path of zero" is in the past and is virtually unknowable, but it provides clues as to how mathematical knowledge in general moved through the ancient world. Zero was perhaps conceived in Sumer, gestated in Babylonia and Greece, born in India, and reared through the medieval Arab world and Renaissance Europe. We are being extra kind to ancient Greece here—we can see zero there if we squint hard—and perhaps ungrateful to Egypt, which may have provided a hidden assist. In short, there may be no trajectory in a strict sense. The various cultures may have acted as particles in a field, with concepts bouncing freely back and forth among them.

The Mayan zero is a challenge to any notions of white superiority. Separated from the Old World by an ocean, the Maya are unlikely to have stolen the concept. The zero was first used in the Mayan civilization somewhere between A.D. 292 and 357, as revealed by the chronological monuments called stelae. The oldest stela without a zero is Stela

29 at Tikal, dedicated in 292, but Stelae 18 and 19 at Uaxactun, dated at 357, contain a zero sign. Somewhere in between, the Maya invented their own zero, free of Greek or other Old World influence. How do we rationalize that these people, acting on their own, had supple enough brains to conceive of what Dantzig calls "one of the greatest single achievements of the human race"?[171]

MORRIS KLINE, the best-known modern historian of mathematics, has characterized Babylonian and Egyptian math as the "scrawling of children." He called the Indian mathematicians "fools." Are students being adequately informed of the contributions of non-Western cultures?

Harvard's Barry Mazur says that math is very difficult to learn and very difficult to teach. It is hard enough finding talented math teachers, let alone requiring that those teachers know a spectrum of Western and non-Western forms of math. "Taking a broad serious view of math," he says, "would require learning several languages. You'd have to spend your life on this stuff to understand it. No modern mathematician has done it." Then he paused. "But we have to do it better than Morris Kline."[172]

3

ASTRONOMY
Sky Watchers and More

THROUGHOUT human existence we have looked to the sky. Were there perhaps moments involving a monolithic slab and a summer solstice sunrise as in Kubrick's Ape Prelude in *2001: A Space Odyssey*? Why not? Astronomical observations predate writing. The human integration of events in the night sky into a larger view of human order seems to verify a hard wiring of the brain to confer upon celestial goings-on a pattern and organization. The movements of heavenly bodies have been recorded, noted, or remarked upon in an endless variety of ways, but the continuum of astronomical observation across cultures has been consistent.

Here in the present we have better equipment. In the opening of his book *Frozen Star,* the astronomer George Greenstein writes, "Out for a stroll one starry night, it struck me how rarely we astronomers ever look at the sky." Data from satellites are piped down, in digital form, to computer screens in windowless offices bathed in fluorescent light. There are more data, but there is certainly less romance. Greenstein noted that ancient astronomers, with cruder equipment, may have had a greater appreciation of the sky as whole. When Carl Sagan gazed upward, in the TV series *Cosmos,* and rhapsodized about billions and billions of stars, comedians mocked him. He was, however, validly pointing out how any human, even those without telescopes, can begin to understand the world by looking at the sky. (Sagan later claimed he had never used the phrase "billions and billions.")

Stargazing isn't what it used to be. The stars are just as bright as they were during Sumerian times, but light pollution from terrestrial sources obscures their brilliance. For many, real sky watching has been replaced by enhanced images of stars in the media. We have become accustomed to "photographs" of galaxies, which appear as large, fluffy, bulging wheels of stars lathered with some sort of creamy galactic frosting. They look good enough to eat. In reality, galaxies are wispy things, barely visibile to the naked eye. The photographs you're familiar with are taken using long time exposures—to make the galaxies look beefier—through a powerful telescope.[1] These are cartoon galaxies, bearing no resemblance to reality, which promote our modern view of the universe as one dense with stars.

All galaxies, including the one we live in, the Milky Way, are primarily empty space with a sparse smattering of stars and interstellar matter. If two galaxies met head-on, they could pass through each other with few collisions.[2] If you shot a rocket randomly through the Milky Way, the chances of it hitting a star would be one in one billion trillion.[3] In fact, in the early 1970s, NASA did shoot unmanned ships *Pioneer 10* and *Pioneer 11* out of the solar system. The spacecraft carried drawings of humans and other messages supposedly for aliens, but this was an exercise only, since NASA knew the rockets had little chance of contacting a star, let alone a planet.[4] Still, the misguided thought the messages were meant for alien eyes and ears.

Only modern earthlings believe in jelly-doughnut galaxies, a universe so congested with matter we can send messages to our extraterrestrial friends. In fact, the night sky reveals more space than stars.

ANCIENT CULTURES were often more realistic in their relationship with the heavens. In recent decades we have come to recognize the astronomical sophistication of ancient non-Western cultures. Otto Neugebauer's 1957 *Exact Sciences in Antiquity* became a foundation text and spurred the beginning of a new multidisciplinary field, archaeoastronomy. Anthony Aveni, a professor of astronomy and anthropology at

Colgate University, defines archaeoastronomy as the study of the practice and use of astronomy among the ancient cultures of the world, taking into account all forms of evidence, written and unwritten.[5] Though archaeoastronomy has only been around since the early 1970s, it already has had considerable success as a tool for interpreting astronomical accomplishments of pre-Renaissance cultures. Revivified by Harvard-Smithsonian astronomer Gerald Hawkins's interpretations of the alignments of Stonehenge—after some early work by Sir Norman Lockyer around the turn of the twentieth century—the field has expanded to include cultures everywhere.[6]

In most ancient cultures in which sky observations were important, astronomers served also as priests. Although the carefully oriented temples and ball courts of the Maya and Aztecs doubled as astronomical observatories, they were also temples and structures for the observance of civic and religious rituals. Using the observatory-temple, the ancient peoples of Mexico and the Andes linked the stars to their lives through omen and prophecy. Although this marriage of astrology and astronomy common to most ancient non-Western cultures has discredited their efforts in the eyes of some scholars, their accomplishments remain.

Astrology was held in high esteem in the West for many years. Johannes Kepler, the founder of planetary astronomy, simultaneously supported himself in part by casting horoscopes, as did his mentor, the Danish nobleman Tycho Brahe, sometimes called Europe's first great astronomical observer.

It was an eclipse of the sun predicted for August 21, 1560, that first interested the fourteen-year-old Tycho Brahe in astronomy and astrology. He was struck by the fact that men could understand the motions of the stars and planets so finely that they could foretell their positions years in advance. Like other astronomers, Tycho was fascinated with the regularity of the universe. If we can predict that Halley's comet will traverse our sky every seventy-five years, is it so far-fetched that Tycho, Kepler, and Galileo considered the possibility that the lives of men could also be charted with similar regularity? In fact, until the sixteenth century, *astrology* was the correct term for the science of studying planets and stars. *Astronomy* was the practice of naming and identifying stars and constellations, a minor associated science of classification, much as taxonomy is to biology. (The suffix *-nomy* means "to arrange.")

A CAVEAT. Aveni brings to our attention the fact that when they looked at the sky, the ancients may have had different things in mind than we do. "All people," he says, "Western and non-Western, before the Enlightenment, tended to approach the sky differently." He conducted a study of Aztec eclipses, and found that the eclipses the Aztecs chose to record for history were those that occurred at the "right times"—that is, those that occurred at fifty-two-year cycles, that dovetailed with their calendar. These were not necessarily the most spectacular eclipses, which attract our modern eye. "We wouldn't think of connecting a volcanic eruption with the death of a president, but they would," says Aveni.

What were ancient peoples looking for? Aveni says, "I think they were looking for events that validated their belief systems, or sometimes caused their belief systems to be altered." If the astronomers were among the ruling elite, as they were in Babylon and in the ancient Maya and Aztec worlds, they would be looking for signs in the cosmos to validate their actions: making war, conducting a battle, having a marriage alliance or a merger of states, and so on. "We usually interpret this in a simplistic way, and say, Oh well, they were like some kind of Hitler who would just dupe the masses. We have to understand that this was a deeply held belief, that you look to the cosmos for signs from the gods to indicate how you should behave, which way you should turn, in much the same way a president looks to his or her cabinet."[7]

Ancient and medieval non-Western astronomy is pretelescope, naked-eye astronomy. Even without scopes, however, the ancient Indians, long before Copernicus, knew that the earth revolved around the sun and, a thousand years before Kepler, knew that the orbits of the planets were elliptical; the Arabs invented the observatory and named most of our popular stars; the Chinese mapped the sky; and the

Amerindians noted important astronomical events with daggers of light or optical snakes that thrill us to this day.

We shall start our survey in the New World. Most New World cultures lacked a written language (exceptions being the Maya, the Aztecs, and possibly the Inca), but they left a rich astronomical heritage.

The New World

It is easy to assess the impact of Old World astronomy on the Western Hemisphere: there wasn't any. Mesoamerican and other New World cultures were "hermetically sealed," as Aveni puts it, from the rest of their sky-watching peers by the Pacific and Atlantic Oceans. While most of Europe languished, Mesoamerican cultures, influencing only one another, synthesized an astronomy package that was sophisticated, complex, precise, and solely their own.

Mesoamerican astronomical interests were inseparable from religious and sociopolitical ones. (Mesoamerica stretches from northwest Mexico through central Guatemala and El Salvador.) As in ancient Mesopotamia, China, India, Greece, and Italy, astronomical gods form the core of the pre-Columbian pantheon. Mesoamerican societies saw the heavenly bodies as gods who influenced their fate and controlled what happened on earth.[8] They also thought if they tried really really hard, they could influence these divinities.

Although hundreds of distinct ethnic groups existed, the presence of shared calendars and a body of astronomical knowledge suggests communication between these groups reaching back further than two thousand years.[9] Some of the later written Maya astronomical accounts are tied to earlier cultures—Epi-Olmec, Mixtec, Zapotec-derived phenomena two thousand years old. As early as the twelfth century B.C., the Olmecs, the progenitors of the Maya, Aztecs, and other peoples of Mesoamerica, were building ceremonial pyramids one hundred feet high, probably for better viewing the celestial events, as well as for ritual purposes.

Shortly before the Christian era, a mysterious civilization began building the city of Teotihuacán in a valley thirty miles from today's Mexico City. Teotihuacán's Pyramid of the Sun stands 210 feet high and 700 feet wide at the base.[10] This culture's preoccupation with sky observation would serve as a template for Mesoamerican civilizations to follow.

The solar zenith plays a central role in all Mesoamerica. On the day (usually June 21) the sun arrives at its summer solstice point—that is, when it crosses the overhead point, or zenith—something special happens at a latitude of 23.5 degrees north. Unlike in more temperate latitudes, in the tropics the sun reaches a true overhead position at noon two times a year. Since most Mesoamerican cities were located south of this latitude, their citizens could observe the sun directly overhead during the time it traveled over their latitude. Zenith passage observations are possible only in the tropics (that is, between 23.5 degrees north and south latitudes) and were unknown to the Spanish conquistadors who descended upon the Yucatán in the sixteenth century.[11] Early Mesoamericans were certainly aware that the zenith varied subtly as they traveled north and south. Early architectural complexes showed orientations keyed to the local horizon position of the sun on the zenith date.[12]

For all Mesoamerican societies the sun was the ruler of time and space. Pre-Columbian peoples fashioned their architecture to integrate time and space. Astronomers used fixed locations in temples and pyramids to track the rising and setting of the sun and other celestial bodies. They marked solar events by placing sets of crossed sticks along lines of sight on buildings' terraces and ramparts. The precise direction of the sun at sunrise was a prime orientation.

A common system of Mesoamerican knowledge included not only prediction of expected solar and lunar eclipses, but also intense observation of the seasonal rising and setting of Venus and, possibly, Jupiter, Mars, and Saturn, as well as the marking of the dates of significant conjunction of planets, the moon, and bright stars and constellations. These events were recorded on monuments from as early as the first century A.D.[13]

Of all the ancient timekeepers, the Mesoamericans (especially the Maya) developed the most complex and intricate calendrical systems. The oldest Mesoamerican calendar inscriptions date from 600 B.C. They devised a 260-day calendar called the sacred count, used for divination, astrology, and religious record keeping. This calendar gave each day a name, much like contemporary days of the week. There were twenty day names, each represented by a unique symbol. The days were numbered from one to thirteen. With twenty day names, after the count of thirteen was reached, the next day was numbered one again. The 260-day sacred-count calendar was in use throughout Mesoamerica for centuries, probably before the beginning of writing. No other cultural groups in the world have used a 260-day calendar. No one knows just when, how, or why the Mesoamericans decided upon a period of 260 days. Their shared geographical location and weather patterns and the agricultural cycles of the northern tropics probably influenced its development. Its design may tie together several astronomical events, such as the configurations of Mars, appearances of Venus, or eclipse seasons. Contemporary Mesoamericans, who still use the 260-day calendar for ritual events, have suggested that the 260-day count is based on the length of the human gestation period.

The planet Venus plays a central role in Mesoamerican culture, especially in the timing of warfare. The Venus warfare cult, recognized at many Mesoamerican sites by the images of a goggle-eyed deity known as Tlaloc, apparently originated in Teotihuacan, and can be traced at least as far back there as the sixth century A.D.[14]

In the centuries between A.D. 200 and 900, the period of the so-called Classic Maya, astronomical, calendrical, and timekeeping powers reached an apex. The Maya took all these elements and pushed them to a level of originality and brilliance. The Maya were probably the most sophisticated astronomers and mathematicians of their era.

If you visit the ruined plazas of the Classic Maya cities, you will still see carved stelae inscribed with effigies and the exploits of kings and queens. You will read of their royal descents from the gods. All of this is offered side by side with complex calendrical dates fixing the time of year of the event and its position in the ubiquitous 260-day ritual calendar. The royal doings are also accompanied by the correct phase of the moon, its position in the zodiac, the count of the days since the time of Maya creation and even since mythic times predating creation—a number of days running into millions. All were precisely cued to each royal personage or occurrence.[15]

During their Classic Period, the Maya developed a Venus calendar accurate to one day in five hundred years, as well as an eclipse-warning table that still functions in the twenty-first century. They created their own zodiac as well as tables to follow Mars, the moon, Venus, and possibly Jupiter and Mercury. To make all this work congruently, they devised a sophisticated mathematics to facilitate the computations. They projected their astronomical tales hundreds of years forward and backward, even to eras preceding the creation of their contemporary version of the universe. Maya astronomy reached a level comparable to that achieved by the Babylonians and surpassing in some ways the Egyptians'.[16] Almost as remarkable as the precision and scope of Mayan astronomy was their drive to elaborate upon it, a preoccupation with celestial accounting that developed into an unparalleled obsession.[17]

Of the thousands of texts in which the Maya recorded their findings, only four survived the Spanish book burnings. It is as if, observed Mayanologist Michael Coe, the only things the future knew of us were based on three prayer books and *Pilgrim's Progress.*[18]

The Dresden Codex (the names of the codices indicate where they were or are kept, thus the European names) is the most beautiful of the Maya folding-screen texts. It is eight inches high and, when folded out from its accordion-like form, eleven feet long. Written on a long strip of bark paper coated with fine stucco or gesso, it is primarily concerned with the 260-day ritual counts divided up in several ways, the divisions being associated with specific gods.[19]

The Madrid and Paris Codices are less perfect in execution. The Paris Codex is very fragmentary but suggests specific timetables for prognostications of astronomical events. The Grolier Codex (named after the Grolier Club in New York City, where it was exhibited in

1971) is also in bad condition but comprises one-half of a twenty-page table concerning the Venus cycle. The radiocarbon date of A.D. 1230 is now considered accurate, thus making it the earliest of the manuscripts by about twenty years.

Venus is the planet of primal religious significance for the Maya, who made extensive calculations of its multiple apparitions. Unlike the Greeks of the Homeric age, however, the Maya knew that the evening and morning stars were the same object. To chart Venus's synodical period (the time it takes for the planet to return to the same position relative to the earth's orbit around the sun) the Maya used the figure of 584 days (the actual figure is 583.92, nearly too close to call). They divided this number into four periods of varying length; Venus as the morning star was one. The second was Venus's disappearance at superior conjunction—the point when the planet is invisible as it passes behind the sun. The third was its reappearance as the evening star; the fourth, its disappearance again at inferior conjunction—when it is obscured by its passage in front of the sun. Venus's first and last visibilities were of great concern, with the first being especially important in the Dresden Codex.

In 1982 Yale University linguist Floyd Lounsbury brought to light how strongly the Maya linked Venus with warfare, demonstrating that war imagery is associated with the first visibility of Venus in the morning and evening sky. The timing indicates that the highly important war events of the Maya clustered in the dry season, the preferred time for waging battle. Scholars have concluded that war was avoided during times when Venus was invisible in superior conjunction.[20] The visibility of Venus as the evening star on December 3, 735, for instance, set off an attack on the southern Peten site of Seibal, in present-day Guatemala, leading to the capture of its ruler the next day. This unfortunate, says Coe, was kept alive for twelve years, finally being sacrificed at a ritual ball game timed for an inferior conjunction of Venus.

The stationary point (at the end of retrograde period) of Jupiter apparently signaled accession to the throne, or inaugural rituals at Palenque. Ball games and associated bloodletting events seem linked with Jupiter's retrograde period. The accession at age forty-nine and apotheosis twenty-one years later of the great Palenque ruler Kan Balam was set by Jupiter's second stage.[21]

Stars were the "eyes of the night" to the Maya. The Pleiades, as they were named in the Old World, were important calendar stars. Astronomers used a window in the Caracol at Chichén Itzá, in Mexico, to view the Pleiades as they set at dusk in late April, and again before the onset of the rains at the time of the first solar zenith in late May. The Maya visualized Scorpius as a scorpion, and temples were oriented toward the setting points of its stars. There was a long-standing association between the period of Orion's conjunction and maize planting.[22]

Aveni and his colleagues determined that the Maya used buildings and doorways for astronomical sightings, especially of Venus. At Uxmal, the tenth-century capital of an ancient city-state in western Yucatán, all the buildings are aligned in the same direction, except the Governor's Palace. There Aveni discovered that a perpendicular measurement taken from the central doorway reaches a solitary mound 3.5 miles away. Venus would have risen precisely above this mound when the planet reached its southerly extreme in A.D. 750.

In 1975, Aveni determined that the building's orientation and sight lines might approximate this southernmost rise of Venus, an event that takes place only every eight years. In the mid-1990s, David Rosenthal, an explorer, photographer, and Mayan enthusiast, spent months attempting to photograph Venus's southernmost rise at the palace, which is richly adorned with Venus gylphs. After much trial with fog and clouds, Rosenthal finally saw the event one morning in January 1997. And he described something more than the astronomical detail:

> Accounts also indicate the Yucatan climate hasn't changed significantly in the last 1,000 years or so, and this is particularly true in sites as far away from urban areas as Uxmal. Chances are the ancient Maya might have been subject to the same problem I'd experienced.
>
> But was it really a problem? The mist-enshrouded early-

> morning horizon seen from a promontory like the Palace of the Governor appears as the shoreline of an endless ocean. This view is very similar to a Mayan cosmological construct where the edge of the world meets an infinite sea, which in turn constitutes the surface of Xibalba, the Underworld. Like other rising celestial objects, Venus emerges from this mysterious realm to sail across the sky. Could it be that my perspective of this brilliant traveler surging free from shadowy darkness was the very one sought and shared by priest-astronomers more than a millennium before?[23]

With his collaborator Sharon Gibbs, Aveni showed that at around A.D. 1000 the entire Caracol, a round tower with windows at Chichén Itzá, was aligned with the northerly extremes of Venus. Another diagonal sight line through the windows matched the planet's setting position when it attained its maximum southerly position.[24] The Classic Period Castillo at Chichén Itzá dramatically expressed the Maya sun-monument-ritual relationship in its orientations and its four staircases of ninety-one steps per side (which when added to the temple platform as the final step totals 365 steps). At sunset at the equinoxes, shadows formed by the nine levels, or stages, of the pyramid create a great slithering snake design all along the serpent balustrade on the north side of the Castillo. Today, thousands of people come to witness this event.

The habit of incorporating the year's beginning at the winter solstice within the architecture plan of a ceremonial center was the first phase in Mesoamerica of astronomically aligning the city as a whole. Built before the birth of Christ and already abandoned for several centuries during the European Dark Ages, Teotihuacán was carefully planned. The fifty-square-mile ceremonial center was laid out in an east-west axis and grid—approximately 15.5 degrees to the east of north and the west of south.

Further, if you could travel back two thousand years and stand by a marker on the Street of the Dead and look over a petroglyph on the western horizon at the right time of year, you'd see the setting of the

Pleiades star cluster. When the Pleiades reappeared in the east after having been invisible in the light of the sun for forty days, they did so on the precise day of the sun's zenith. Here, says Aveni, was a highly visible, convenient timing mechanism to signal the start of the new year. Tying the sun to stars was different from beginning the solar calendar by marking the sun's northern- or southernmost passage. "The Pleiades, being both prominent and in the right place at the right time, became the new celestial timer of choice" to the astronomers of Teotihaucán.[25]

The Aztecs believed they were the children of the Teotihuacanos, whom they considered gods. When the Aztecs built their capitol, Tenochtitlán, around A.D. 1325,[26] it was with Teotihuacán in mind. The great Templo Mayor in Tenochtitlán was positioned so that the rays of the dawning sun on the spring equinox (usually March 21) would fall in the notch between the twin temples, the shrines of Tlaloc and Huitzilopochtli, on top of the flat pyramid. Measuring the temple ruins, Aveni found that it is skewed nearly 7 degrees south of true east to match the sun's path over the elevated twin temples on equinox day.[27]

LIKE THE MESOAMERICANS, the Incas in ancient Peru (A.D. 1200 to 1532) created an astronomical system extraordinary in its vast organizational structure. The Incas encoded their calendar in architecture, as did the Maya, but they also built a unique system based on the topography of the city, the empire, and the Andean landscape itself.

This astronomy evolved from the Inca *ceque* system, an organizational device for the recording of time.[28] The *ceque* (which means "ray") system grew out of the Inca capitol. Cuzco lies in latitude 13.5 south, at the confluence of two rivers in a 10,500-foot-high mountain valley. The Incas may have called it Tahuantinsuyu, or "the Four Quarters of the Universe." Cuzco was ground zero of the *ceque* system, which Aveni describes as a giant cosmogram, "a mnemonic map built into Cuzco's natural and manmade topography." A semiabstract, conceptual system, it consisted of a number of imaginary radial lines, the *ceques,* grouped like spokes on a wheel, radiating out from Cuzco and extending to the ends of the empire. The wheel's hub, the system's epicenter, was the Coricancha, Cuzco's sacred temple of ancestor worship.[29] The *ceque* system unified Inca ideas about religion, social organization, calendar, astronomy, and hydrology.[30]

There were a total of forty-one *ceque* lines. Each was traceable by the line of small shrines or sacred natural spots, called *huacas,* leading outward from the Coricancha across the landscape. There were 328 in all. Each *ceque* had a kin or social group designated responsible for maintaining the *huacas,* which were located at intervals along each line.[31] Stone pillars or other landmarks along the visible horizon marked the positions of important celestial objects. By direct sighting of celestial rises and settings over particular *huacas,* the Incan astronomers kept accurate records of important dates of the seasonal year without needing to write them down.[32] The land itself served as codices.

The stone pillars marked the sun's passage at the middle of August, which signaled the beginning of the planting season. But the timing of the planting season occurred at slightly different times at higher or lower mountain elevations. The Incas placed the *huacas* at different elevations to record the sunrise at different times, so that planting would begin on the optimal date for that altitude.[33]

According to anthropologist R. Tom Zuidema, who conducted extensive fieldwork on Andean astronomy, the *ceque* mapping system was not only a directional scheme incorporating astronomical events happening at the horizon; it was also a seasonal calendar, with each *huaca* representing a day in the year, and a cluster of *huacas* representing a lunar month.[34] Zuidema and others saw the *ceque* system as a macrocosmic version of the quipu, the all-purpose Incan accounting device. The quipu was a set of colored strings attached to a main cord. Knots on each string represented the numerical equivalent of the items being counted. Zuidema and others perceived the *ceque* system as a giant quipu overlying Cuzco; the *ceque* lines were the cords; the *huacas,* the knots.[35]

Although the Incas rose to empire in less than a century, before they were destroyed by the Spanish invasion, there are precursors to the *ceque* system in the artifacts of previous cultures. Investigating the famous Nasca lines on the desert coast of Peru, Aveni detected the ancient traces of the *ceque* system. Built by the Nasca people in the first centuries A.D., the lines consist of some geometrical figures and animals, but primarily of straight lines etched on the desert. There are approximately eight hundred of them, some several miles long, emanating from sixty-two focal points. The whole organized pattern resembles an assemblage of *ceque* systems tied together in a network spread across one hundred square miles. Aveni's research strongly suggests that these were pathways, probably walked over by participants in a rain ritual, a kind of rain dance. Most begin and end at water sources. Sun watching may have played a role, too. A significant number of the lines point to the place where the sun rises during the season of the year when water begins to run in rivers and underground canals.[36]

Zuidema theorized that the Incas relied on a series of stone-pillar celestial markers, but for many years there was no evidence to back this up. Throughout the 1980s, however, University of Chicago archaeologist Brian Bauer and Lawrence Livermore Laboratory astrophysicist David Dearborn and their colleagues searched around Cuzco for the pillars described by sixteenth-century Spanish chroniclers who wrote that the structures were large enough to be seen against the setting sun at a distance of nine miles. One such group of pillars marked where the sun sets at the June solstice, which is the northernmost point at which the sun crosses the horizon. A combination of postconquest looting and recent urban growth in the Cuzco valley has destroyed the area where the Cuzco pillars once stood. According to Bauer, many scholars of Latin American antiquity believe that the Incas built the large pillars to record the sun's horizon location at the June and December solstices, but archaeologists had not yet found physical evidence of the pillars and there had been no detailed investigation into the organization of the solstice rituals, though this is the thrust of Zuidema's current research.[37]

During a survey of pre-Hispanic sites on the Island of the Sun, in Lake Titicaca (on the Peru-Bolivia border), Bauer, Dearborn, and others discovered the remains of two stone pillars. They also found a large platform area just outside the walls of a sanctuary on the island. The team's archaeological and astronomical research, which they presented in a 1998 issue of *Latin American Antiquity,* suggests that the Incas used the site to support the ruling elites' claim to power through elaborate solar rituals. In the early fifteenth century the Inca empire expanded into the Lake Titicaca region and usurped the Island of the Sun from local control. The island and a sacred rock, which locals believed was the birthplace of the sun, had been the focus of worship for centuries. Under the Incas it became one of the most important pilgrimage centers in South America.

The team's research indicates that on the June solstice, the Incan king and the high priests of the empire assembled in a small plaza beside the sacred rock to witness the setting of the sun between the stone pillars. Their findings also indicate that, as the elites paid homage to the sun from within the sanctuary, pilgrims observed the event from a second platform outside the sanctuary wall. From the perspective of the pilgrims, the sun set between the stone pillars and directly over the ruling elite, who called themselves the children of the sun. "While both groups participated in solar worship, the non-elites simultaneously offered respect to the sun and the children of that deity. This physical segregation emphasized that the Inca alone had direct access to the powers of the sun," Bauer wrote.[38]

David Dearborn's study of the Incas began in the early 1980s, at Machu Picchu, in Peru. He and his colleagues found archaeological and ethnohistorical evidence to support the use of certain structures at Machu Picchu and other monuments as working observatories where the Incas monitored the sun's motion. In one of Machu Picchu's most exquisitely masoned buildings, often called the Torreón, Dearborn discovered that one window centered on the June solstice sunrise. At another of the Torreón's windows, to the southeast, when you sit on the floor with your back to the room's altar, you can see the stars in the tail of Scorpius—a constellation in the Andes sometimes known as Collca, the storehouse—rising through the window. In the Incan era these stars would have been rising when the June solstice sun set. Anthropologist

Gary Urton noted that the name of Collca was also given to the Pleiades. In Incan times, beginning about a month before the June solstice, these two groups of stars were on opposite sides of the sky. The Pleiades appeared in the morning sky, rising in the winter solstice window.[39]

LESS OBVIOUS than the astronomy of the Mesoamericas and the Incas is the sky-watching skill of the North American peoples. Yet clearly they developed a body of knowledge about the nighttime skies. One striking phenomenon is the "sun dagger." Its interpretation is dubious, according to some researchers, but we would be remiss in not mentioning it.

On June 29, 1977, Anna Sofaer, an artist studying ancient Anasazi rock art, was dangling on the side of Fajada Butte, 405 feet above the floor of Chaco Canyon, New Mexico. (The Anasazi were cliff-dwelling Native Americans of the Southwest.) She had climbed to this precarious spot to have a look at a pair of spiral petroglyphs known to be sheltered behind three slabs of sandstone, propped upright against a decorated wall. Near noon, Sofaer saw the shade beneath the slabs pierced by what she called a "dagger" of sunlight. It nearly bisected the larger, footwide spiral, taking almost twelve minutes to pass through it. Knowing the summer solstice had just passed, Sofaer thought the location might have been designed to mark it. She returned to the same high perch at monthly intervals, as well as at the equinoxes and solstices.

With other experts, Sofaer became convinced the Fajada petroglyphs were an accurate and precise calendar marker. She and other researchers observed that on the next summer solstice, sunlight first slipped between the middle and the right slabs about an hour before noon. A spot of light glowed at the upper rim of the larger spiral, grew into a thin dagger, and sliced down through the carved turns, splitting them through the center. Eighteen minutes after the first appearance, the light vanished. Throughout the year, a spot of sunlight appeared near the small spiral and also turned into a dagger. At the winter solstice two light daggers framed the large spiral for forty-nine minutes at noontime, when sunlight entered between the three uprights.

We cannot know with certainty if the entire structure was constructed intentionally for astronomical purposes or whether the rock formations simply had fallen there and been used serendipitously. Astronomers believe that this is an early form of Anasazi sun clock (dating to around A.D. 1000) that functioned as an approximate solstice marker. Whether the the sun daggers were intended to function precisely is open to question, as is a more recent theory that the spirals designated the limits of the shadow cast by the 18.6-year oscillation of the rising full moon and were used as part of a scheme for predicting eclipses.[40]

There are similar examples of the Anasazi technique for marking solstices in the Four Corners region of the Southwest. At Colorado's Hovenweep National Monument, the ancient lost people occupied many hidden canyons. In one canyon, in what is called the Holly House group, below something called the "sun room," an inner wall of rock corridor is decorated with many petroglyphs, including a set of concentric circles and two spirals. On mornings near the summer solstice, sunlight penetrates the crack between the overhang and a block of stone that forms the opposite corridor wall. At that point, two daggers of light appear on the decorated panel. Both extend horizontally across the south wall, and the left dagger cuts through the spirals. The right dagger bisects the rings. As the morning progresses, the dagger points meet in a symphony of light playing across the rock.[41]

In the early 1950s British astronomer Fred Hoyle suggested that records of the great 1054 supernova, a bright exploding star, could appear concurrent with a crescent moon in the rock art of the U.S. Southwest.[42] Soon thereafter Mount Wilson astronomer William C. Miller found possible supernova depictions. One Chaco Canyon petroglyph in particular has become renowned as a record of the 1054 supernova. In 1975, astronomers John Brandt of the University of New Mexico at Albuquerque and Ray Williamson of George Washington University made calculations that place this extraordinarily bright starry event next to a crescent moon on the morning of July 5, 1054, more or less the way it is portrayed on the petroglyph. Some critics say that simpler explanations exist. It is far more likely, they say, that the petroglyph represents the crescent moon sliding by Venus as the morning or evening star.[43] The controversy continues.

After Sofaer's rediscovery of the sun daggers at Fajada Butte, interest in Anasazi astronomy burgeoned until, according to astronomer Von Del Chamberlain of the Hansen Planetarium in Salt Lake City, Utah, "A host of people were ready to leave no piece of rock art untouched by sunlight, moonlight, starlight, infrared light, or the absence of any of these. The quest for light and shadow casting on rock art seems close to being a religion with devotees scrambling over the rocks at solstice, equinox, and more recently cross-quarter dates to watch in awe as photons beam or shadows touch the enchanting figures left by ancient peoples."[44]

It would be a mistake to discount the Native Americans' quests for astronomical knowledge merely because there was little written transmission, and because the Europeans who did write about it were biased. Indigenous peoples tended to integrate their knowledge in different epistemological sets than did the Europeans, who dismissed valid information as native superstition. Renaming a sacred spot "Devil's Tower," notes archaeoastronomer Paula Giese, is all too common across the continent. In fact, she adds, if you see such religiously pejorative names attached to geological features, you can be fairly certain the site was once sacred to some indigenous peoples.[45]

THE PREHISTORIC PEOPLES of eastern North America often built earthworks, constructing thousands of mounds and pyramids, probably in the service of astronomy as well as politics and religion. One of the greatest remaining is the so-called Monk's Mound (named after a nearby nineteenth-century Trappist monastery) found near what is now Cahokia, Illinois, eight miles east of downtown St. Louis. The mound is located by the confluence of two of the continent's mightiest rivers, the Mississippi and Missouri. Archaeologists posit that to build it, the Cahokians hauled fifty-five-pound loads of dirt on their backs from nearby pits. They must have done this 14.7 million times over three centuries to construct one rectangular platform on top of another until the twenty-two-million-cubic-foot mound was finished, in around A.D. 1000.

Somewhat more speculative is "Woodhenge," a reconstructed circle of forty-eight wooden posts west of Monk's Mound. It's called Woodhenge because of its general functional similarity ("*verrry* general," says Aveni) to Britain's Stonehenge, the circle of large stones, erected around 1500 to 2000 B.C. on the Salisbury Plain, that some think was used for astronomical purposes. Woodhenge's circle may have served a calendrical purpose, since a pole at the center, when aligned with the circle's eastermost post at the front of Monk's Mound, marks the equinoxes.[46]

To the Cahokians, the sun, rather than the moon, was of primary importance. By following the sun's annual path along the horizon, the rulers of this vast economic hub could regulate the seasonal flow of goods and services and plan ceremonies and holidays to take place in front of the mound.[47] It is now protected as Cahokia Mound State Historic Site.

In other places this same "Mississippian" culture would manifest these characteristic astronomically organized earthen designs, often truncated pyramid mounds having possible calendrical functions. The Mississippians lived into the sixteenth century and vanished in the wake of Hernando de Soto's epidemic-spreading advance. Most of their monumental geometric forms followed them into oblivion. (Epidemiologists are not sure what disease de Soto spread except that it was probably carried by the herd of pigs the Spaniards took with them for food.)[48]

We know less about the early dwellers of the Great Plains than about their contemporaries to the south and their relatives in Mesoamerica. "I am surprised we know anything about them at all considering how few there were and over what a great area they ranged," comments John A. Eddy.[49] Eddy, a solar astronomer at the High Altitude Observatory of the National Center for Atmospheric Research in Boulder, discovered a possible reason why the Plains hunter-gatherers built monuments of stone circles or medicine wheels, about fifty of which are known to exist today on the Great Plains, the east ridges of the Rockies, and the grassy plains of Canada. Eddy's analysis lends credibility to the astronomical sophistication of the Great Plains Indians. His interpretation is widely disputed, by Aveni and others.[50] I would be remiss in ignoring Eddy's findings, but I present them as controversial.

In the 1970s Eddy began his astronomical analysis of the wheels, focusing on the Big Horn Medicine Wheel, located in Wyoming at 9,642 feet elevation on a windswept flank of Medicine Mountain. The wheel is a collection of small cairns (conical piles of rocks) and spokes, with a central cairn two feet in diameter and two feet high. The ring itself is no higher than the scattered rocks that define it, but its largest diameter is eighty-seven feet—only slightly smaller than the large circle of Stonehenge in Britain.[51]

Radiating from the central cairn to the rim are twenty-eight spokes made of piles of boulders. The spokes terminate with cairns. Eddy discovered that the view from one cairn at the end of a spoke, across the center cairn, and on to a low ridge to the northeast aligned with the summer-solstice sunrise. He hypothesized that other spoke-cairn relationships would reveal a consistent set of alignments, including three whose lines oriented on the rising points of four stars: Aldebaran, Rigel, Fomalhaut, and Sirius. In centuries past, these stars rose heliacally, just flashing as they came up within a day or two of the summer solstice from A.D. 1600 to 1800. (Precession has shifted the stars now, with respect to the solstices.) Today experts do not agree on whether the Bighorn Medicine Wheel was designed to give precise determination of information about the summer solstice[52] or if the wheel served only as a ceremonial place—or both. Some even call the medicine wheels early solar-star analog computers.[53] The spot, revered as sacred by many Indian peoples, is designated as Medicine Wheel National Historic Landmark within Bighorn National Forest and is visited by as many as seventy thousand tourists during the summer months, when the site is accessible.

The existence of the medicine wheels is evidence that the nomadic peoples of the Great Plains had a deep interest in the night sky of blazing stars. They also must have had a need for navigational tools provided by the celestial bodies to guide them on journeys through the often featureless expanses of the plains.[54] In 1977, Eddy verified that another ruin, 425 miles north of Bighorn in Saskatchewan's Moose Mountain Park, had the same basic plan as the Wyoming wheel. The alignments were the same. The oldest parts of the Moose Mountain wheels may be two thousand years old, part of a millenniums-old shamanic tradition.[55] The similarity of the wheels' patterns and proposed use, says Eddy, suggests that a Plains people may have used a sky calendar for at least a thousand years, and that the summer-solstice dawn stars were an important part of an abiding lore.[56] "The problem is," says Aveni, "that there are dozens of these wheels. Eddy only picked two!"[57]

Perhaps the most extensive record of constellations in North America comes from the Pawnee of central Kansas and Nebraska. A star map painted on buckskin, now on display at Chicago's Field Museum, contains hundreds of star symbols. The map, possibly more than three hundred years old, depicts stars of various magnitudes by simple crosses of different sizes. A band of stars denotes the Milky Way, with the winter constellations shown at the left and the summer heavens at the right. Constellations include Lyra, Ursa Major and Minor, Coma Berenices, and Andromeda, all indicated much as we see them today.[58]

The Dipper bowls and handles and Cassiopeia are recognizable. Corona Borealis (called "Chiefs in Council" by the Pawnee) seems especially prominent and exaggerated. Painted on the opposite side of the Milky Way are the Pleiades. Farther off center is Orion's belt. The only surviving artifact of its kind, the star map was clearly a sacred text of enormous power, as well as a mnemonic device for remembering the many stories that go with the appearances of various constellations.[59]

Although there may have been European influences in the making of the star map, the Pawnee were keenly interested in celestial configurations. They arranged their permanent villages in a prescribed order, with four subvillages formed around a central group placed as if at the corners of a great square. At the west end of an imaginary line through the center of the square was the fifth village, with a shrine derived from the positions of the star of the west, or the evening star. At the opposite end was a village with the shrine of the star of the east, or morning star. Around this basic grouping they put seven other villages so that the arrangement on the Nebraska landscape mirrored the pattern formed by the patron stars in the sky.

The star cult was also reflected in the plan and construction of the Skidi Pawnee earth lodge, the usual type of habitation at permanent villages. The circular floor, from twenty to fifty feet in diameter, symbolized the earth; its domed roof-wall superstructure was a microcosm of the sky. Around the central fire pit—which held a small bit of the sun, the fire—the superstructure was supported by four large posts set more or less in the semicardinal directions (northwest, southwest, southeast, northeast). The Pawnee sometimes painted these posts in colors coded to the directions. The lodge's tunnel-like doorway opened toward the east, so that the rising sun could shine upon the household altar.[60]

Von Del Chamberlain, of the Hansen Planetarium, who has analyzed the structure of the Pawnee lodge, sees it as a functioning calendar. The path of the shaft of noonday sunlight entering the lodge's smoke hole would change with the year's course, extending only partway down the wall at the winter solstice and all the way to the floor by mid-February. By the time the Pawnees abandoned the lodges for their summer tipis, the solar image would have migrated to a position close to the center of the lodge.

An observer could have briefly glimpsed the star groups the Pawnee recognized if he had sat against the wall along the lodge's axis of symmetry just before sunrise in late July, then again just after sunset around the time of the winter solstice. The Corona Borealis (the Pawnee "Chiefs in Council") would enter the smoke hole directly opposite in time to the Pleiades. This may explain their opposed location in space on the Pawnee buckskin star map. Pawnee star charts served as a kind of calendar, and the stars were positioned so as to express their relationships over a year's time, rather than being a snapshot of the heavens at any one moment. The spatial positioning of Pleiades and Corona B. was actually temporal, vaguely like the face of a clock. Chamberlain thinks that rather than being solely astronomical observatories or calendrical time slots, the lodge and its apertures were living astronomical schoolrooms, through which visually dramatic star and sun scenes could enhance stories and moral tales.[61]

Oceania

Aside from the Vikings, the Oceanic peoples were arguably the most accomplished mariners in the world before advanced instrumentation made long-distance navigation a relatively sure thing. Long before Columbus, two-way voyaging was common throughout the many scattered islands and across the great expanse of the Pacific Triangle, bounded by the Easter Islands, the Hawaiian Islands, and New Zealand. For Oceanic peoples such as the Polynesians and Micronesians, celestial bodies served as accurate navigational instruments.[62]

The ability of the Pacific peoples to reach out over many atolls and make voyages over vast stretches of open ocean using minimal references required an expert knowledge of navigation and astronomy. In the Gilbert Islands of Micronesia, there was no word for *astronomer;* if you wanted an expert in the stars you asked for a *tiaborau,* a navigator.[63] There may have been no one more esteemed in a local society.[64]

Polynesia's location facilitated its people's remarkable celestial observations and seafaring acumen. Like the astronomical civilizations of Central America, the Polynesians lived close to the equator. Near-equatorial latitudes offer a much more symmetrically partitioned sky than latitudes farther away from the equator. In the tropics, an observer sees the motion of heavenly objects as heading straight up in the east and straight down in the west. The observer seems to be at the center of things, with the north and south hemispheres behaving identically.[65] Consequently, using celestial objects as navigational agents was much easier for equatorial sea voyagers, and they developed highly efficient star compasses. Theirs is a more orderly sky than the one seen by people who live farther north or south, where the stars appear to move westerly on a slant either to the south or north, depending on whether the viewer is in the northern or southern hemisphere.

The Gilbert Islands, at a latitude 3 degrees south, lie almost on top of the equator. So these islanders, like most Polynesians, observed the east-to-west-moving stars rising and setting in almost vertical east-to-west directions, and divided the sky into symmetrical boxes according to the cardinal directions. This gave them a method for describing the location of a star or constellation in terms of its position within one of these imaginary boxes.[66]

The transmission of knowledge about the star charts was a crucial part of Oceanic culture. Instruction started young. The following journal entry from the missionary vessel *Southern Cross,* traveling in the late 1800s, describes three boys from the southwest Pacific Santa Cruz Islands. The oldest boy was

> teaching the names of various stars to his younger companions, and [I] was surprised at the number [of stars] he knew by name. Moreover, at any time of night or day, in whatsoever direction we might happen to be steering, these boys, even the youngest of the three, a lad of ten or twelve, would be able to point to where his home lay; this I have found them able to do many hundreds of miles to the south of the Santa Cruz group.[67]

Even today on some islands there remain fixed stone structures, sometimes called stone canoes, aligned with the constellations. These stone canoes served as navigational trainers, like flight simulators. Each pair of stones on the canoe was aligned with the place where certain stars would appear or disappear on the sea horizon at different times during the night. In August on one island, for instance, the bright star Regulus lined up with one stone pair at sunset, while at midnight Arcturus gave the same bearing. A student would sit between the stones, face in one of the cardinal directions, and memorize the constellations he saw and the bearings they indicated.

Europeans sailing the Pacific were impressed with the orientation ability of Pacific navigators, although explorers such as Captain James Cook, in the late eighteenth century, never realized the extent to which

Polynesians used celestial navigation. His crew recorded an encounter with Tupaia, a Tahitian navigator who accurately demonstrated the position of numerous island chains from the Marquesas to Fiji totally by memory, an area greater than the span of the Atlantic Ocean and containing multitudes of islands. Tupaia led Cook to many islands unknown to Europeans.[68]

The Polynesians used the fixed rising and setting points of stars to establish north, south, east, west, and every direction in between.[69] Navigators used zenith stars, bright stars they recognized as passing the direct overhead position, or zenith, from specific islands. The declination—the angular distance north or south of the equator—of the zenith star was equal to the latitude with which it was associated. So navigators could associate every island with one or more of its own zenithal guide stars. For example, Sirius is the zenithal guide star for the Fiji Islands, at latitude 17 degrees south; Rigel, the zenith star of the Solomons, at latitude 7 degrees south; Altair, for the Carolines, at 9 degrees north.[70] If a navigator knew the zenith stars of different latitudes, he could tell what latitude he was on by observing which star passed directly overhead at night.

The islanders also used pairs of stars that rise or set at the same time as clues to latitude. Star pairs rise and set together only at specific latitudes. For example, when Sirius and Pollux set together, the observer is at the latitude of Tahiti, 17 degrees south. As the observer moves north or south of that latitude, one star will begin to rise or set before or after the other star. This strategy was easier to use with setting, rather than rising, star pairs because the navigator could watch the pair as it sank toward the horizon instead of trying to anticipate its appearance.[71]

Traversing the larger Pacific Ocean did not seem insurmountable to the Polynesians. They connected Asia to the Pacific Islands at the very least. Those who were sent on land-discovering missions had to master the navigational skills to find their way home. To many contemporary people of the mid-Pacific, their ancestors were true "Vikings of the Sunrise," a name coined by a Maori ethnologist, Te Rangi Hiroa (also called Sir Peter Buck, 1880–1951). Polynesians, he thought, developed "highways" on the ocean that were charted in the heavens above. And lacking the fear of falling off the edge of a "flat" earth, they were further encouraged in their desire to sail far off into the horizon, to the Americas.

Te Rangi Hiroa surmised that Polynesian navigators had already sailed to the Americas—as recorded in oral history, chants, songs, and traditional lore—centuries before Columbus. The fact that the Polynesians had settled every inhabitable island from Hawaii to New Zealand to Easter Island centuries before Europeans arrived is evidence enough of their ability to accomplish the voyages. Most hard evidence for Polynesian-American contacts has vanished, however, except one, in the form of a tuber: the sweet potato (*Ipomoea batatas*).

Botanists have determined that this staple food crop, common to all of the Polynesian islands, is native to South America. Either Polynesians made round-trip sailings and returned with sweet potatoes or American Indians brought them to the Pacific. Somehow the sweet potato was transferred from South America to Polynesia between A.D. 400 and 700. (Experts say birds could not have transported it.) Linguistic evidence has established the Peruvian and Ecuadorian *kumar* as the root word for *kumara, kumala,* and *'uala,* varieties of names for the sweet potato in Polynesia.[72] There may, of course, be other explanations. We cite this evidence cautiously.

The Old World

It is tempting to say that Old World ancient astronomy was more advanced than that in the New World because it introduced instrumentation to the science of star watching. There were no telescopes, of course—that innovation belongs to the West—but Chinese and Islamic astronomers did develop elaborate metal sighting devices to chart the heavens. There is more to astronomy than hardware, however. First of all, as we've seen, New World astronomers used architecture, *huacas,* windows and doorways, pillars, houses, mountains, sun daggers, and other natural structures to delineate the movements of celestial phenomena; in some ways, such methods are more creative than quadrants and the like.

Beyond instrumentation, the ancient Old World astronomers' biggest contribution was applying mathematics to the skies, setting up a rigorous basis for astronomy based on nothing more than the naked eye and a grasp of logic.

Mesopotamia

Mesopotamian astronomy constitutes one of the earliest systematic, scientific treatments of the physical world. Ancient astronomers, seeking to forecast the future by means of the heavens, had developed a complex system of arithmetic progressions and methods of approximation by the fourth century B.C. Since they could not see what lay ahead in human life, they became adept at predicting celestial events. The mass of observations they collected and their mathematical methods were crucial contributions to the later flowering of astronomy among the Indians and Muslims as well as the Greeks.

For more than two thousand years, the efforts of Mesopotamian astronomers lay forgotten under the ruins of palaces and ziggurats in what is now mainly Iraq. All that was known of the subject came from a few passages in the Bible and reports of Greek and Roman writers. But the reports were tantalizing. The Roman scholar Pliny the Elder, for instance, wrote that the Babylonians inscribed observations of the stars on baked-clay tablets for 720,000 years, a number doubled several centuries later by a Greek philosopher, Simplicius, to the astounding figure of 1,440,000 years.[73]

In the mid–nineteenth century, archaeologists began to unearth thousands of these tablets inscribed with cuneiform writing in Mesopotamia. One hundred years later, an estimated half million tablets were in museums around the world. During the brief cease-fires during the conflicts with Iran, international teams of archaeologists ran to the Iraqi fields and dug at an unprecedented rate to find more. At the site of the ancient city of Sippar, just southwest of Baghdad, for example, excavators discovered a library from the late Babylonian empire containing a huge cache of astronomical records and mathematical exercises. But when Iraq invaded Kuwait, all archaeological activity ceased, and the tablets were supposedly shelved in Baghdad somewhere, to join all but a few of the hundred or so tablets that have been translated to date.[74] What we know about ancient Near Eastern astronomy may constitute just the beginning of the story.

The fraction of translated texts reveals the presence of an astronomy in Mesopotamia that goes back at least as far as the eighteenth century B.C. The Sumerians, who invented the cuneiform writing system shortly before 3000 B.C., were the first to catalog the brightest stars, outline a rudimentary set of zodiacal constellations, note the movements of the five visible planets (Mercury, Venus, Mars, Jupiter, and Saturn), and chart the motions of the sun and moon against the background of the constellations. They gave names to the constellations, appellations still familiar in some cases—Scorpio, Taurus, Leo. The zodiac with these names was in use throughout the rises and collapses of Mesopotamian empires, down to the last days of Babylonian astronomy at the beginning of the Christian era.[75]

The Sumerians may have been the world's first people to develop a calendar based entirely on the recurrence of the complete, or synodic, phases of the moon and to use the moon's synodic periods as the basis of a twelve-month, 360-day year. To keep the lunar-year calendar in step with the solar year, they intercalated an extra month every now and then, probably when the royal astrologers realized the calendar had fallen severely out of step with the seasons. The official decision to intercalate a month was taken by King Hammurabi of Babylon (1792–1750 B.C.). The later Babylonians knew that the 360-day year did not match the lunisolar year, and may not have used the solar year much.[76]

Throughout the long history of Mesopotamia, timekeeping centered on one primary problem—knowing when the new moon crescent would first be visible at rising time each period. Predicting the positions of the moon phase, or period, and these periods' correspondence to the periods of the planets, evolved into the later Babylonians' celestial obsession. Their solutions to these problems would in turn evolve into a science and a method of scientific thinking we use today.

But first they needed a moon-based calendar. Around 1000 B.C., the Assyrians spelled out the rules of this calendar, called MUL.APIN (the Plow), the first compendium of its kind. Each month began precisely at sunset with the first visible crescent of the new moon.[77] MUL.APIN had its roots in a calendrical scheme developed even earlier, during the Ur II period, around the twenty-first century B.C.—and perhaps even earlier than that, around 2900 B.C. Thus its design—which we use in part today—could have been in place at the earliest moments of civilization. The calendar specifies twelve months to a year, thirty days to a month, and 360 days to a year and is the ancestor of our division of the circumference of a circle into 360 degrees.

Around 500 B.C., the Babylonians set in place their final zodiacal system of twelve signs in 30-degree intervals. It was a reference system for the position of the moon and planets and the fundamental system for Babylonian mathematical astronomy to come.[78] The earliest extant Babylonian texts were written in astrological format during the reign of King Ammizaduqa (1702–1682 B.C.). The great Enuma Anu Enlil ("When the gods Anu and Enlil . . ."), a Babylonian equivalent of the Mayan Dresden Codex and other codices, may have roots in the lunar-eclipse omens from the dynasty of Akkad and Ur, late in the third millennium B.C.[79]

Centuries' worth of Venus observations were incorporated in Enuma Anu Enlil, particularly in the so-called Venus Tablet 63, more popularly known as the Venus Tablet of King Ammizaduga.[80] In each written statement, Venus (Ishtar) was said to disappear on a particular day of a given month and return on another. Today's investigators, using computer software packages, have zeroed in on 1581 B.C. as the most likely choice for the start of the Venus synodic run. One of the Enuma Anu Enlil's primary ephemerides, or databases, made it possible to forecast lunar eclipses and record the intervals between successive eclipses.[81] In all astronomy-centric cultures, eclipses were the first celestial phenomena from which predictions were derived. And understandably so, for the eclipses were stressful events, terrifyingly inexplicable as long as the conditions for their occurrences were not understood. In the violent world of Mesopotamia—with its continuous wars—where there was a need to make predictions and hedge against upheavals, both natural and man-made, prognostications must have seemed somewhat more crucial than even the need for stock market analysts' forecasting today.

As in other schools of early astronomy, divination was a driving force in the development of Mesopotamian celestial science—although there is debate among experts about how much the early diviners bifurcated into the separate disciplines of astrology and astronomy, especially during the ascendancy of Babylonian mathematical astronomy. Divination established considerable motivation for the development of a predictive astronomy, but the content of the mathematical astronomy that emerged may not be justified solely on the basis of the needs of omens.[82] Even if the motivation for mathematical astronomy had been astrological, the level of sophistication that emerged, in terms of its predictive range and conceptual grasp of the celestial phenomena, far exceeded anything reflected in the omen literature.[83]

The observations upon which the Enuma Anu Enlil was built hint at the mathematical astronomy to come, especially in the rising and setting phenomena of the moon and planets. Present in the Enuma is already a mathematical function describing the amount of time the moon is visible over the course of a month, as well as another elaborating a variation of this function over a year's time. Both functions give values in time-degrees (1 time-degree = $\frac{1}{360}$ of a day = 4 minutes), reflecting the use of this unit in the seventeenth century B.C. Both computational devices use proper sexagesimal place-value notation to express these values.[84]

The usefulness of the Babylonian sexagesimal numerical system prevails even today. Although the use of base 60 is its most conspicuous feature, it was not essential for the success of the system. The real advantage of the Babylonian number system in astronomy and elsewhere is this place-value notation. Its invention, Otto Neugebauer says, can be compared to the invention of the alphabet.[85] Place-value notation allowed for the development of an algebraic arithmetic mode.

From the outset, the Babylonians would treat elementary geometrical problems in algebraic form. They chose to account for celestial motions in a primarily temporal mode, as opposed to the Greek spatial, or geometric, one. Thus Babylonian algebraic and place-value notations became the foundation of a theoretical astronomy of mathematical character. This astronomy reduced empirical data to the minimum. It took celestial phenomena of a rather complicated character and found simple mathematical functions whose combination described the phenomena cleverly and elegantly.[86]

From the eighth century B.C. on, the Babylonian compulsion to pile up records of astronomical observations went full steam. The origins of computation began in these copious records of the motions of moon, planets, and sun. Ptolemy remarked that the earliest observations available to him came from the reign of King Nabonassar (747–734 B.C.), and he used eclipse records from that reign in his own computations. Around that time Babylonian astronomers starting keeping daily, monthly, and annual "diaries" of observations (fragments of which survive in the British Museum). The diaries typically contain, for each month: time intervals for the setting and rising of the sun and the moon in the middle of the month; descriptions of lunar and solar eclipses; and dates on which the moon approached the so-called Normal Stars. These were a group of thirty-one stars in the zodiac belt the Babylonians used as reference points for the movement of the moon and planets.[87]

The goal of Babylonian astronomy was to be able to compute, starting from a few empirical elements, the positions of the celestial bodies for any given moment. Over centuries of observation and recordings, patterns began to reveal themselves. The huge mass of collected data provided the astronomers with fairly accurate average values for the times of movements for the moon, eclipses, and planetary events. Once they had such averages, they could make short-term predictions by calculation methods we today would term linear extrapolation. This forecasting method, according to Anthony Aveni,[88] was based on a simple sequence:

PLACE + SPATIAL INTERVAL = FUTURE SPACE

or

TIME + TEMPORAL INTERVAL = FUTURE TIME

The first formula charts the future place in the sky where an event ought to be observed; the second, the future time when a phenomenon should take place. Aveni demonstrates how the Babylonians might, for example, use their tables to compute where the moon would be next month in Aries. The first place in the constellation of Aries they'd mark as 2 degrees 02′ 06″ 20′ ″ (the ′ ″, which stands for 1/60 of a second, is no longer used in our geometry). To that line the astronomer would add the interval: 28 degrees 50′ 39″ 18′ ″. The sum would be 30 degrees 52′ 45″ 38′ ″, the distance into Aries where the conjunction would occur.[89]

By the fifth century B.C., Babylonian astronomers had begun experimenting with these radically new techniques for predicting celestial phenomena. Purely mathematical in nature and rational in approach, they entailed separating data into components describable by mathematical functions and combined to predict the events in question.[90] By this time, the astronomers had realized that they must account for the fact that the apparent motions of the sun and moon from west to east around the zodiac do not have constant speeds. These objects appear to move with increasing speed for half of each revolution to a definite maximum, and then decrease in speed to the former minimum. The astronomers worked to represent this cycle arithmetically—by giving the moon, for example, a fixed speed for its motion during the first half of its cycle and a different fixed speed for the other half. The mathematical tables that resulted from this effort represent Babylon's principal contributions to the science of astronomy.

Some historians of science think that early mathematical methods were developed to a certain level by a single, unidentified man. He conceived a new idea that rapidly led to a systematic method of long-range prediction. This idea, now familiar to every scientist, consists of considering a complicated periodic event as the result of a number of smaller periodic effects, each of a character simpler than the actual phenomenon. "The whole method probably originated in the theory of the moon, where we find it at its highest perfection," writes Neugebauer.

The Babylonians could easily calculate the movements of new moons if the sun and moon each moved with constant velocity. Perhaps they assumed this to be the case and used average values for this ideal movement: that gave them average positions for the new moons. The actual movement deviates from this average but oscillates around it periodically. Some treated these deviations as new periodic phenomena and, for the sake of easier mathematical treatment, considered them as linearly increasing and decreasing.[91]

The astronomer refined the mathematical method by representing the speed of the moon as a factor that increases linearly from the minimum to the maximum during half of its revolution, then decreases from maximum back to the minimum at the end of the cycle. Using contemporary graphic descriptions of this model, one would approximate the moon's speed plotted against the sequence of months as a zigzag function made up of alternating sets of sloping straight lines. The ancient astronomer did it with numbers.[92]

This unnamed astronomer also realized that one could treat additional deviations in orbits using a similar method. Thus, starting with average positions, he applied the corrections required by the object's periodic tables that led to a close description of the actual facts. What we have here, says Neugebauer, is "the nucleus, the idea of 'perturbations,' which is so fundamental to all phases of the development of celestial mechanics, whence it spread into every branch of exact science."[93] It's not clear when and by whom this idea was first employed, but the consistency and uniformity of its application in some lunar texts point to an invention by a single person. A few of these tablets, which originated in the cities of Babylon and Uruk, on the Euphrates River, bear the names of Naburiannu, who lived in about 491 B.C., and Kidinnu (about

379 B.C.), astrologers who may have invented these systems of calculation.[94]

The lunar computational theories are known by scholars today as System A and System B. Each consists of a set of arithmetic functions, including the so-called zigzag function, tabulated in columns in lunar ephemerides (tables showing the daily position of the moon) and in auxiliary tables by which are calculated the times, dates, and magnitude of eclipses. The powerful theory of System A was exceeded, six hundred years later, only by the *Almagest,* Ptolemy's second-century A.D. thirteen-volume work detailing the classic mathematical and astronomical accomplishments of the Greeks. These mathematical constructs of Babylonian astronomy invite the question of whether their invention was not a transforming event in the creation of science as we know it.[95]

In System A, one admiring expert, John Britton, says, remarkable accuracy pervades the theory, including an evident mastery of all aspects of the properties and behavior of linear zigzag functions, an affinity for algebraic formulation, and a disciplined sense of rigor governing all aspects of its construction. Beyond these, there is an aesthetic sensibility in the structure of the theory, expressed in an evident preference for symmetry and simplicity. In general, System A favors simple numbers, but with apparent care that this does not compromise its fundamental accuracy. Finally, Britton says, there is an air of privacy in it, where one senses that the subtleties of its structure are not intended to be seen, hidden as they are under the cloak of several additional sexagesimal places. "It's a pity we lack its author's name," Britton adds.[96]

Babylonian mathematical theory was not quite so ambitious in planetary as in lunar astronomy. It was not developed to the same degree of refinement, and probably could not have been without better instruments. Nevertheless, planetary positions were calculated, with both their eastward and retrograde motions represented, visibilities and disappearances computable. Planetary theory may have been intended as an approximation to phenomena too complex and irregular to compute with absolute precision. The theory may have been based upon a number of deliberate approximations for the purpose of computation.[97]

Astronomers lingered on in Babylon's temple of Bel into the first century A.D. By then a substantial part of their tradition had passed to Greek and probably Indian astronomers. Babylonian influence on Greek astronomy, as reflected in the *Almagest,* included the names of many constellations; the zodiacal reference system; the degree as the basic unit of angular measure; observations, especially of eclipses, going back to the beginning of the reign of King Nabonassar in 747 B.C.; and fundamental parameters including the value for the mean synodic month.[98]

The destruction of the Mesopotamian tradition would have been complete in medieval Europe had not Babylonian astronomy, via Greek distribution, found a new and interesting development among Indian astronomers. Then, when the Arab conquest reached India, this ancient science saw a triumphant revival everywhere in the Muslim world, and prepared the staging grounds for astronomy in the Renaissance.[99]

Egypt

The Ancient Egyptians were not noted for their astronomy, but they were highly practical people, and possibly it was their realistic approach to sky watching and timekeeping that has given us two big contributions: the 365-day year and the division of a day and night into a cycle of twenty-four units.

Perhaps no calendar-keeping institution has continued longer than theirs. After its uninterrupted run during all of Egyptian history, the Greek astronomers adopted the Egyptian calendar for their calculations. Ptolemy based certain tables in the *Almagest* on Egyptian years. Even Copernicus in 1543, in *De revolutionibus orbium coelestium,* used it, explaining simply that astronomers are practical people, and because the main requisite for every time-measuring unit is constancy, the Egyptian calendar is an ideal tool.[100] The calendar consisted of twelve months of thirty days each, with five additional days at the end and no intercalations at all. It is no wonder, remarks Neugebauer, that Hellenistic astronomers preferred this system to the Babylonian lunar calendar with its irregularly changing months combined with a complicated cyclic intercalation, not to mention the "chaos of the Greek and Roman calendars."[101]

Why did the Egyptian astronomers impose upon the citizenry a calendar with no respect for sun and moon? Was it the sheer primitiveness of their observations? Probably there was a more important cycle in their lives. To the early Nile people, the annual flooding of the great river, "the Inundation," seemed to occur fairly predictably around the time of Sirius (Sothis). The brightest star in their firmament, Sirius reappeared in the east after disappearing in conjunction. Then it would appear for a few minutes before the light of the day, having been obscured for seventy days by the sun. This familiar heliacal rising occurred close to the summer solstice (early June in around 4500 B.C.). Thus the revered event Peret Sepdet (The Going Forth of Sothis) was a cross-reference, a double time check, to reset the year clock and restart the count of the months.[102] The calendar was an agricultural as much as or even more than an astronomical one. Neugebauer showed that a simple recording of the variable dates of the Nile's inundations led to an average interval of 365 days. Only after two or three hundred years could this "Nile calendar" no longer be considered correct.[103]

At the same time, the Egyptians did observe the real lunar cycle, which had a well-defined religious significance. The two calendars—the lunar and the thirty-day-month civil calendar—coexisted, as dual calendars did in much of Mesopotamian history. As Neugebauer points out, the behavior of the moon is so complicated that it was not until the last centuries of Babylonian history, around 500 B.C., that astronomers figured out a satisfactory and sufficiently accurate program to predict the length of the lunar month for any appreciable time. That is, only a highly developed celestial mechanics could make it worthwhile to run an empire on moon time. Organized societies, furthermore, need to be able to determine future dates, regardless of the moon's phase. A simplified calendar is equally practical for keeping track of the past, because it eliminates the need for keeping exact records of the actual length of each month. A thirty-day month met the requirements for running a country as big as Egypt, even as in Babylon the simplified calendar versions served the needs of rent collectors and merchants dealing in accounts receivable. The thirty-day month was not an attempt to approximate reality but constituted a way of expressing time in round numbers.[104]

The beginnings of the system of night-day divisions probably started with the decans (called thus by the Greeks—groups of stars that denoted ten-day periods). The Egyptians, creating a twelve-unit division of the period of total darkness, developed the decans for time measurement at night.[105] This method may have originated when the Egyptians, always on the lookout for Sirius, had trouble distinguishing one bright predawn star from another on similar horizon points. If one bright star was part of three or four other fainter ones that formed a distinctive pattern, rising a few minutes before Sirius, then the astronomers could forewarn of the superstar's rising. So they identified the fainter stars as attendants, the brighter Sirius as gatekeeper, and the place of the star's appearance as the gate. As each succeeding star group dominated the horizon gate, the rising group would be associated with specific myths, characterizing the order of the sequence. Later, when the Egyptians wrote these sequences down, they organized them as twelve star groups emerging from the underworld, an underworld with twelve gates.[106]

The decans made their appearance in drawings and tables on the insides of pharaonic coffin lids in the Ninth through Twelfth Dynasties. The coffin artists painted decanal "star clocks"—covering a year at 10-day intervals—on the sarcophagi lids to aid the dead one's journey through the underworld and up into the sky to join the immortal stars. By 1100 B.C., the Egyptians had drawn up a list of decanal stars and denoted the night hours by combining only those stars that resembled Sirius.[107]

For one week, the star clocks marked off twelve hours of the night by the rising of a specific star or set of stars. The next week, the stars changed position by sliding over one hour; that is, those stars that rose to mark the first hour of the first week would then mark the second hour of the second week, and so on. Each decan passed out of the clock at the end of 120 days, or one-third of a year.[108] Thus, given the star table and the appearance of the night sky at any moment, one could tell the time by noting the tabular position of a specific star for a specific date. The star clocks did not account for the fact that 365 days do not return the sun to the same star, so every four years, the tables would be one day in error. After 120 years, they'd be off by a whole month. The Egyptians evidently tried to solve this problem by shifting the stars' names by the appropriate amounts to reset the clock with the civil calendar. But they had abandoned this procedure by the time of the New Kingdom (1550–1070 B.C.).[109]

The decanal hours were neither constant nor sixty minutes. But each decan had to serve for ten days as the indicator of its hour, and these hours could not be a part of twilight. Again, this is a simple twelve-unit scheme that worked for all seasons of the year. And a symmetry of night and day, of upper and nether worlds, suggested a similar twelve-unit division for the day. But it was not until the Hellenistic period that Babylonian time reckoning, with its sexagesimal division, combined with this Egyptian norm of two phases of twelve hours and led to twenty-four "equinoctial hours" of sixty minutes each.[110]

During the New Kingdom, astronomy was characterized by pictures of priests seated before star grids in the tombs of Ramses VI, VII, and IX. They represent the final stage of telling time with stars. Rather than noting rising stars, the new procedure involved any stars transiting the meridian (highest point) and several adjacent latitudinal lines. These astronomers may have been using water clocks to chart the transiting stars. The first direct evidence of water-clock usage in Egypt comes from an inscription of a prince, Amenemhet, around 1520 B.C. The water clock looked like a vase, having a scale on the inside marking hours and a hole into which fit a finely bored plug; the diameter of the bore was no larger than that of a hypodermic needle. The clock was filled with water, which escaped through this small-bore outlet.[111]

Above all in the world, the Egyptians manifested their celestial awareness in their pyramids. The Old Kingdom (2613–2125 B.C.) is sometimes called the Age of Pyramids, with the Fourth Dynasty pyramids at Giza reflecting the religio-astronomical zeal of Old Kingdom pharaohs. The Egyptians were inspired to shape the pyramids by imitating the way clouds and dust scattered sunlight into broad swaths that formed stairways to heaven. Indeed, they considered the pyramids to be stone pathways to the immortals—the northern circumpolar stars. The Egyptians called these stars *ikhemu-sek,* "the ones not knowing destruction," since these stars never set on the Egyptians' world.

The entrances to the pyramids all face north, and their corridors slope downward at such an angle that the north circumpolar stars could be seen from them. The three Giza pyramids are oriented diagonally and offset with respect to each other so that their north faces did not block each other's view of these circumpolar stars, especially Alpha Draconis (Thuban), the pole star of those times. The Giza orientation was also governed by the fact that the Egyptians believed the entrance to the underworld lay due west, a point on the horizon where the sun set into the mouth of the sky goddess, Nut, on the spring equinox. The pharaoh was expected to pass safely through the underworld before joining immortal gods.[112]

The Great Pyramid at Giza has inspired various astronomical interpretations involving alignments of interior corridors and shadows cast by its profile. Whatever else may be said about it, there is no doubt the pyramid is aligned fairly precisely and that the four sides of its huge base (covering more than thirteen acres) run north, south, east, and west. The worst agreement of any side with exact cardinal orientation is on the east, and even there the misalignment from true north-south is only 5.5 arc minutes. Preserving this accuracy on the immense scale of the pyramid means not "twisting" the sides at higher levels. The Egyptians' success emphasizes the concerns they had for the four cardinal directions.[113]

How the Egyptians aligned the Giza pyramids so nicely despite their less than sophisticated astronomy has been a puzzle. Recently, a British Egyptologist announced that she has solved it. The best estimate of the age of the royal tombs, roughly forty-five thousand years old, is based on chronologies of the period and the reign of kings, and is accurate only to within one hundred years.

Reporting in *Nature* in November 2000, Kate Spence, of the University of Cambridge, estimated that the building of the pyramids began between 2485 and 2375 B.C. and that two stars helped the engineers align them to true north. The Egyptians were trying to find true north, but they did not have a star marking the pole. So they used two stars, Kochab in the Little Dipper and Mizar in the Big Dipper, to find the pole. "It is on a line between those two stars," Spence said. "You measure when the two stars are basically on top of one another and if you line them up with a plumb line that will give you true north." According to astronomical data, 2467 B.C. is the year in which the line that goes between the two stars passes exactly through the trajectory of the pole. "If they had started building on that date we would have a pyramid which is absolutely aligned to north," Spence added. "But they seemed to have started work about 11 years before that, which means it is still a few minutes off north."[114] Aveni says the jury is still out on this theory, and notes that the Egyptians were "pretty good engineers" and that the alignment problem may not be so difficult.[115]

Sun (Ra) worship reached its zenith during the Fifth Dynasty (ca. 2750–2400 B.C.) when six kings built immense temples to honor him. Ra temples had special designs making it easier to measure night hours for predicting sunrise. King Userkaf's temple, the first of the six, was associated with a series of stars containing Denab as its brightest, the star from which Ra was "born." Astronomers, or "Overseers of the House," could have stood on the roof to monitor the axial crossing of stars used as hour markers. They employed an instrument called the *bay,* a palm rib with a notch cut into one end. This was also used with the *merkhet,* to ascertain the orientation of the building itself and the axial line on the roof. Similar tools were probably used to determine the orientation of the pyramids and to maintain that orientation as construction progressed.[116]

Over fifteen hundred years later, the Royal Tombs of the New Kingdom were cut into the cliffs in the Valley of Kings at Luxor. There, numerous representations of the "Northern Group," a figurative portrayal of circumpolar constellations, were painted on corridor ceilings in the tomb of Ramses VI, a Twentieth Dynasty pharaoh. The finest version of this theme appears overhead in the burial chamber of the tomb of Seti I, of the Nineteenth Dynasty, who ruled to about 1292 B.C. It encompasses most of the northern heavens.[117] After the rise of the Ptolemys (323 B.C. through 30 B.C.), Greek and Babylonian influences were visible in temple construction and astronomy, and there is nothing purely Egyptian to be found—except for the enduring 365-day-year and twenty-four-hour-day calendars.

India

According to David Pingree, a Brown University historian of mathematics who has carried out an extensive survey of the literature on Indian astronomy, there exist "at present in India and outside of it some million manuscripts[118] on the various aspects of *jyotihshastra*—texts on astronomy, mathematics, astrology, and divination. These vast numbers of manuscripts have been neither cataloged nor translated and constitute a territory that remains remarkably unexplored."[119] Yet Indian astronomy, perhaps more than any other, has served as the crossroads and catalyst between the past and future of the science.

Many fundamental ideas of Indian astronomy were introduced from other cultures, and Indian astronomy is part of broad currents of knowledge, theory, and practice that crisscrossed the major civilizations of Eurasia between the late second millennium B.C. and the nineteenth century in the West. The headwaters of this great multicultural stream was Mesopotamia, but within India the tradition was reconfigured, later by sophisticated mathematics, to fit Indian social and intellectual patterns.[120] Indian mathematical innovations had a profound effect on neighboring cultures. Trigonometry and analemma (a system of ways to reduce problems in three dimensions to a plane), for instance, greatly influenced Islamic astronomy and its heirs in western Europe. In serving as a conduit for incoming ideas and a catalyst for influencing others, India played a pivotal role in the development of astronomical sciences.[121]

The first references to astronomy in India are to be found in the

Rig-Veda, an oral religious, moral, and speculative epic written down around 2000 B.C. Vedic Aryans deified the sun (Surya), stars, and comets. As was common across so many cultures, astronomy in India was interwoven with astrology and divination. Indians integrated the sun, moon, and planets into the determination of human fortunes.

The Vedas recognized the sun as the source of light and warmth, the source of life, the center of creation, and the center of the spheres. This perception may have planted a seed, leading Indians thinkers to entertain the idea of heliocentricity long before some Greeks thought of it. An ancient Sanskrit couplet also contemplates the idea of multiple suns: "*Sarva Dishanaam, Suryaham Suryaha, Surya.*" Roughly translated this means, "There are suns in all directions, the night sky being full of them," suggesting that early sky watchers may have realized that the visible stars are similar in kind to the sun.[122] A hymn of the Rig-Veda, the Taittriya Brahmana, extols *nakshatravidya* (*nakshatra* means stars; *vidya,* knowledge).[123]

As in so much of Indian astronomy, even the details of this old ritual knowledge are a blend of native and exotic knowledge. Some of the hymns of the Rig-Veda are clearly related to the content of MUL.APIN, the eleventh-century B.C. Mesopotamian text. MUL.APIN gives the dates of the heliacal rising of the constellations in terms of an "ideal calendar" of twelve thirty-day months and a 360-day year. A late Rig-Veda hymn refers to the same calendar. MUL.APIN describes the oscillation of the rising point of the sun along the eastern horizon at the solstices. The same oscillation is described in a Vedic hymn, the Areyabrahmana.[124]

In the fifth century B.C. in the Indus Valley, Indians developed a water clock for calendrical purposes. The clock's operation was governed by a mathematical linear zigzag function, with the ratio of the longest to the shortest day of the year being 3 to 2. This, too, is a Babylonian import. The Mesopotamian omen texts of the Enuma Anu Enlil were also imported about 400 B.C. Embedded in them are theories of planetary motion for use in making rough predictions of the dates of the occurrences of the ominous planetary happenings, such things as first and last visibilities, retrogressions, and conjunctions with constellations.

The Mesopotamians' first attempts at constructing math models are here too, such as dividing the ecliptic for each planet into various arcs, wherein each arc has a given velocity and employs standard intervals of time. A more advanced form of Babylonian planetary theory, fully developed around 300 B.C., is reflected in later Indian texts. This material first passed through Greek intermediaries in the form of astrological and astronomical treatises, sometime between the second and the fourth centuries A.D.[125]

AMONG THE EARLY Indian texts are the *siddhantas,* treatises on astronomy and math, which were written in hymnal poetic form, probably both as a mnemonic device and because art and science, as in much of the non-Western world, were never considered to exist on the opposite side of a conceptual ravine but to enhance the spirit of each other. Of the eighteen early *siddhantas,* only five survive as extracts, including the major early textbook on Hindu astronomy, the *Surya-Siddhanta,* written around 400 A.D. The *Surya-Siddhanta* contains many things, including a method for finding the times of planetary ascension of arc of the ecliptic, a fundamental problem of ancient Mesopotamian trigonometry.[126] Yet the *Surya-Siddhanta* contains ancient Indian doctrines, too, such as the conception of the strings of air that push and pull the planets in their irregular motion,[127] which is an early intimation, albeit vague, of the force of gravitation. The Sanskrit word for gravitation is *gurutvakarshan. Akarshan* means to be attracted. From early times, the language itself reflected the idea that the character of this force was attraction.

Some scholars argue that these early texts show that early Indian astronomers flirted with, at least, heliocentrism and a theory of gravitation one thousand years earlier than these concepts were articulated by Copernicus, Galileo, and Newton. For example: "He [the sun] is denominated by the golden wombed, the blessed; as being the generator." The sun is also referred to as "the supreme source of light upon the border of darkness—he revolves, brings beings into being; the creator of creatures." The *Surya-Siddhanta,* furthermore, credits the Vedas with establishing the sun "within the egg as grandfather of all worlds; he himself then revolves causing existence."[128] Well, it's a stretch. There is no inverse-square law here, as Newton articulated. George Saliba, of Columbia University, calls Indian gravitation an anachronism. Aveni says it doesn't spell gravitation to him either.

Around 425 A.D. the *Paitamahasiddhanta* expressed geometric models of terrestrial and celestial spheres and mechanics to account for planetary motions. The text takes the basic flat-earth model and converts it into a spherical universe. Here two epicycles (rather than the Ptolemaic single one) exert pulls on each planet that dislodge it from its mean longitude, causing the resulting motion to be discontinuous rather than uniform.[129]

The epicycle, the construct that so dominated pre-Copernican astronomy, is a Greek geometric device that was universalized for hundreds of years by Ptolemy. One way of explaining an epicycle is to see it as a circle in which a planet moves and which has a center that is itself carried around on the circumference of a larger circle. That is, variations in the distance of a planet from the earth could be explained by assuming a planet moved in a circle, the epicycle, whose center traveled around another circle, the deferent, centered on the earth.

Hipparchus proposed an amended concept, suggesting that the planets moved in eccentric circles around the earth, with their orbital centers situated some distance from the earth's center. Hipparchus explained the apparent motions of the sun in terms of a fixed circular orbit eccentric to the earth, using epicycles to describe the orbits of the planets. Ptolemy appropriated both epicycles and eccentric orbits in creating his celestial model. By Ptolemy's time it took around forty-one circles to account for all the goings on in the heavens.[130] So it was in this light, centuries later, that Indian astronomers manipulated the epicycle with great success (erroneously, of course), inventing various new algorithms to compute these complex processes within processes in order to explain how the planets revolve on their axes as they revolve around the earth.[131]

A school of Paitamaha called Brahmapaksa had a wide influence outside India, beginning with the adaptation of at least some of its mathematics by astronomer-mathematicians in Iran in about 450.[132] Around this time, individual astronomers began to enter the stage of Indian history. The Brahmapaksa school was influential not only in Iran but on the young Indian astronomer Aryabhata, born in today's state of Kerala. In 499 he presented a treatise on mathematics and astronomy, the *Aryabhatiya.* The *Aryabhatiya* is a summary of Hindu mathematics up to his time, including astronomy, spherical trigonometry, arithmetic, algebra, and plane trigonometry. In it one of Aryabhata's main goals was to simplify the ever more complex computational mathematics of Indian astronomy. He had a practical purpose for this: to fix the Hindu calendar for easier forecasting of eclipses and movements of celestial bodies.

In the process, the *Aryabhatiya* presented a new treatment of the position of the planets in space. It proposed that the apparent rotation of the heavens was due to the axial rotation of the earth, a view of the solar system that future commentators could not bring themselves to follow. In fact, most later editors changed the text to save Aryabhata from what they thought were gross errors.

A revolutionary thinker in many areas, Aryabhata gave the radius of the planetary orbits in terms of the radius of the earth-sun orbit—that is, their orbits as basically their periods of rotation around the sun. He explained that the glow of the moon and planets was the result of reflected sunlight. And with incredible astuteness, he conceptualized the orbits of the planets as ellipses, a thousand years before Kepler reluctantly (he originally preferred circles) came to the same conclusion. Aryabhata wrote that the cause of lunar eclipses was the shadow of the earth, despite the prevailing belief that eclipses were caused by a demon called Rahu. His value for the length of the year at 365 days, six hours, twelve minutes, and thirty seconds, however, is a slight overestimate; the true value is fewer than 365 days and 6 hours. Another astronomer, Bhaskara I, writing a commentary on the *Aryabhatiya* about a hundred years later, had this to say:

> Aryabhata is the master who, after reaching the furthest shores and plumbing the inmost depths of the sea of ultimate knowledge of mathematics, kinematics and spherics, handed over the three sciences to the learned world.[133]

The *Aryabhatiya* was translated into Latin in the thirteenth century. Through this translation European mathematicians eventually learned methods for calculating the squares of triangles and the volumes of spheres, as well as square and cube roots. Explanations about the cause of eclipses and the sun being the source of moonlight may not have caused much excitement in Europe when astronomers there finally read the treatise, since by then they had learned of these things through the investigations of Copernicus and Galileo. But Aryabhata had conceptualized these ideas a thousand years before the Europeans.[134]

Fifty years after the *Aryabhatiya,* the philosopher, astronomer, and mathematician Varahamihira wrote the *Pancasiddhantika* ("Five Treatises"), a compendium of Greek, Egyptian, Roman, and Indian astronomy. Varahamihira's knowledge of Western astronomy was thorough. In these five sections, his huge work progresses through Indian astronomy and culminates in two analyses of Western astronomy, showing calculations based on Greek and Alexandrian reckoning and even giving complete Ptolemaic mathematical charts and tables.

Included in it is the Indian transformation of the Hellenistic astronomer Hipparchus's tables of sines into a table of chords, and the first application of periodic theorems to problems in spherical trigonometry—a field, says Pingree, "in which Indian astronomers were brilliantly innovative."[135] Varahamihira, like Aryabhata before him, contemplated the idea that the earth was spherical in shape. He believed that there could be an attractive force keeping bodies stuck to the earth. If Varahamihiri believed in gravity, it is safe to assume that he posited also a general gravitational force.[136] (Gravitation is the general force between masses; gravity is the effect of the gravitation on the surface of a planet.)

In 628, Brahmagupta, the last and arguably the most accomplished of the ancient Indian astronomers, set forth his astronomical system in characteristic *siddhanta* verse form in the *Brahmasphutasiddhanta* (The Opening of the Universe).[137] Brahmagupta became the head of the astronomical observatory at Ujjain, the foremost mathematical center of ancient India, where great mathematicians such as Varahamihira had worked and built up a strong school of mathematical astronomy.

The *Brahmasphutasiddhanta* contains twenty-five chapters, the first ten of which are arranged by topics such as true longitudes of the planets, lunar eclipses, solar eclipses, risings and settings, the moon's crescent, the moon's shadow, conjunctions of the planets with each other, and conjunctions of the planets with the fixed stars. The other fifteen chapters seem to form a second work—a major addendum to the original treatise, including reworkings of the previous treatises on astronomy and mathematics and additional works on algebra, on the gnomon (a sundial-like object), on meters, on the sphere, on instruments and versified tables.[138] A large part of the *Brahmasphutasiddhanta* was translated into Arabic in the early 770s and became the basis of various studies by the astronomer Ya'qub ibn Tariq. In 1126 it was translated into Latin. This translation, along with other associated texts translated from Arabic, provided the basis for the Indo-Arab stage of Western astronomy.[139]

The culmination of southern Indian astronomy was the tradition begun by Madhava in Kerala right before 1400. Madhava was renowned for his derivation of the infinite series for pi and the power series for trigonometric functions. His pupil Paramesvara attempted to correct solar and lunar parameters by conducting a long series of eclipse observations between 1393 and 1432. In these observations he used an astrolabe, an instrument devised to measure the positions of heavenly bodies, to determine the angle of altitude of the eclipsed body and, possibly, the time of the phases of the eclipses. This is remarkable, says Pingree, since the astrolabe had been introduced in a translation or adaptation of a Persian text only in 1370, and far away in northern India.[140]

Observation played a minor role in Indian astronomy until observational instruments were introduced from Islamic and Ptolemaic traditions in the late fourteenth century. This new Indo-Muslim coalition of empirical astronomy culminated in the immense masonry observatories erected by Jayasimha in 1730, when they were already obsolete. Generally, Indian astronomers took both theoretical models and equations from outside sources and adapted them to their traditions and needs. Those needs were the computation of their complex calendars, timekeeping, the casting of horoscopes, prediction of solar and lunar eclipses, and conjunctions of planets with fixed stars or with each other. To these ends, they employed a sophisticated mathematics of approximation and developed elaborate arrangements of tables.

The Indians advanced astronomy by mathematics rather than by deductions elicited from nature. Some of these mathematical innovations had a profound effect on neighboring cultures—as for instance, trigonometry and analemma on Islamic astronomy—and in medieval western Europe. In its reception of external ideas and its influences on others, India played a pivotal role in the development of astronomical sciences in the rest of the world.[141]

Islam

Soon after the prophet Muhammad's death in 632, Muslims had established a commonwealth stretching from Spain to Central Asia. With conquest they brought an Arab folk astronomy that mixed with local knowledge, especially the mathematical traditions of Indian, Persian, and Greek astronomy, which they mastered and adapted to their needs. Early Islamic astronomy was a potpourri, but by the tenth century it had acquired distinctive characteristics of its own.[142] From then through the fifteenth century Muslim scholars were unequaled in astronomy. Astrology, too, was part of the package, and remained so; even the greatest Islamic astronomers, such as al-Biruni, practiced the occult art.[143] In the deep background was the legacy of ancient Mesopotamia, partly intact over the thousands of years.[144]

The earliest astronomical documents in Arabic may have been written in Sind in Afghanistan (now in Pakistan) an area conquered by Muslims by the seventh century. They consisted of text and tables called *zij,* after a Pahlavi (old scholarly Persian) word meaning "cord" or "thread" and, by extension, "the warp of a fabric."[145] Around 771, an Indian political mission arrived from Sind to the court of al-Mansur, the caliph of Baghdad. The group included a scholar versed in astronomy who brought along the famous *Brahmasphutasiddhanta.* The caliph ordered it translated into Arabic, and the resulting *Zij al-Sindhind al-kabir* was the springboard for a series of *zij*s by great Islamic astronomers writing through the tenth century. The Sindhi tradition flourished all the way to Andalusia, Spain, and as a result, the influence of Indian and Islamic astronomy spread from Morocco to England in the late Middle Ages.[146]

In their later form, the *zij*s consisted of several hundred pages of text and tables. Aspects of mathematical astronomy one could find in a typical *zij* included trigonometry; spherical astronomy; solar, lunar, and planetary equations; lunar and planetary latitudes; planetary stations; parallax; solar and planetary visibility; mathematical geography (lists of cities with geographical coordinates) determining the direction of Mecca; uranometry (tables of fixed stars with coordinates); and, not least, mathematical astrology.[147] In one *zij,* the famous Egyptian astronomer Ibn Yunus describes forty planetary conjunctions and thirty lunar eclipses. Using modern knowledge of the positions of the planets, we find that Yunus is exactly right.[148]

Although religion was not the entire driving force spurring the growth of astronomy in Islam—a tolerant, multiracial, highly literate society, with a predominant language, Arabic, also fostered it—sacred matters played a big part. Islam needed a way to figure out how to orient all sacred structures, as well as daily worshipers, precisely toward Mecca. Celestial mapping sprang from this requirement for establishing the holy coordinates and the right direction, or *qibla,* toward the Kaaba, the shrine of Mecca toward which all Muslims face five times a day in prayer.

But which way is Mecca? In early times, religious authorities probably determined the *qibla* through sighting by celestial bodies, such as the star of Bethlehem, that lay in the general direction taken by pilgrims walking to Mecca. The Kaaba itself is aligned to specific directions; its major (south) axis is positioned toward the rising of the star Canopus; its minor axis, or east and west facades, aligns to sunrise at the summer solstice and sunset at the winter solstice.[149] An expert at a distant shrine had to devise schemes to face the segment of the Kaaba corresponding to his location, as if he were actually in front of that segment of the Kaaba perimeter.[150]

By the ninth century, astronomers were using trigonometric and other computational devices to determine the *qibla* from geographical coordinates. The puzzle was easily turned into a problem of spherical astronomy by considering the zeniths of the localities involved. In the astronomer al-Biruni's treatise on mathematical geography, for example, the goal was determining the *qibla* at Ghazni, in Afghanistan.

In the ninth century, a great patron of science, the Abbasid Caliph al-Ma'mun, gathered astronomers in Baghdad to create the House of Wisdom (Bait al-Hikmah). There the astronomers conducted observations of the sun and moon, directed toward determining the local latitude and longitude in order to establish the *qibla*. They collected some of the best results into a *zij* called "Tested" (*al-Mumtahan*). Not until the eighteenth century and the invention of the marine chronometer, was it possible to measure longitudinal differences correctly. By then it became obvious that most medieval coordinates were incorrect. Even *qiblas* derived by correct mathematical procedures but based on these coordinates were off by a few degrees. Nonetheless, determining the *qibla* was one of the most advanced problems faced by Muslim astronomers, and the solutions they found were of great sophistication.[151]

Along with sacred direction came sacred time. Like the ancient Babylonians, the Islamic calendar begins with the first sighting of the crescent following the new moon in the west. The precise determination of the beginning and end of months has been especially important for setting the time for Ramadan, the sacred month of fasting. Astronomers devised a wide variety of ways to find Ramadan's start. Al-Khwarizmi, for instance, compiled a table of the minimum ecliptic elongations of the sun and moon for each zodiacal sign, computed for the latitude of Baghdad. (Today there is often even more confusion than in medieval times about the start of Ramadan, since the crescent moon may be seen in some locations around the world at one time and not in others.)[152]

There was also a pressing need to know what time it was in order to fix the five prayer times each day. Again, the astronomical and mathematical disciplines were called into service; this application has virtually no parallel in the science of Greece or medieval Europe. It was an unrivaled and increasingly sophisticated effort, as an enormous body of recorded observations and calculations testifies.[153]

Until the ninth century experts determined prayer times by shadows and lunar mansions. After that, they used tables to calculate the times, correlating shadow lengths and height of the sun to indicate the lengths of the intervals between prayers. Such tables enabled the *muwaqqit,* the official timekeepers employed by the mosques, to inform muezzins of the time for each of the five prayers so they could summon the faithful.[154] The tables proliferated throughout Islam and began to evolve into increasingly accurate and comprehensive data sources as naked-eye observations and calculation were replaced with more sophisticated sundials, quadrants, astrolabes, and compasses.[155]

It was convenient to have the tables displaying prayer times for each day of the year. By the mid–tenth century Islamic astronomers had compiled two tables displaying time of day as a function of solar meridian altitude computed for Baghdad. These Islamic timekeeping devices became an obsession of artisans in the thirteenth and fourteenth centuries. A table for Damascus shows twelve functions relating to timekeeping tabulated for each day of the year. Another table had over 400,000 entries. (Today the tables Muslims use to regulate prayers are published in newspapers, pocket diaries, calendars, and on the Internet. The muezzins' calls are recorded and amplified by loudspeakers.)[156]

The analysis of time and timekeeping inspired Muslim investigators into more complicated zones of abstraction, such as analyses of shadows. In the eleventh century al-Biruni wrote a work on shade and shadows, strange phenomena involving shadows, gnomonics, the history of the tangent and secant functions, applications of the shadow functions to the astrolabe and to other instruments, shadow observations for the solution of various astronomical problems, and the shadow-determined times of Muslim prayers.[157]

Al-Biruni introduced techniques to measure the earth and distances on it using triangulation. He found the radius of the earth to be 6,339.6 kilometers (3,930.6 miles), a value not obtained in the West until the sixteenth century. One of his *zijs* contains a table giving the coordinates of six hundred places, almost all of which he had direct knowledge of. Not all, however, were measured by al-Biruni himself, some being taken from a similar table given by al-Khwarizmi. (Al-Biruni seems to have realized that for places given by both al-Khwarizmi and Ptolemy, the value obtained by al-Khwarizmi was the more accurate.) Al-Biruni also wrote treatises on the astrolabe and devised a mechanical calendar. He made observations on the speed of light, stating that its velocity is immense compared with that of sound.[158]

THE ARABIC FASCINATION with mechanical devices led to the development of the first serious collection of astronomical instruments designed to acquire precise data about time and the motion and position of celestial objects. The instruments devised in the Arab world during the medieval period were far more complex and more ornate and provided far more precise information than any predecessors.[159]

The astrolabe was a particular favorite of Islamic astronomers. Meaning "star taker," it was a source of precise data gathering for collections of tables across the southern Mediterranean world from the tenth through fifteenth centuries.[160] Invented by the Greeks in the second century B.C., it was enhanced or, some say, perfected by Muslims. The compact, often small, device functioned like a sophisticated engineer's slide rule. With its replaceable plates it could be calibrated for use in different geographical locations, and manipulated to provide many types of year-round celestial and timekeeping data, terrestrial measurements, and astrological information. It combined the sighting properties of the telescope and the figuring capacity of a little analog computer.[161]

The viewer looked through a pair of sighting holes at opposite ends of a rod mounted on a circular plate. The astrolabe's face was outfitted with a series of star-map plates that could be removed and substituted like CDs, one for each latitude. The disks consisted of a flat stereographic projection of the sky onto the celestial equator (the extension of the earth's geographic equator onto the sky). The central hole marked the position on the celestial sphere approximated by the pole-star. On top of this another plate gave basic coordinates in horizon, equatorial, and ecliptic systems.[162]

Once the astronomer sighted the target object through the peepholes, he read its positions off a rotatable ruler. Another circle on the flip side of the star clock served to fix the hour of night, day of the month, and position of the object sighted on the zodiac. Craftsmen fashioned the instrument's pointers in the form of dragon's teeth, serpent's tongues, or other animal parts. Made of brass and filigree, the astrolabe presents a relationship between science, art, and nature possible among the world's cultures then but scarcely today.[163]

Muslims developed other instruments, including the spherical astrolabe. Treatises on it were written from the tenth through sixteenth centuries. Al-Khwarizmi and other astronomers introduced features such as shadow squares and trigonometric grids on the instrument's back, azimuth curves on plates for different latitudes, and a universal plate of horizons. One spherical astrolabe, dated 1329, represents the culmination of Islamic astrolabe making and has no equal in sophistication among instruments from the European Renaissance. Whereas standard astrolabes require a different plate for each latitude, this one, of Ibn al-Sarraj, has plates that serve all latitudes. Its various components can be used to solve all the problems of spherical astronomy for any latitude.[164]

The Muslims developed the armillary sphere, named for the many concentric armillae, or "bracelets," that compose it. Armillary spheres create a physical representation of features of the celestial sphere, such as the circles of the equator, the horizon, the meridian, and tropical and polar circles.[165]

Another set of instruments taken to new precision were quadrants. The astrolabic quadrant was a simplified astrolabe, shaped like a 90-degree pie segment, that could be used to solve all standard problems of spherical astronomy, especially those involved in mapping features of the celestial sphere. Developed by Muslims in Egypt in the eleventh through twelfth centuries, it had by the sixteenth century replaced the astrolabe everywhere in the world except Persia and India.[166] The quadrant, sextant, and octant, for measuring altitude, along with the early form of the surveyor's transit, or theodolite, are tools highway crews use for surveying today. Keep in mind that the medieval sextant and other tools were not as sophisticated as today's sextant, which has a built-in telescope and a precision scale for ticking off degrees.

THE DEMAND for precise data in preparing calendric prayer tables gave rise to observatories. Another force behind their construction was the ever present thirst for astrologic forecasting. Observatories were usually established or sponsored by caliphs and other rulers to serve their interest in astrology.[167] As part of his "House of Wisdom" projects, al-Ma'mun, the Abbasid caliph (813–833) had observatories built at Baghdad and Damascus.[168] These centers attracted eminent scholars and celebrity astronomers, who served as magnets to pull in ever brighter students from all Muslim regions. The great mathematician-astronomer al-Khwarizmi, for instance, took part in the celebrated Houses of Wisdom at Baghdad and Cairo during the reign of al-Ma'mun. Thabit ibn Qurrah observed at al-Ma'mun's establishment at Baghdad. In the eleventh century, Ibn Yunus led an observatory team at Cairo. At a court observatory at Ghazni in Afghanistan, al-Biruni provided data that formed the basis for the most significant *zij*s of Islamic astronomy.[169]

In 1259 the mathematician Nasir al-Din al-Tusi founded an observatory with a large professional staff at Maragha in Persia, where observers had the use of three sorts of astrolabes: planispheric, linear, and spherical. These were substantially the same as the instruments used later by European astronomers until the invention of the telescope.[170] The observatory of Ulugh Beg at Samarkand, in what is now Uzbekistan, was built between 1420 and 1437. Ulugh Beg equipped this three-story observatory with the best and most accurate instruments available, including a sextant made of marble. He also procured a quadrant so large that part of the ground was removed to allow it to fit inside the observatory.[171] Beg was assassinated in 1449, leading to a catastrophic neglect of the observatory, which lost its position as a leading center of astronomy. There is little doubt that the organization of this observatory and the instruments employed there influenced Tycho Brahe's famous sixteenth-century observatories at Uraniborg and Stjerneborg.[172]

Tycho Brahe has always been held up to students in the West as the master of pretelescope instrumentation. In reality, al-Ma'mum built a lavish observatory in 829 and furnished it with a fifty-six-foot-radius stone sextant and a twenty-foot-radius quadrant, a quadrant larger than Tycho's famous instrument built seven centuries later. Beg's sextants ranged up to 180 feet in radius, and the margin of error of his instruments was reportedly as good as, if not better than, that of Tycho's more than a century later. The Arabs' theoretical work was also superior. Because a cannonball fired east did not go farther than one fired west, Tycho decided that the earth did not rotate, thus pushing astronomy backward by several centuries.[173] He also developed an erroneous model of the solar system that stalled, at least for a while, the acceptance of the Copernican view.[174]

At all these observatories, astronomers recomputed and refined Ptolemy's coordinates for the stars and, eventually, revised Ptolemy's second-century A.D. catalog of stars. This catalog, which gave the positions of 1,022 stars, classified, as they are today, by magnitude, or brightness, was heavily corrected, notably by the tenth-century astronomer Abd al-Rahman al-Sufi (Azophi), whose *Book of the Fixed Stars* is the earliest illustrated astronomical manuscript known. It is still considered an important work for the study of proper motions and long period variables of stars. Al-Sufi was the first astronomer to describe the "nebulosity" of the nebula in Andromeda in his atlas of stars. (The copy in the Bodleian Library, the work of the author's son, is dated 1009, and the author expressly states that he traced the drawings from a celestial globe.)[175] These observatories mapped the skies in unprecedented detail, thereby providing an invaluable framework for observations carried out by later generations in both Islam and the West.[176]

It has been claimed that the observatory—as an organized and specialized institution—was born in Islam, and was passed on in a rather highly developed state to Europe. Muslim observatories, perhaps, are the first to meet the essential conditions for observatory qua observatory: large precision instruments, fixed and specialized locations, royal patronage, and a number of specialized scientists and astronomers working in collaboration.[177]

For centuries the main theoretical work Muslim astronomers did at an observatory centered on simplifying the Ptolemaic model and bringing it into line with the Aristotelian model, which postulated uniform circular orbits for the planets. The *Almagest* was translated into Arabic at least five times in the eighth and ninth centuries, and these texts were all available in the twelfth century, when they were used by Ibn al-Salah for his critique of Ptolemy's star catalog. Muslim astronomers' versions of the *Almagest* not only contained reformulations and paraphrases but also corrected and criticized it. Within their general allegiance to Ptolemy's cosmos, however, Muslim astronomers began to express their growing awareness about the divide between the Greek's model of the universe and their observed reality. The translations gave rise to a series of discussions on whole texts or parts, many of them critical. One by al-Haytham (circa 1025) was actually entitled "Doubts about Ptolemy."[178]

Earlier Greek and Indian astronomers had struggled to make sense of phenomena such as precession of the equinoxes and planets' retrograde motion. Ptolemy had refined the epicycle-deferent mechanism by adding the device of an equant, an eccentric, or off-center, point around which orbits the large circle, or deferent, on which is centered the epicycle indicating the planet's path. This equant was supposed to make sense of the apparent approach, recession, and backing up of a planet. It represented the most sophisticated attempt to square what the eye could observe with the way the theory stated a planet must move.[179]

Muslim astronomers eventually came to object in particular to the way Ptolemy's epicyclic motions violated the principle of uniformity of motion, a principle central to Greek and Indian concepts of all heavenly bodies, as well as a cornerstone of Ptolemy's system. This objection ultimately brought about a reform of planetary astronomy, making modifications to Ptolemy's planetary models. One school of dissent reached its fullest expression in the thirteenth century, notably with the Persian al-Tusi and his colleagues.

In his major treatise, *Memoir on Astronomy* (*Al-Tadhkira fi'ilm al'hay'a*), al-Tusi devised a new model of lunar motion, essentially different from Ptolemy's, abolishing the eccentric, among other things. In his model, al-Tusi further invented a theorem that occurred again 250 years later in Copernicus's *De revolutionibus.* This was the famous Tusi couple, devised to overcome objections to the notion of an equant. The Tusi couple resolved linear motion into the sum of two circular motions, with the aim of removing all parts of Ptolemy's system that were not based on the principle of uniform circular motion. (See chapter 1.) It presented a hypothetical model of epicyclic motion involving a combination of motions, each of which was uniform with respect to its own center.[180]

A century later, in Damascus, Ibn al-Shatir, who served as *muwaqqit* at the Great Mosque, developed a model based on the Tusi couple. But it wasn't until the 1950s, when this work was rediscovered, that scholar Edward S. Kennedy noticed that the solar, lunar, and planetary models al-Shatir proposed in his book *The Final Quest Concerning the Rectification of Principles* (*Nahayat al-su*) were quite different from those of Ptolemy. Al-Shatir had, in fact, put forth the details of what he thought was a true theoretical formulation of a set of planetary models describing planetary motions; and he'd intended them as alternatives to the Ptolemaic models. Indeed, they were mathematically identical to those of Copernicus, writing some 150 years later.[181]

Europeans learned of Islamic astronomy by way of Spain, but because of political turmoil and communication problems, the most current writings were not always available. This is why Europeans discovered two major works of the Muslim astronomers al-Khwarizmi and al-Battani at a time when these works were no longer widely used in

Islam. And it explains why so few works ever reached Europe at all. On the other hand, some early eastern Islamic research, later forgotten in the East, was transmitted to Spain and westward. These contributions have been taken as European developments because evidence to the contrary was not obvious. For example, the horary quadrant, with a movable cursor, which was invented in ninth-century Baghdad and forgotten soon after, became a favorite in medieval Europe.[182]

China

Chinese astronomy resembles most other premodern sky technologies in that it was driven by divination. Yet Chinese astronomy differed from all others. It was run solely by a government bureaucracy and based on a worldview that said the ruler was "emperor under all heaven"—a divine appointment. Yet the connection between celestial events and human fate was perhaps even more profound. The link was not just between heavenly deities and the emperor; the earth, the emperor, and the entire cosmos were bound together in one gigantic entity, a superorganism in which the five elements, or "phases"—fire, air, wood, earth, and water—were in constant interaction as they sought their affinities with one another.[183]

Yet in China as elsewhere, portent astrology called for careful and regular observations of celestial events. The cosmic importance of every omen in the sky demanded that its results be noted down in detail. As a consequence, the Chinese possess the longest unbroken run of astronomical records in the world, observations of considerable importance to modern astronomers, whose research requires data about long-term celestial events.[184]

China developed astronomy very early in its history. Evidence goes back five thousand years. The ancients wrote stars-laden texts in many forms—on wine jugs, tortoise shells, and silk. The earliest records from archaeological sites in Qinghai Province consist of ceramic fragments on which are painted images of rayed sun disks and moon crescents.[185] A piece of bone found to be thirty-five hundred years old contains writing showing that the Chinese already knew the length of the year to be 365¼ days. There is evidence of star observation from before the twenty-first century B.C.[186]

The first recorded astronomical inscriptions date from the sixteenth to nineteenth centuries B.C. in the Shang kingdom of Henan Province. These artifacts are examples of an astronomical-divination system, technically called scapulimancy, a technique going back to Neolithic times. Selecting an ox or deer shoulder blade (scapula) or a tortoise shell, the diviners then dried, polished, and drilled the material with holes. They inserted a hot metal brand into one hole and examined the pattern of resulting cracks in the bone or shell. The diviner noted both the prognostication and later results on the cracked material.[187]

The oracle bones' existence was lost to the modern world until 1899, when a scholar from Peking became ill and sent his valet to a drugshop for medicines. One ingredient in the potion the pharmacist sent him was labeled "dragon's bones." The scholar realized it was bone chips with words inscribed on them in ancient Chinese—oracle bones.

During the following decades the bones were traced to a field near An-yang, around three hundred miles southwest of Beijing. During the 1920s and '30s, some twenty-five thousand oracle bones were excavated there, from what may have been a palace archive. At least 135,000 more pieces have been excavated since, forming a treasury of information going back to Shang times. This vast library recorded on the bone texts has enabled modern historians of astronomy to backtrack regularly occurring celestial events with computers to match sky phenomena inscribed millennia ago.[188]

Recently, NASA astronomers used fourteenth-century B.C. oracle bones to help determine how much the earth's rotation is slowing down. Based on analysis of the tortoiseshell inscriptions, Kevin Pang and his colleagues at the Jet Propulsion Laboratory at Pasadena reported they had fixed the exact date and path of a solar eclipse seen in China in 1302 B.C. That, in turn, led them to calculate that the length of each day was 47⁄1,000 of a second shorter in 1302 B.C. than it is today.[189]

A cache of five thousand pieces of oracle bones excavated in An-yang in 1972 yielded a series of divinations of sky events. The Chinese astronomical historian Zhang Peiyu found that six dates recorded in the inscriptions matched perfectly with a series of solar eclipses visible from the Henan area in the twelfth century B.C., half a millennium earlier than records of such events obtained from Babylonia or Egypt.[190] Other Shang bones yielded inscriptions of lunar eclipses.

A reconstruction of another bone recording from around the same time revealed the observation of a supernova. The supernova inscription, perhaps the most ancient extant record of a nova sighting, says, in part, "On the seventh day of the month . . . a great new star appeared in company with Antares." The Chinese called these supernovas "guest stars."[191] Thus the Chinese knew well what they were observing when, in June 1054 (A.D.), a star in the constellation Taurus blew. Chinese sky watchers reported it to be as bright as Venus, apparent during daylight, and visible for twenty-three days. The remnant of this explosion can be seen today and is called the Crab Nebula. (The Greeks have no record of the supernova.) Experts today have compiled detailed descriptions of supernova explosions that coincide with contemporary X-ray and radio sources.[192]

In the Greek-influenced West, the sun and heavens were supposed to be immaculate. But Chinese astronomers saw spots on the sun. The earliest surviving record of a sunspot observation is by the astronomer Kan Te in the fourth century B.C. Kan Te assumed that these spots were eclipses that began at the center of the sun and spread outward. Although he was wrong, he recognized the spots for what they were—solar phenomena.

The next documentation of sunspots was in 165 B.C., when it was reported the Chinese character *wang* appeared in the sun—shaped like a cross with a bar across the top and bottom. It is accepted as the world's earliest precisely dated sunspot. The West's earliest reference to sunspots is Einhard's *Life of Charlemagne,* around A.D. 807.[193] Joseph Needham found 112 instances of sunspots recorded in Chinese histories between 28 B.C. and A.D. 1638. In other Chinese books he found hundreds more notices, "but no one has had time or stamina to collect them into a body," comments sinologist Robert Temple. Nonetheless, the sunspot records constitute the oldest continuous series of such observations. And again, these are of great use to modern astronomers. Sunspot cycles, for instance, affect the earth's ionosphere and weather (magnetic storms are related to sunspots). Analyzing available records, Japanese astronomer Shigeru Kanda reports he has detected a 975-year sunspot cycle. If so, it may have significant implications for weather cycles.[194]

The Chinese were also careful observers of comets. They computed the approximate orbits of about forty comet trajectories with such precision that many of their trajectories could be drawn on star charts simply from reading ancient texts. They were interested in the precise position and direction of the tail of each comet.

In the year 240 B.C., astronomers officially documented the appearance of a comet today known as Halley's. Another comet recorded in 467 B.C. is also thought to be Halley's. In the 600s A.D., they observed that comets shine by reflected light like the moon. They noted that comet tails always pointed away from the sun, suggesting this phenomenon was the result of a solar "energy." Today it's known that this cometic tail direction is caused by the force of "solar wind," the sun's radiation. It wasn't much of a stretch, says Temple, for the Chinese to formulate the idea of solar wind. It is congenial with their cosmological assumptions, Chinese literature being filled with references to the *ch'i* of the sun's radiation. *Ch'i,* translated as something like the "emanative or radiative force," comes from the sun. To Chinese astronomers it would have been obvious that the sun's *ch'i* was strong enough to blow the tails of comets as if in a strong wind. The Chinese conceived of space as being full of strong forces.[195]

As a consequence of the emperor's divine connection with the cosmos, it became traditional after important changes of rulership, and always after a switch to a new dynasty, for a fresh calendar to be drawn up. This custom was well established by Han times (206 B.C. to A.D. 220) and led to some forty new calendars made up between early Han and the beginning of the Ming dynasty in 1368.[196]

According to the theory of monarchy, the ruling dynasty remained fit to rule because of the accord the emperor maintained with the heavenly order. His special status in the order of nature allowed him to maintain a parallel order in the political realm, for the state was a microcosm. If the emperor lacked virtue or was careless in his duties, disorderly phenomena would appear in the sky as a warning of potential political disaster. Thus astronomers had to incorporate as many phenomena as possible in a "correct" calendar. The calendar, issued in the emperor's name, became part of the trappings of power that demonstrated his dynasty's right to rule, a function, writes sinologist Nathan Sivin, "not entirely different from that of economic indicators in a modern nation."[197]

The importance of astronomical observing in this world of extreme politics, then, made secrecy absolutely necessary. Because the data could be easily manipulated, it could be dangerous in the hands of someone trying to undermine the current dynasty. It was therefore state policy that the proper place to do astronomy was the imperial court. In certain periods it was illegal to do it elsewhere.[198] With this information virtually classified as top secret, the astronomer became a high-level administrative functionary in a country that developed the most elaborate bureaucracy in the ancient world. The databases resided in a state observatory deep within the bowels of the palace.[199]

If not the greatest astronomical mathematicians, the Chinese were the greatest star charters before the Renaissance. Their earliest star chart goes back to at least 2000 B.C., to a carving on a cliff at Jiangjunya in Jiangsu Province. The carving contains many stars, as well as human and animal heads. There are disks indicating the sun in seasonal positions and where a number of bright stars and the moon appear over the seasons. This bright region is recognizable as the Milky Way by its position and appearance; the Milky Way displays gaps and divisions that are depicted on the carving.[200]

China, being in the Northern Hemisphere, fixed itself on the northern circumpolar stars, both for orientation and to express its concept of divine rulership. The circumpolar stars in the higher latitudes are raised quite high in the sky as they rotate about the pole, so the fixity of the polar axis became an apt metaphor for the divine right of emperors. The pivot point about which this rotation occurs is known as the north celestial pole. The emperors were clever to adopt the stars of the north, such as Cassiopeia and Cepheus. These stars are located near the celestial pole, so that in the temperate latitudes of most of China they are visible eternally in the sky, never hidden by the horizon.[201]

The first catalogs of star positions appear to have been drawn up by Shi Shen, Gan De, and Wu Xian, the earliest notable astronomer in China, who worked between 370 and 270 B.C., two centuries before Hipparchus. Together their lists enumerated 1,464 stars grouped into 284 constellations. (The West made bigger groups, with only 88 constellations.) In A.D. 310, during the Western Chin dynasty (A.D. 265–317), this early work was collated by the astronomer royal Qian Luozhi, who cast a bronze celestial globe with the stars on it colored in red, black, and white to distinguish the listings of the three astronomers. As early as the Han dynasty, astronomers prepared star charts. Carvings and reliefs show individual constellations or asterisms depicted as dots or small circles connected by lines to delineate the constellation itself. This ball-and-link convention did not appear in the West until the late nineteenth century.[202]

Star maps need a means of specifying positions of heavenly bodies with reference to one another. The science of mapmaking took a leap forward in the second century B.C. when Chang Heng invented what's now called quantitative cartography. Chang, the inventor of the seismograph and a leading scientist, applied a grid system to maps so that positions, distances, and itineraries could be calculated and analyzed. Chang Heng's own works are lost, although an official history of the Han dynasty stated, "He cast a network of coordinates about heaven and earth, and reckoned on the basis of it."[203] Copies of these maps were never made, since the information on them was too dangerous to risk its falling into the wrong hands. Meanwhile, in Europe, mapmaking had degenerated under the influence of religion, says Robert Temple, "to a point scarcely credible."[204]

Drawing actual charts of the sky means finding a way to depict positions as if one is drawing a map. Preparing maps also involves the problem of mapping the curved surface of the celestial sphere on a flat surface, just as mapping the near-spherical surface of the earth requires the use of map projection. This is made more difficult if the sky is seen as a dome curving above one's head. In both China and the West, projection goes back a long way for mapping the earth. But for mapping the stars, Chang Heng was first, drawing up in Han times a chart that was a "Mercator" projection.

Mercator projection was "invented" in Europe by the Flemish mathematician and geographer Gerhard Kremer, a.k.a. Gerardus Mercator, and first published in 1568. But this projection system had been used by the Chinese centuries before Mercator. The projection works by means of a cylinder. If one inserts a transparent globe of the earth (or other celestial sphere) in the center of a hollow cylinder and turns on a light inside the globe, the features of the sphere's surface will be thrown, or projected, onto this cylinder, and will reflect a certain distortion. The higher up and lower down from the sphere's center, or equator, the more the features are distorted. Virtually useless for land travel, this projection has the odd property that a navigational course drawn on it will come out as a straight line, whereas with other maps such courses are arcs.[205]

The oldest surviving projection chart depicting the whole of the visible sky is painted on paper and now resides in the British Library. Dating from about A.D. 940, it comes from Dunhuang in Gansu Province and gives a flat representation of Qian Luozhi's (the astronomer royal's) tricolored chart, working from his celestial globe. It presents the celestial globe as projected onto a surface by the cylindrical projection technique, displaying over 1,350 stars in thirteen sections. One section is a planisphere—that is, in a kind of Mercator projection it depicts the circle of the sphere on a flat map centered on the north pole. The remaining twelve are flat maps centered on the celestial equator.[206]

A century later, in 1094, Su Sung published further Mercator-style map projections in his book *New Design for a Mechanized Armillary Sphere and Celestial Globe.* One map had a straight line running across the middle as the equator and an arc above it, the ecliptic. The rectangular boxes of the lunar mansions are clearly seen, with the stars near the equator being more tightly packed together and those near the poles spread farther apart.[207]

The evolution of Chinese instrumentation parallels that of the West. Notched jade disks and cylindrical sighting tubes date back to the fifth century B.C., and probably functioned as means of computing rudimentary celestial cycles. The Chinese used the gnomon as far back as 1500 B.C. Just as the Chinese had begun to standardize weights, measures, and other practical details in the sixth century B.C., and more extensively in the next three centuries, they standardized the gnomon. In addition to timekeeping, they used the gnomon to determine the terrestrial distance corresponding to an arc of the meridian. Such a determination of this north-south line was vital for precise calendar making, because precision calendars required measuring the latitude of those stations where the relevant observations were made. The gnomon was also significant in mapmaking and in the Chinese fascination with determining the size of the earth—nearly a millennium before Eratosthenes![208]

Between A.D. 721 and 725, under the auspices of the Buddhist astronomer and mathematician Yi Xing and the Chinese astronomer royal Nangong Yue, Chinese scholars set out to do this. To measure the size of the earth, they selected nine locations covering the prodigious distance of more than 3,500 kilometers (2,175 miles) on a nearly north-south axis. They made simultaneous measurements of shadows at the summer and winter solstices at all nine stations. The main outcome of this feat: they determined that the distance on earth corresponding to 1 degree latitude was 155 kilometers (97 miles). This is larger than today's value of 111 kilometers (69 miles), but far more accurate than previous attempts. Indeed, they found that the variation on shadow length with changed latitude was four times the value previously thought. There was no piece of research like it carried out anywhere else during the Middle Ages. In making his tabulations, Yi Xing used "tangent tables." This was thought to have been a Muslim invention of the ninth century, but it turns out that the Chinese discovered the use of tangents and tabulated them at least one hundred years earlier.[209]

The Muslim big-observatory concept came to China in the thirteenth century, and during the Mongol dynasty, in 1276, the astronomer Guo Shoujing built a giant gnomon called Tower of the Winds. It was an all-purpose observatory with the tower itself serving as a gnomon. A horizontal rod in an aperture at roof level—about forty feet above the ground—cast a shadow on the long low wall extending northward below. A chamber at the top was designed for watching stars, while the inner rooms of the tower housed a water-driven clock and an armillary sphere.[210]

Modern astronomical observatories derive not from the European tradition but from the Chinese. Modern telescopes are oriented and mounted in the equatorial system, which in China goes back at least to 2400 B.C. Equatorial mounting takes the equator as the horizontal circle around the side of the instrument, and the pole as the top point. Europeans originally followed the Greek-Indian-Muslim tradition in which the two circles that were important were the horizontal and the ecliptic, the circle described by the sun's motion in the sky that is in the same plane as the earth's orbit around the sun. This tradition more or less ignored the equator. China, meanwhile, largely ignored the horizon and the ecliptic. In the seventeenth century, European astronomers came to realize that the Chinese equatorial system was more convenient and showed greater promise. It was adopted by Tycho Brahe and his successors and remains the basis of astronomy today.[211]

The Chinese, furthermore, had the skills to build the precision observational instruments to display this system. Having invented cast iron, they built large astronomical instruments of bronze and iron that took the form of armillary spheres—huge metal rings precisely graduated with the degrees of a circle.

Different rings representing different sky circles were joined together at the two points where they crossed each other. Always with an emphasis on the meridians, one ring would represent the equator, another the sky-circle meridian passing directly overhead and through the celestial pole. These devices had sighting tubes through which astronomers could observe specific stars. The astronomer could move the sighting tube along the equator ring until he found a star. Then he counted the number of degrees marked on that ring back to the meridian ring, which stood up from it at 90 degrees. As soon as he counted the degrees, he could detect the exact position of the star along the equator and tell which sky segment it was in. These instruments aided astronomers in drawing star maps with great precision.[212]

The earliest known instrument of this type was built in 104 B.C., and the instrumentation became increasingly complex until the thirteenth century. Ken Shou-Ch'ang introduced the first permanently mounted equatorial armillary ring in 52 B.C. and in A.D. 84 Fu An and Chia Kmuei added a second ring to show the ecliptic. Chang Heng, the mapmaker, added a ring for the meridian in A.D. 125, as well as one for the horizon. But Chang Heng was not yet satisfied. He made an armillary sphere that rotated by water pressure in about A.D. 132. He used a wheel powered by a constant pressure head of water in a water clock to rotate the instrument slowly. This instrument was a tremendous tool for demonstrating and computing the movements of heavenly bodies.[213]

An advanced version of the armillary sphere is the torquetum, first invented by Arabs sometime between A.D. 1000 and 1200. (Some credit al-Tusi for the invention.) Here all the various rings are no longer nested together in a single sphere but are mounted at various different parts of a set of struts in a way more efficient than that allowed by the constraints of a single sphere. In 1270 Kuo Shou-Ching made a metal torquetum called the "simplified instrument." It was purely equatorial, with all the Arab ecliptic components left out. It survives today at the Purple Mountain Observatory in Nanking. It was moved there from its home site at Linfen in Shanxi during the Ming dynasty when government officials no longer understood that the difference of 3¾ degrees in latitude caused by the move would render it useless.[214] Needham called this "simplified instrument" the precursor of all equatorial-mounted telescopes. Needham believed that some knowledge of it eventually reached Tycho Brahe in Denmark three centuries later and led to Brahe's taking up equatorial astronomy for his instruments. Actually, an equatorial mounting of the kind devised by Guo Shoujing wasn't constructed in the West until 1791, when it was used for a telescope made in England, and thus its design became known as the "English mounting."[215]

Astronomy was the first real science practiced by the world's ancient cultures. It was primarily observational (rather than experimental), but it meets most criteria for what a science should be. Next we will continue on to an associated discipline, cosmology. I write "discipline" rather than "science" because, as you shall see, it's not clear what cosmology is. Cosmology is dependent on astronomy, extrapolating its data to a worldview. Which is not to say that astronomers always agree with the tales spun by their cosmology colleagues.

4

COSMOLOGY
That Old-Time Religion

ISAAC Asimov was prolific, having written more than three hundred books, on topics from biochemistry and physics to Shakespeare and the Bible. His best work came early. In 1941, he published "Nightfall," a story about a doomed civilization on the planet Lagash, which doesn't revolve around a single sun, as the earth does, but is held in the gravitational field of six separate suns. (Asimov does not work out the orbit of the planet—a seven-body problem!—but let's not fault him for that.) The result is that the inhabitants of Lagash are bathed in constant sunlight. Unaware of the night sky, its astronomers extrapolate that their universe comprises only a few dozen stars. These are mysterious lights barely visible against the light of the six suns. Those who see the stars as important are considered cultists. Still, there is an uneasy feeling on Lagash. Archaeologists have found the remains of nine previous cultures, each of which had reached a technological sophistication equal to the present culture and had then vanished. The geological strata indicate that each civilization had lasted about two thousand years.

At the end of the story, we find out the terrible truth: every 2,049 years all six suns set, and night falls. The Lagashians are terrified of the dark and the cold. They start fires, and the culture perishes. Anthony Peratt, who has served as a physicist at Los Alamos National Laboratory and the Department of Energy, points out that the Lagashians are destroyed by more than fire. The appearance of the night sky and countless stars destroys their cosmology; it shakes their faith and the philosophical basis of their society, which then crumbles.[1]

Cosmology is the study of the universe as a whole, its history and origin. It is usually (but not always) based on astronomy, along with religious and social beliefs. George P. Murdock, an anthropologist, listed sixty-eight civilizations that have fashioned cosmologies. Some had little science and scant astronomy. As soon as we humans identify a handful of stars, we construct a picture of the whole universe. Barbara C. Sproul, director of the program in religion at Hunter College, City University of New York, takes issue with Murdock's figure of sixty-eight civilizations: "*All* civilizations have cosmologies of some sort that say how reality is structured. By 'reality' they mean *their* universe, which might only be the neighborhood, but it is as far as they can see." As we shall see, our present universe may not be much larger.

If one demands that I justify cosmology as a science, I will find myself struggling to answer. The root of *cosmos* refers to a word that encompasses *everything.* How can you have a science that depends on knowing everything? We don't even know the size of the universe. Yet here we are. I can say this: cosmology is interesting; cosmology is important. Because it is so intertwined with general societal beliefs and attitudes, cosmology is a clue to the collective psychology of a civilization. There is usually some science in there, too.

Many cultures, I suspect, would react as the Lagashians did when their cosmology collapsed. Our psyches crumbles when our cosmologies crack. As we shall see, even in the modern age (perhaps *especially* in the modern age), we panic when our favorite model, the big bang, comes under fire.

In 1966, when the cosmologist Edward Harrison accepted a teaching post at the University of Massachusetts, he was handed the red book, a manual for faculty members. It explained what a university was, and what it wasn't, citing two courses that it said one wouldn't find in a curriculum of higher education: witchcraft and cosmology.

Decades later, Harrison tells this as a funny story, but at the time he was not amused, and he demanded that the university remove the aspersion toward cosmology from the red book. Still, he concedes that cosmology may not be a science,[2] and he is one of the few in his field who has spelled out carefully what cosmology is and what it isn't. In the first sentence of *Masks of the Universe,* Harrison writes: "The universe in which we live, or think we live, is mostly a world of our own making." An ancient Greek agrees. About the universe, Socrates said, "I know nothing except the fact of my ignorance."[3]

Harrison says that the real Universe (capital U) is unknown; it is everything, and we will never know what it is in its own right, independent of our changing opinions. There are, however, universes (lowercase), which are our models of the Universe, and cosmology is a study of these universes. "A universe is a mask fitted on the face of the unknown Universe," says Harrison.[4]

Cosmologists have always struggled with ambitious goals and too little data. The medieval European Christian universe was a comforting and static one: humans at the center; heavens populated by spirits; a sphere of fixed stars; beyond it, the primum mobile, a sphere maintained in constant motion by divine will; and, finally, the empyrean, a realm of pure fire where God lives. Western medieval cosmologists provided their followers with purpose and place in the Christian universe. It was primarily an Aristotelian model. The Arabs supplied the primum mobile.

Today we have exiled angels and fixed stars from our universe. Our leading cosmological model is the big bang (or, as cosmologists write it, Big Bang). With its fiery explosions, wormholes, white dwarfs, red giants, and black holes, the big bang universe satisfies our Lucasfilm sensibilities. It also features an abrupt beginning to match our creation myths and is constantly expanding. The big bang is the biggest-budget universe ever, with mind-boggling numbers to dazzle us—a technique pioneered by fifth-century A.D. Indian cosmologists, the first to estimate the age of the earth at more than 4 billion years. We are not at the physical center of the big bang universe, and there is no God, yet it is an anthropocentric model. The huge numbers—the comparative strengths of the four forces, the surplus of matter over antimatter, and soon—are balanced delicately to result in the evolution of intelligent life. That is, the known forces—electromagnetism, the strong and weak nuclear forces, and gravitation—are in such proportion to one another as to allow us (humans) to evolve and exist. The tiny imbalance of matter over antimatter—for every 100 million quark-antiquark pairs at the big bang there was one extra quark—makes possible a universe with matter (us included). Had there been equal amounts of matter and antimatter, they would have annihilated each other, leaving a universe of pure radiation.[5]

A human-constructed universe ends up with humans, at least mathematically, at its center. This is not to mock the big bang but to emphasize that it encompasses modern interests and beliefs as much as it does astronomy, and in that sense varies little from past models of the universe.

Harrison and I discussed the work of two physicists, Fred Adams and Greg Laughlin, who wrote a book in which they accept the big bang's precepts so thoroughly that they predict the course of the universe for the next googol years. (A googol is 10^{100}, or a 1 with a hundred zeroes after it.) Such extravagant extrapolations assume a rock-hard confidence in one's cosmology, that we have accurately charted all the stars and the forces that move them.

"The notion that we have lifted all the veils," said Harrison, "can't be true. We would leave no new universes for our descendants to discover."[6] Today we believe that past cultures were mistaken in their cosmologies; we ignore in doing so that those people believed in their universes just as firmly as we believe in our big bang universe.

"In the Babylonian universe," Harrison writes, "the flowers danced and fluttered in the breeze, the sun rose and set, the moon waxed and waned, the jeweled lights in the night sky traveled the heavens, and a rock was a rock, and a tree a tree. But the nature of these things and their meaning was other than what we now think. The lifestyles and modes of thought of the Babylonians, so unlike our own, were in harmony with the Babylonian universe."[7]

We don't know what the Babylonians thought about members of

their society who disagreed with the consensus cosmology. Were they tolerated? Tortured? We know in our own time that scientists who do not endorse the big bang are committing professional suicide. Past universes are dismissed as religion. Harrison notes, "Our universe is the only rational universe. Ones that came before us are mythologies. Contemporaries who disagree with our cosmology are crackpots." He calls cosmology "that new-time religion."

That said, the big bang universe is a fine theory, and it seems superior to the alternatives. Let's take a brief look at the hypothesis to give us a standard for comparison as we travel back in time to the Sumerians, ancient Amerindians, and others.

During the first thirty years of the twentieth century, there were two schools of thought concerning the cosmos. Some astronomers felt that the universe was small and static. They believed the Milky Way, in which our solar system is located, was the entire universe, and that the faint pinwheel clouds called nebulae that they saw through their telescopes in the far distance were insubstantial gaseous bodies. This "one big galaxy" theory was challenged by those who subscribed to a competing "island universes" theory, a term coined by the German philosopher Immanuel Kant. This hypothesis held that the Milky Way was but one small spiral in a great ocean of similar island universes.[8]

The champion of the "one big galaxy" cosmos was an astronomer from Missouri, Harlow Shapley, who debated all comers on the topic. In the winter of 1921, Shapley was at the eyepiece of the one-hundred-inch Mount Wilson telescope in California. He had been taking photographs of the Andromeda Nebula, which, if he was correct, was composed of nothing but gases. His night assistant at the time was Milton Humason.[9] Humason had dropped out of school at fourteen to become a bellhop, then a mule driver, and had then landed a job as a janitor at the Mount Wilson Observatory. He became adept at making astrographic plates of the heavens and was eventually promoted to assistant astronomer.[10]

On the night in question, Humason was "blinking the plates" of Andromeda on the stereocomparator, which compares two different plates taken at different times to reveal new features. What he saw was startling. The plates contained images of Cepheid variables, stars used as reference points, from beyond the Milky Way. This was evidence that the nebulae were galaxies of stars. Confronted with the evidence, Shapley patiently explained to Humason how the universe comprises but one galaxy, took out his handkerchief, and wiped the plates clean.[11] Humason was acting like an astronomer. Shapley was acting in the role of the cosmologist.

Fast-forward a year and a half to the fall of 1923. Shapley has left for Harvard, and another Missouri-born astronomer, Edwin Hubble, is at the eyepiece of the Wilson hundred-incher. Hubble and Shapley didn't care for each other. In fact, few people cared for Hubble. Still, he was a great astronomer. On the night of October 5, he locked onto a spiral arm of Andromeda and exposed a plate that, upon later examination, revealed a Cepheid variable. Cepheids pulsate, and their periods of pulsation are related to their absolute brightness—their "wattage." Detect a Cepheid, measure its period, you know how bright it really is. This gives you a measure of its distance. Shapley's putative universe was about 300,000 light-years in diameter. Hubble measured the Cepheid as being a million light-years away. The Andromeda Nebula was actually a galaxy full of millions of stars. Hubble had expanded the size of the universe a hundredfold.[12]

In 1929, Hubble struck again, this time with an even greater discovery. He measured the red shifts of the nebulae. (By now, we knew they were galaxies, but Hubble preferred the old term, *nebulae.*) Light stretches out or scrunches up as it moves through space; if the light source is moving toward the observer, the spectral lines are shifted toward the blue, or "blueshifted"; if moving away, they are "redshifted." This is analogous to a train whistle changing its pitch as it approaches and then departs from the listener. This Doppler shift can reveal the velocity and direction of a light source.

What Hubble found was startling. It changed our worldview, even that of Albert Einstein, who had believed in a static universe. No matter where Hubble pointed the Mount Wilson telescope, he found the nebulae were all running away from him—and fast! They moved at a significant fraction of the speed of light. The almost universal conclusion was that the universe is blowing up as we speak. Hubble's law, or the law of redshifts, holds that the farther away a nebula, the faster it is moving: double the distance and the speed doubles; triple the distance and the speed triples, and soon. Think of yourself as a student in a desk in a lecture hall, said Arthur Eddington. The student a foot away from you must move only one foot to double the distance. The student twenty feet away must move twenty feet in the same time period. Something similar is happening to the universe.

Hubble's work exposed two critical facts: 1) The universe is bigger than anyone had thought; and 2) It's getting bigger all the time. His discoveries fueled cosmology for the rest of the century. The Russian mathematician Alexander Friedmann and the Belgian physicist Georges Lemaître had devised expanding-universe theories in the 1920s, and Hubble's data gave them credibility. The ultimate extrapolations of Hubble's work today are the various big bang theories by the Russian-American physicist George Gamow and others.

Deriving a big bang creation from Hubble's redshift work is straightforward. Say you had a movie of the life of the universe. At this point in time, the galaxies are rushing away from one another. So if you ran the projector backward, as it were, the most distant galaxies, moving faster, would close in on the nearer ones, and finally the universe would rush together into a very small volume. That was the beginning. All of the universe was packed into a tiny space and exploded;[13] we're still exploding. We can't go back to the big bang—now estimated at about twelve to fifteen billion years ago—but the scenario is certainly logical.

Only consider this: the big bang hypothesis may never have been stronger than just after Hubble's redshift discovery. In those early days, the big bang expanding universe was simple, elegant, and easy to understand. This was before astronomers sought additional evidence for the theory, sometimes with unfortunate results. During the next several decades, multiple holes were bored in the model. Newspapers and magazines are continually celebrating "triumphs" of the big bang. In reality, most such triumphs have been devices to patch the gaps in the evidence. What was once an aesthetic paradigm now looks like a Rube Goldberg device. Let's examine a few big bang problems and their putative solutions.

One major glitch is isotropy, the fact that no matter where in the sky you point a telescope, you see similar patterns of stars, galaxies, and dust. Using the COBE (cosmic background explorer) satellite, astrophysicists find that the universe is the same temperature no matter what part of the sky they measure. (The temperature can be extrapolated by measuring the wavelengths of radiation.) These temperatures match with a precision of .01 percent. Scientists don't think this is a coincidence. They think that all parts of the sky must have been in contact with one another.[14]

Imagine yourself inside a giant popover. In every direction, you see yellowish pastry, billowy stalactites of baked dough jutting from the sides—and all at 350 degrees F. You figure all the dough had been in contact before baking and expanding, and you would be right. Same deal with the universe. Except that astrophysicists, running the film backward, as it were, found that the various parts of the sky couldn't have ever been in contact. Let's say our popover has been baking for an hour, and the sides are five inches apart, but we know that the dough expands at only two inches per hour. If we go backward in time, each side would shrink two inches inward—so they could never have been together. We have the same problem with our universe. Even if the universe had expanded at the speed of light, the numbers don't work out. The skies, rushing toward one another in the backward film, never come together.[15] (Another popular pastry metaphor is raisin-bread dough, with the raisins representing galaxies.)

Alan Guth, a particle physicist at MIT, saved the universe. In 1980 he proposed the theory of inflation, or the inflationary universe, which solves the isotropy problem, and more! Guth found a loophole in Einstein's theory of relativity, which normally holds that nothing can go faster than the speed of light. Guth figured that space itself, not being an object, would be immune from the cosmic speed limit and therefore could expand at superluminal speeds. This, he said, happened a tiny fraction of a second after the big bang. The specklike universe, about the size of a proton, was filled with a bizarre explosive force, expanding to the size of a golf ball. Not big, but all this happened in 10^{-33} (.000000000000000000000000000000001) seconds. This solves isotropy. The universe expanded faster than we thought. Different parts of the sky, like the crust of the popover, were once combined.

This magical inflation solved other problems, too. The *flatness* of the big bang universe has always troubled scientists. The universe had three choices: closed, open, or flat. If there were lots of mass, the universe would have curved inward and imploded quickly, because the mass would make for a hefty gravitational pull. The big crunch, and hence closed universe. With sparse matter, we'd get an open universe, or persistent runaway expansion. The universe would have burned itself out by now. What we have is a Goldilocks universe—flat, just right. It expands but doesn't burn out. This is the most unlikely of all universes. The universe's flatness at the age of one second has to be almost perfect.[16] Scientists don't like such coincidences. It smacks of an intelligent plan. It smacks of God.

Inflation solves this theoretical and theological crisis. A closed universe is curved inward, like the surface of a sphere. An open universe is curved outward, like the surface of a saddle. A flat universe is flat. With inflation, it can curve either way because inflation takes a segment of any curve, inward or outward, and makes it appear flat by stretching it humongously. In short, inflation makes any universe flat.[17]

Then there's the lumpiness problem. A big bang should have produced a homogeneous, smooth universe. Yet we have galaxies, lumps. Gravitation alone is not strong enough to cause these. Again, inflation to the rescue. Spooky quantum fluctuations can lead to the lumps. Inflation enlarges them so they can become galaxies.

If you read the above very quickly, it all makes sense. Very little of the phenomena described, however, has been seen through a telescope or replicated in the laboratory. There is sparse evidence for inflation except for the fact that, without it, we have to walk away from our comfortable big bang cosmology.

Besides the many pieces of evidence that point away from the creation of our universe by a spontaneous, unaided big bang, something even more suspicious troubles scientists. Our present universe is unlikely. The big bang, if it happened, could have randomly resulted in any number of universes. The world we now live in was a long shot.

Of course, most events are unlikely. As the statistician's koan goes: "Everything is impossible, yet something happens." Billion-in-one events happen constantly. A golf ball hit over a course containing a billion blades of grass must land on one of them. In the case of our universe, however, it was a hole in one into the wind. We live in the unlikeliest of universes, in which the flatness of mass, the relationship of the electromagnetic force to gravitation, the weak nuclear force and the strong nuclear force, along with the unlikely presence of carbon and other bizarre coincidences, make our world, in the words of George Greenstein, "amazingly hospitable to life." Greenstein, a physicist and professor of astronomy at Amherst College, cites the presence of carbon, the atom of biological life, as puzzling. Given a big bang, a universe composed entirely of lighter elements, hydrogen and helium, is far more likely.[18]

Greenstein has compiled a list of unlikely coincidences that have occurred to ensure life (as we know it): the vast distances between stars, without which they would be in frequent collision (even near collisions would hurl planets out of orbit); the peculiar conditions in the cores of red-giant stars (like our sun) that stimulate the creation of heavy elements; the exact equal but opposite electrical charges on the electron and proton that allow matter to form; the slight excess in weight of the neutron over the proton, which allows stars to shine for a long time; the smoothness and uniform temperature of the big bang.[19]

A solution of sorts is offered by the British philosopher John Leslie. Suppose, he says, you are sentenced to stand before a firing squad composed of twelve crack marksmen. They fire. But you find yourself unscathed. Do you say, "Hmmm. Curious," and walk away? No, something awesome has happened. There had to be a plan.[20] Our universe is like that firing squad. This analogy has led many people, including the British theoretical physicist John Polkinghorne, to posit a simple solution: God.[21]

Hope for the atheists also occurs to Leslie. Suppose, he says, that there were millions of firing squads executing people continuously worldwide. Now it seems almost inevitable that at least one squad will miss the mark. This is the interpretation that most big bangers apply to the universe. It is called the many worlds hypothesis. If you live in a zillion-to-one universe and there are a zillion other universes with a random distribution of qualities, then it's not so odd that one is just right.

So where are these extra universes? Doesn't science require better evidence than statistical logic? Cosmologists say these universes are cut off from our own in space and time and are unobservable.[22] Proof of their existence is the very fact that we can't find them! How do you argue with that? The many worlds hypothesis has been promulgated by cosmologists for a couple of decades, though it is hardly original to them and Leslie. In 1779, for instance, the British philosopher David Hume pondered whether the present universe is the final product after a series of mistakes. Numerous universes, wrote Hume, "might have been botched and bungled throughout an eternity ere this system was struck out."[23]

The firing-squad analogy can thus lead one either to believe in a Grand Designer or in many worlds, depending on one's predilection. Of course, there is no physical evidence for either. Rocky Kolb, one of the country's most respected cosmologists and the director of the astrophysics group at the Fermi National Accelerator Laboratory (Fermilab), in Batavia, Illinois, sees a hard distinction between the cosmologies of the past, such as those of India and Islam, and modern cosmology: "The western view of the universe has been grounded," he writes, "in science rather than religion or philosophy."[24] This isn't entirely true, as we shall see. The big bang universe embodies the philosophy of the scientific community as much as the medieval Christian universe embodied the

Church's, or Amerindian cosmologies Native American social constructs.

NOT ALL MODERN scientists subscribe to the big bang. There are at least two other major cosmologies with intelligent disciples today, the steady state universe and the plasma universe.

Steady state. Cosmologists Fred Hoyle, Thomas Gold, and Hermann Bondi have no argument with the concept that the universe is expanding, but they don't subscribe to the idea that everything emerged from a primordial singularity. Instead, they argue, as the galaxies move farther apart, new galaxies form in the empty space left behind at a rate that makes the universe seem unchanged. Because there are so many stars and galaxies, you would still find roughly the same number of them ten billion years ago as you do now, despite the new ones. Unlike the big bang, the steady state universe is not evolving. It violates the law of thermodynamics, which states that the total amount of energy in a closed system never changes. On the other hand, so does the big bang, which produces an entire universe out of nothing.

Plasma cosmology. This theory explains the universe in terms of its plasma, an ionized gas of electrons and positive ions. Most of the universe is, in fact, plasma. The beauty of plasma cosmology is that it relies on electromagnetism to explain the structure of the universe, and electromagnetism is 10^{41} times stronger than gravitation, the force used to explain the structure of the big bang universe. The plasma approach gets rid of lumpiness problems, because there's plenty of electromagnetic energy to do anything you want. The weird thing is that plasma cosmology does not explain the solar system nearly so well as a gravitation-based theory, as big bangers are quick to point out. Locally (in the solar system) gravitation more than adequately explains large masses (the planets) revolving around an even greater mass (the sun). In the universe as a whole, in which charged plasma rules, electromagnetism takes on a greater role, explaining the lumpy structure of the universe more obviously than gravitation can. Each theory has its strengths and its Achilles' heel.

Edwin Hubble never swallowed the big bang hypothesis, even though his redshift discovery started the whole thing. According to his protégé Allan Sandage, Hubble wasn't interested in theory, or in "worlds that could be." He took "what the universe gives you."[25] Hubble found the big bang theory pointless and never accepted that his discovery was necessarily proof of an expanding universe.[26] Ever the scientist, he said he was measuring redshifts. There are explanations for redshifts other than a big bang.

Creation stories fall fairly into categories: the primal earth or water mother who spontaneously creates the universe from her body; the sole male progenitor, like the Judeo-Christian God, who creates the universe from thought or word; the "world parents," whose procreative union gives birth to the cosmos; the world egg and world tree; the cosmic serpent.[27] All these refer to cosmological theories of creation out of nothing (e.g., the big bang), creation out of chaos (plasma cosmology), a universe without beginning or end (steady state), or cycles of cosmic birth and destruction (alternating universes).

Let us examine cosmologies of four pre-Western cultures: Mesopotamia, Mayan Mesoamerica, Oceania, and India. Mesopotamia has long since become dust; Mayan civilization has severely diminished, but remnants of it exist (more than 2 million Yucatecans call themselves Mayan, speaking twenty-nine dialects); there are also remnants of the original Oceanic cultures; and Indian Hindu culture still thrives in modern form. Mesoamerican, Mesopotamian, and Indian societies represent complex, literate civilizations, while the oral traditions of Oceania are dismissed by some with the negative buzzword *primitive.* Yet all four cosmologies have similarities: an initial separation of primal elements, a cosmos made of successive levels, and a divine lineage that inevitably leads back to human genealogy. Most cosmologies, including our own, contain inconsistencies and contradictions. The Moscow-born physicist Andrei Linde says Americans are too fixated on consistency. "In Russia," he told a meeting of cosmologists, "when we dig a tunnel, we put a team of workers on each side of the mountain. If they meet in the middle, we have a beautiful tunnel. If they don't meet—two beautiful tunnels."

The table gives a simplified preview of some upcoming cosmologies.

CULTURE	TYPE OF UNIVERSE	CREATION MYTH
Mesopotamia: Sumer	plasma	universe split from primal waters of mother goddess
Mesopotamia: Babylonia	plasma and big bang	universe formed from corpse of mother goddess, killed by grandson
India: Hindu	alternating universes	universes dreamed in (and out) of existence
	big bang; plasma	universe hatched from golden egg
India: Jainist	steady state	universe always existed
Oceania: Maiana	big bang, inflationary	universe bursts from swelling on god's head
Oceania: Tahiti	plasma	universe formed from shells
Oceania: Mangaia	big bang, inflationary	universe grows from coconut root
Mayan Mesoamerica	alternating universes; many worlds hypothesis	gods take four tries to create the universe

Mesopotamia

Mesopotamian cosmologies, which reflect the civilizations of Sumer, Babylonia, and Assyria, have come down to us in many forms. The most complete is the Babylonian Enuma elish, which sets down how the universe was created. This account of the story comes from Barbara C. Sproul, whose extensive study of creation myths is chronicled in her book *Primal Myths: Creating the World,* and Hebrew University Assyriologist Wayne Horowitz, whose decade of research and translation is set forth in his book *Mesopotamian Cosmic Geography.* The text of the Enuma elish was probably composed either around 2000 B.C., directly after the fall of the earlier Sumerian civilization and the rise of Babylonian power, or seven hundred years later during an upsurge of Babylonian nationalism under Nebuchadnezzar I.[28]

The male progenitor Apsu (manifested as "sweet" underground drinking water) and the primal mother Tiamat (manifested as brackish or salt water) are found lying side by side in the primal murk. Out of their union comes slime and silt, Lahmu and Lahamu, who then beget the horizons, who beget the sky god, Anu, and the god of water and earth, Ea. Ea then begets the beautiful, manly god of the sun and creator, Marduk, whose mission is to annihilate Tiamat. Tiamat's temper has suddenly gone bad, and she becomes a chaos monster[29] armed with "the Worm, the Dragon, . . . the Mad Dog, the Man Scorpion, the Howling Storm."[30]

Marduk mounts a chariot and dispatches Tiamat using winds, storms, and floods, then sets about rebuilding the universe with the precision of a good carpenter: "Then [Marduk] relaxed examining her corpse. . . . / He split her in half like a dried fish. / Then he set half of her up and made the Heavens as a roof. / He stretched out her skin and assigned a guard. / He ordered them not to let her waters escape."[31]

After measuring Apsu's underground ocean of drinking water, Marduk walks the heavens (i.e., Tiamat's upper half) and declares it similar in dimension and nature to Apsu—that is, watery. Marduk divides cosmic space further into the earth's surface and an area where winds and storms occur, corresponding to the atmosphere. The highest heavens are set aside for the House of Anu, Marduk's sky-god uncle. Marduk's great-great-grandfather, Apsu, is demoted from godhood to become merely the watery location of the House of Ea, Marduk's father. The middle heavens are given to Enlil, an air god still in favor from ancient Sumer.[32]

After properly housing his family, Marduk "set up the constellations. / He fixed the year, drew the boundary-lines. / Set up three stars for each of the 12 months." These stars/constellations (the "Plough Star," Piscis Austrinus, and a third unnamed body) are made in the respective images of Enlil, Ea, and Anu, and follow three east-west divisions of the heavens. They and the sun and moon, all created from Tiamat's belly, are set in "the Heights," somewhat below the House of Anu.[33] Marduk's star—Jupiter[34]—guides all the stars of the sky through their orbits with help from Enlil's and Ea's stars.[35]

Marduk then makes the earth out of Tiamat's head, forms the Tigris and Euphrates from her eyes, and plugs up her nostrils. Finally he creates the bonds that hold earth and heaven together out of Tiamat's tail and crotch.[36] Looking through a contemporary lens, one could speculate that this is an acknowledgment of how female procreative power holds the bonds of human life together.

Marduk thus emerges as the shining hero of Babylon, the power of the Tigris-Euphrates river valley. On earth, Marduk replaces the brackish sea or salt water of Tiamat, from which nothing could grow, with Apsu's "sweet water" of underground springs, upon which Mesopotamian agriculture depended. In the heavens, he sets up a correspondence between Babylon and the highest region, the House of Anu, a direct tie between the human and divine powers. It should not be surprising that the Enuma elish was written, according to Barbara Sproul, "to praise Marduk, the main god of Babylon; to explain his rise from a . . . local deity to the head of the whole pantheon; and to honor Babylon itself as the most preeminent city."[37] Anthony Aveni adds, "That this all takes place at the shore of the Persian Gulf where the Trigris and Euphrates (sweet water) intersect makes the tale highly geographically relevant."[38]

Going back in time to Sumer, we find a different creation story. The male Apsu is absent. Creation begins with the goddess Nammu, mother of the gods, who creates the heaven and earth all by herself. Enlil, the air god, is born from heaven and earth. She splits earth from heaven using a pickax. According to E. O. James, a professor of religion at the University of London, "It appears that the cosmos was conceived as a product of the union of . . . water, air and earth, which Nammu brought forth."

The story does not deal with the mechanics, but it has a beguiling simplicity that is lacking in the later Babylonian version,[39] which more closely resembles current big bang theory, with its inflation clauses and other hedges.

In Sumerian texts, Nammu is also the cosmic eternal sea out of which all matter emerges in order. To Ewa Wasilewska, a University of Utah anthropologist and archaeologist who has done extensive fieldwork in the Middle East, this order is inherent in the primal water and permits a limited but evolutionary progression of creation. Nammu's Sumerian sign is the same as one used for *absu,* which in Sumerian times meant both "sweet water sea" and "the watery deep." Hence Nammu is the historical source of both the pure water of Apsu or Ea and the salt water of Tiamat.

Nammu's importance is reflected in the Sumerian division of the heavens. As they charted the east-west paths of the stars like the Babylonians,[40] the Sumerians also divided the sky into three north-south regions corresponding to different city states: Nippur was north and center, associated with Enlil; Uruk to the west and Elam to the east were both centrally sited and under the influence of Anu. But Eridu, Sumer's oldest city-state, occupied the southeast and was associated with Nammu. Hence the worship of Nammu reflects the most long-standing cosmology and the oldest political entity in Mesopotamia.[41]

Yet as Sumer waned and Babylon ascended, Nammu descended from being the Sumerian goddess of creation and order to a Babylonian chaos monster. She lost the fertile, life-sustaining quality of "sweet water" to Apsu, and was left with the undrinkable salt or "bitter" (probably alkaline) waters.[42] According to Wasilewska, Marduk became the top god only by murdering the "female principle, Tiamat, the mother of all, personification of the salt waters."[43]

The Sumerian and Babylonian tales exhibit contradictions and reflect with candor the cultural contexts of the time. But certain parallels can be drawn between modern big bang cosmology and the ancients. Totally apart from the constant revisions, there are the separations of primal matter into polarities. In the Sumerian version, the primal water splits spontaneously into heaven and earth first, and then later anthropomorphically, when Enlil, the air god, separates earth from heaven with a pickax. Hence, creation comes into being from nothing, without cause.

India

In *Cosmos,* Carl Sagan describes several ancient creation myths, which are, he writes, "a tribute to human audacity." While calling the big bang "our modern scientific myth," he points to a critical difference in that "science is self-questioning, and that we can perform experiments and observations to test our ideas."[44]

Yet clearly Sagan is drawn to the Indian cyclic cosmology, where Brahma, the great creator god, dreams the universe into being. According to the religious scholar Mircea Eliade, during each Brahma day, 4.32 billion years to be exact, the universe putters along. But at the onset of the Brahma evening, the god tires of it all, yawns, and falls into a deep sleep. The universe vanishes, dissolving the three material realms of the earth, the sun, and the heavens, which contain the moon, the planets, and the North Star. (Four higher realms are not destroyed in this cycle.) The night slips by; then Brahma starts to dream again, and another universe shakes into being.[45]

The cycle of creation and destruction continues forever, manifested in the Hindu deity Shiva, Lord of the Dance, who holds the drum that sounds the universe's creation in his right hand and the flame that, billions of years later, will destroy the universe in his left. Meanwhile, Brahma is but one of untold numbers of other gods dreaming their own universes.[46]

The 8.64 billion years that mark a full day-and-night cycle in Brahma's life is about half the modern estimate for the age of the universe. The ancient Hindus believed that each Brahma day and each Brahma night lasted a kalpa, 4.32 billion years, with 72,000 kalpas equaling a Brahma century,[47] 311,040 billion years in all. That the Hindus could conceive of the universe in terms of billions (rather than in terms of the thousands of years prevalent in early Western culture and religious doctrine) was, according to Sagan, "no doubt by accident."[48] Yes, it's possible that they were just lucky.

But the similarities between Indian and modern cosmology do not seem accidental. Perhaps ideas of creation from nothing, or alternating cycles of creation and destruction are hardwired in the human psyche. Certainly Shiva's percussive drumbeat suggests the sudden energetic impulse that could have propelled the big bang. And if, as some theorists have proposed, the big bang is merely the prelude to the big crunch, and the universe is caught in an infinite cycle of expansion and contraction, then ancient Indian cosmology is clearly cutting edge compared to the one-directional vision of the big bang. The infinite number of Hindu universes is currently called the many worlds hypothesis, which is no less undocumentable nor unthinkable.

That's the simple part. If one delves deeper into ancient Indian cosmology, the complexities multiply. For one, Brahma is merely the active male manifestation of Brahman, the undifferentiated world-soul,[49] which exists whether or not there is a universe to exist in. Brahma, according to Eliade, is himself subject to the wheels of time and dies at the end of a Brahma century. All seven levels of the universe then dissolve into a feminine primal nothingness—the "rootless root of the universe."[50] Depending on the tradition, only Vishnu, the sustainer god, and/or Shiva, the destroyer, survive. Some stories show the male Shiva and the female primal nothingness reunited in the "bliss of brahman," a "nonprocreative" state.[51] When the bliss is over, Brahma reemerges (sometimes from Vishnu's navel) like an unwanted baby.[52]

Then there is Prajapati, a god who hails from the second millennium B.C., who fashions the universe from his own body in a sacrificial act of both creation and destruction.[53] In the thirteenth-century B.C. Rig-Veda, Prajapati is the divine life force, born from the infinite waters as a golden embryo, and yet one who also fathers himself.[54] Later, in the Brahmanas sacred literature of 1000 to 700 B.C., a golden egg floats on the primal waters, waiting for something to happen. Then Prajapati emerges and becomes the universe and all its forces, the maker of gods, and the "Lord of Brahman," a sort of first force to give the self-created

Brahman a nudge. After forming the watery chaos, the stars, sun, earth, and finally man and the animals,[55] Prajapati dismembers himself and uses his bodily parts to complete the creation,[56] saying "I will reproduce myself, I will become many."[57]

Contradictory? Yes, but there's a repeating theme here: out of Nothing comes Being, out of Being comes a dizzying multiplicity of forms, all faces of the same One: the egg makes the chicken and the chicken makes the egg. Metaphorically, the egg is the infant universe, which suddenly hatches to give birth to all forms of the universe. The big bang.

Another eighth-century B.C. myth from the Chandogya Upanishad leaves the gods out for the most part: "In the beginning this world was merely non-being. It was existent. It developed. It turned into an egg. It lay for the period of a year. It was split asunder. One of the two eggshell-parts became silver, one gold." From the silver, the earth formed, from the gold, the sky. Like the cosmic plasma that split into matter and energy, all successive forms of existence arise from this basic split between silver and gold, earth and sky.[58]

But in the spirit of modern scientific self-questioning, the myth's author is not content with this theory. It violates some internal logic, so he revises it in the next passage: "How," he asks, "from Non-being could Being be produced?" Unable to answer this any better than modern big bang cosmologists can, the ancient cosmologist proposes a theory: "On the contrary, my dear, in the beginning this world was just Being." From being comes heat, from heat comes water, from water (rain) comes food. And so the universe wills itself into being with no god and no beginning at all.[59]

A later variant on the Prajapati-egg myths ties together Hindu time cycles, Brahman, and Brahma in a sort of a grand unified field theory. In addition to the multibillion-year kalpa cycle, smaller cycles described world ages. Many Indian texts divide the kalpas into four ages, but a cycle described in the second-century B.C. Laws of Manu gives fourteen ages in a kalpa, each governed by Manu, a forefather of humanity. Here Brahman, by thought alone, creates both the primal waters and the egg that eventually gives birth to himself. In the egg, Brahman (now split into Brahma, First Man, and the offspring of primal waters)[60] divides the egg in half, again using only thought, and forms the sky and earth, a realm between them, the cardinal directions, and the oceans.[61] Like the big bang, each revision adds another layer of complexity to the origin of the universe.

A discussion of Indian cosmology would be incomplete without looking at the social contexts. Now let us add *Brahmin* to *Brahma* and *Brahman*. All stem from the root *brh*, meaning "to be strong."[62] But whereas *Brahma* and *Brahman* both refer to divinities, *Brahmin* refers to human beings. The Indian Brahmins were the high priests of the Vedic Aryans who conquered the darker-skinned Dravidians starting in 2500 B.C. According to a study by Albert Schweitzer, *Brahman* also meant "sacred power," and the Brahmins validated their position in society through a connection to this divine power.[63] Vedic literature of 800 B.C. first mentions Brahma, who seems to arise "as a process of apotheosization of the *brahma* priest," according to another Indian scholar. Mirroring Brahma's watch over the universe until 72,000 kalpas have completed their cycle and the universe is destroyed in fire, the Brahmins watched over sacrificial rites that ensured the continuation of society until its decreed destruction.[64]

Jainism, a competing sect in India with origins from the fifth century B.C., did not buy into the Brahma(n) story, despite its attraction to the upper classes. (Perhaps Jainists had tired of neighbors who considered themselves just slightly beneath the gods.) In any case, nowhere is the assertion of a steady state universe as explicitly stated as in Jainist texts. The objections to Hindu cosmology written by Jinasena, a ninth-century A.D. teacher, echo the objections heard today about the big bang (substitute "big bang" for "Creator" or "God"):

> Some foolish men declare that Creator made the world. . . .
>
> If God created the world, where was he before creation?
> If you say he was transcendent then, and needed no support,
> where is he now?

> No single being had the skill to make this world—
> For how can an immaterial god create that which is material?
>
> How could God have made the world without any raw material?
> If you say he made this first, and then the world, you are faced with an endless regression.
>
> If you declare that this raw material arose naturally you fall into another fallacy.
> For the whole universe might thus have been its own creator, and have arisen equally naturally.

Jinasena goes along in this vein for some length and finishes by saying, "Know that the world is uncreated, as time itself is, without beginning and end. . . . / Uncreated and indestructible, it endures under the compulsion of its own nature."[65]

Oceania

Here's another big bang story, this time coming from Maiana Island, in the South Pacific Gilbert Islands:

> Na Arean . . . sat alone in space as "a cloud that floats in nothingness." He slept not, for there was no sleep; he hungered not, for as yet there was no hunger. So he remained for a great while, until a thought came into his mind. He said to himself. "I will make a thing." So he made water in his left hand, and dabbled it with his right until it was muddy; then he rolled the mud flat and sat upon it. As he sat, a great swelling grew in his forehead, until on the third day it burst, and a little man sprang forth.[66]

Na Arean makes the initial separation of the universe into its fundamental dualities (matter/energy or land/water) before inflation swells like an inflated ego to create its crowning glory, mankind.

The cosmologies of Oceanic peoples are as various as the islands they inhabit. Barbara Sproul relates sixteen different stories that come from twelve different island cultures.[67] Yet there are many commonalities: the initial separation of the universe into two parts, for example, or the cosmos emerging from a common taproot.

The late London School of Economics anthropologist Alfred Gell's study on Polynesian cosmology and ritual identifies a unifying theme in which the universe begins in an amorphous state that is neither existence nor nonexistence, from which the creator god separates out the two fundamental dualities that propel the unfolding universe. In Polynesia this was the split between *po,* darkness, night, the ocean depths, death, and the world of the gods, and *ao,* light, day, life, and the world of humans. This split, basic to many world cosmologies, is then embroidered by the different islands to produce some complicated cosmologies.[68]

In Tahiti, the creator god is called Ta'aroa, meaning "the sever-er."[69] After splitting *po* from *ao,* Ta'aroa creates the rest of the universe from a series of shells that cover him like an egg. The first shell he turns over to form the sky; the second he pulls off himself like a molting crab to make the earth; then, like Tiamat in Sumer and Prajapati in India, he dismembers himself (i.e., destroys the original homogeneity) to create mountain ranges from his backbone, clouds from his internal organs, the "fatness" of the earth from his flesh, shelled creatures from his finger- and toenails, and red sky and rainbows from his blood.[70] As the ethnographer Teuira Henry set down the myth in 1928:

> As Ta'aroa has crusts, shells, so has everything a shell. The sky is a shell . . . endless space in which the gods placed the sun, the moon, the Sporades, and the constellations. . . . The earth is a shell to the stones, the water, and plants. . . . Man's shell is woman because it is by her that he comes into the world and woman's shell is woman because she is born of woman. One cannot enumerate the shells of all the things that this world produces.[71]

The notion of heavenly realms rising in shell-like levels is found in many Oceanic cultures. In nearby Hawaii, nine levels ascended from the earth, with the three highest containing the sun, moon, and stars and being solid. The intervening levels were collectively "the space in which things hang or swing," and included the realm of clouds.[72]

The Tahitian story above also reflects a common ancient belief of "As is above, so is below": what rules the gods and the heavens manifests itself on the level of everyday human life, and all things are connected to the divine. What works on the scale of the cosmos should also work down to the smallest particle of matter. We believe this today, but we have yet to formulate a theory of quantum gravity, combining our rules for particles with gravitation.

Other cultures from Fiji to Hawaii used roots as metaphors for both cosmological and human origins. To island cultures surrounded by ocean, to be rooted must have been seen as the same as survival; in numerous Polynesian islands, life and land were conceived as evolving from the ocean depths below,[73] just as roots come up from below to support plants. The Fiji islanders had a serpent creator, Ndengei, who was king of the "root-gods," so called because "they were there first . . . rooted in Fiji before there was any Polynesian or European influence."[74] When Ndengei slept he created night; when he tossed and turned in his sleep, he caused earthquakes; when he woke he created the day.[75] He is kind of an animate root, snakelike in motion, rootlike in shape.

On Mangaia Islands, part of the Cook Islands, you find another version of this cosmology. According to a myth recorded by mythologist Charles Long, the Mangaian universe is a hollow coconut that narrows down through a long tap root to a point representing the origin of all things, "the-root-of-all-existence." Inside, at the narrow end of the shell, is a woman (a.k.a. Great Parent, Great Mother),[76] called "The-very-beginning." According to Long, "such is the narrowness of her territory that her knees and chin touch, no other position being possible."[77] In modern terms, here is the infant universe squished down to the size of a point before it suddenly inflates outward—that is, expands from tapered root to spherical coconut.

From the Great Mother comes children. Each child lives at a different level in the coconut's interior, representing both temporal stages of Mangaian ancestry and spatial worlds at different stages of geologic evolution—for instance, "deep ocean" or "the hollow gray rocks." "The thin land" directly under the coconut's top is home to the immediate mythological ancestors of the Mangaians, the First Parents.[78] Above the coconut, ten heavens ascend in a series of domes and cause the movement of the sun, moon, and stars. To the Mangaians, the sun and moon occupy the first dome, rising through a hole in the coconut to the east and setting through an opposing hole in the west.

The First Parents also climbed out onto the coconut's surface through the hole used by the sun and moon.[79] But if the universe is a coconut, what came before the root? In this Mangaian myth, we are not told, but many Oceanian myths go back to a void and/or primal chaos. A Maori chant relates three stages to the creation. First comes a disembodied thought, then night, then light:

From the nothing the begetting
From the nothing the increase
. . . The power of increasing
The living breath;
It dwelt with the empty space,
And produced the atmosphere which is above us
. . . The great firmament above us dwelt with the early dawn
And the moon sprung forth;
The atmosphere above us dwelt with the heat,
And thence proceeded the sun
. . . Then the Heavens became light.[80]

The Maori universe is propelled by physical forces, apparently divested of divine influence and emerging out of nothing. Like modern Western cosmology, however, the exact moment of creation hinges on a contradiction.

A much longer chant from the Tuamotu Islands (originally called

Havaiki) starts with Kiho, "the source of sources," sitting in a void, which nonetheless has spatial qualities:

> Beneath the foundations of Havaiki which was called the Black-gleamless-realm-of Havaiki,
> Dwelling there below Kiho had no parents . . . there was none but him; he was not the root, he was the stability.

In fact, like land solidifying after a volcanic eruption, this state of nonexistence grows increasingly solid, described as eleven kinds of "foundation," the "columnar-support-of-land," "the Upholding-house-of-the-heavenly-regions." Then, following many world cosmologies, Kiho thinks the universe into being: "May I be eloquent of my indwelling occult knowledge; may I be expressive of my outpouring eloquence . . . that all assembled beings shall give ear!"[81]

His thoughts stir up "the inner-urge" of water and land, the two primal components. After a long list of land features that are currently nonexistent (the stratification, the viscosity, the multiplicity), Kiho finally bursts into action with earthquakes and flaming eyes. He is, in fact, a volcano, rearranging the water and land to his liking. Each movement of his body, now floating on the ocean surface, gives rise to eight sky realms and eight land realms, paired along connected planes of existence.[82]

From the standpoint of the big bang, the Oceanian amorphousness that precedes the universe seems closer to modern ideas of cosmic plasma than to a creation out of nothing.[83] The latter idea is really outside of Polynesian thinking, according to Gell, who sees creator gods as forces of separation. Speaking specifically of the Tahitian cosmology, he says, "What the god does is to articulate, or differentiate the world into its distinct components and qualities, but the substance of the newly-articulated cosmos remains what it always was, nothing other than the god himself."[84]

Nonetheless, the Western concept of a universe exploding out of "nothing," as opposed to Oceanian ideas of a primal chaos that is neither "nothing" nor "something," seems to boil down to interesting semantics rather than a fundamental difference in outlook. In the beginning everything was gooey, murky, and unknown.

Mayan Mesoamerica

Isolated from Old World cultures, the Mayan civilization in what is currently southern Mexico and Guatemala emerged around the birth of Christ, flourished, and abruptly and mysteriously disappeared. Other than stone pyramids and steles carved with elaborate glyphs, their story is preserved in a few codices, including the Quiché Mayan book of creation, the Popol Vuh. Mayan cosmology bears many resemblances to other cultures nonetheless: to India in the intermeshing cycles of destruction and creation and the huge time frames in which these cycles are set; to ancient Mesopotamia in the meticulous tracking of heavenly bodies that are manifestations of the gods; to modern cosmology in the gods' careful experimentation and revision, and in the equally ruthless condemnation of outmoded theories.

Before humankind, the Mayan universe unfolds fairly seamlessly. Like many cosmologies, it starts with a primal sea. The Popol Vuh begins: "Now it still ripples, now it still murmurs . . . it still sighs . . . and it is empty under the sky." Translator Dennis Tedlock calls the scene a kind of "white noise"—the sound preceding sound. Only the gods of the sea and earth, collectively called Heart of Lake and Heart of Sea, are present: Maker, Modeler, Bearer, Begetter, and Sovereign Plumed Serpent.[85] Joining them are Heart of Sky and the primal sky gods, called Hurricane, Newborn Thunderbolt, and Sudden Thunderbolt. After conferring, the water and sky gods agree to create land and life in a sequence that resembles twentieth-century biology's "primordial soup": an earth covered by ocean and subject to violent lightning, which helps produce the first amino acids. Thus the cosmic separations occur, the first being the preexistent separation of the water and sky gods, the second being the gods' active separation of earth from water, and sky from earth.[86]

Next the sun, moon, and stars are sown. The ancient Mayans conceived of this as "sowing" and also "dawning," because they connected the planting of seeds, which push up from underground to grow, with the dawning of the celestial bodies, which they believed traveled through the underworld before rising in the east.[87]

Ancient Mayan art pictured a two-headed serpent as the sky, with symbols of Venus—which rises just before dawn—on one end and the sun on the other. Mayan cosmogony depicts an earth whose base is a reptilian earth monster and a sky supported by crocodile and jaguar pillars. Each evening, the sun is consumed by the earth monster and moves backward underground, to rise each morning in the east.

Hence, Venus and the sun (cast in the Popol Vuh as twin boys) emerge each dawn one after the other, just as human twins do at birth. According to Anthony Aveni, "The sinuous image of a two-headed sky serpent offers a graphic depiction of the way the imaginary line connecting Venus above the horizon with the sun below can be followed through time."[88] Venus rises as the "front end of the cosmic monster emerging from the underworld."[89] As the day progresses, the two bodies move across the sky, to set one after the other at dusk. The Popol Vuh chronicles this celestial orbit with the twins' battle with Zipacna, a crocodile earthquake monster, and with their descent into the underworld, the realm of One Death and Seven Deaths. After a series of contests, the twins emerge to be reborn at day.[90]

Taken all together, we have a reptilian monster underneath the earth, a sky serpent overarching the heavens, and crocodile pillars connecting the two. As a total conjecture, perhaps as food passes through the body of a giant serpent (making a large bulge as it does), so the Mayans saw the sun and stars passing through the great serpentine orbits above and below the earthly plane.

However, complications and trouble seem to begin with man, and give rise to the Mayan version of the many worlds hypothesis. According to the Popol Vuh, the gods first create birds, deer, jaguars, and snakes to watch over the forest and to bring offerings to their creators. But the animals cannot praise the gods; they cannot speak at all, and when the gods realize this, they decree that animals are only good for one thing: to be eaten.[91]

So the gods try again. This time they fashion a human being out of clay. But the clay is soft and won't hold together. "It won't last," the mason and sculptor gods say then. "It seems to be dwindling away. So let it just dwindle. It can't walk and it can't multiply, so let it be merely a thought." And the gods abandon their creation.[92]

In the third creation, the gods decide they need something more solid. So they make creatures of wood, who are, well, wooden. These prototypes look like people, talk like people, and reproduce like people, but they don't have feelings, don't think, and, worst of all, fail to remember their creators. (They don't pray to their gods.) The wooden men are busy populating the earth when the gods destroy them by flood; by the Gouger of Faces; by Sudden Bloodletter, who cuts their heads off; by Crunching Jaguar, who eats them; and by their own grinding stones, which pulverize them. Like a plasma cosmologist caught at a big bang conference, there's not much left after this. And so ends the third creation.[93]

But the gods are empiricists, and they learn by experimentation, collaboration, and trial and error, what Aveni calls a "successive approximation process of universe building." In the fourth and final creation and after much conferring, they choose corn for flesh, water for blood, grease for fat. The result is the first true humans, who talk and praise their creators. There is one flaw: the humans are too smart. As the Popol Vuh relates, "Perfectly they saw, perfectly they knew everything under the sky, whenever they looked. . . . As they looked, their knowledge became intense." No one likes competition, so the gods clouded human knowledge such that people "were blinded as the face of a mirror is breathed upon. . . . And such was the loss of . . . understanding, along with the means of knowing everything."[94]

Three tries at the creation fail before a universe emerges that will sustain human life. (The botched attempts are reminiscent of the eighteenth-century cosmology of David Hume, cited earlier.) And so the current world emerges, though it, too, will be destroyed at the end of its era. The Mayans, like the Indians, conceived of great intermeshing time cycles that spawned creations and destructions as easily as a tree unfurls leaves and later drops them.

ODDLY, the dates of the fourth and last Mayan creation mesh rather nicely with those of the fourth and last Indian cycle: August 13, 3114 B.C., and February 5, 3112 B.C., for the Mayan, according to Linda Schele, and February 17–18, 3102 B.C., for the Indian, according to Aveni.[95] In India, these dates lined up with a planetary conjunction in Aries. In Mayan mythology, the two dates represent two acts by the gods to create the universe. On August 13, 3114, they laid the cosmic hearth by wheeling the three stars of Orion's belt to the center of the sky; two years later, on Feburary 5, they lifted up the cosmic tree, which is the Milky Way. As in India, both days corresponded to astronomical events. Schele, an epigraphist and professor of art history at the University of Texas, who sees Mayan myths as "star maps," states that on August 13, 3114 B.C., the three stars in Orion occupied the center of the dawn sky. The Great Nebula (M42), unknown to Europeans until 1610, can be seen between these stars, and was called by the Mayans the smoke of cosmic cooking.[96] A year later, the gods planted the cosmic tree, manifested as the Milky Way, which connected the thirteen layers of heaven to the seven layers of the underworld. According to Schele, "In 3112 B.C. . . . in the morning of February 5, the entire Milky Way rose out of the eastern horizon, until at dawn it stretched north to south across the sky."[97] Aveni agrees with the first interpretation, but is dubious about the February 5 Milky Way assertions.

In the minds of Mayan priests, these heavenly events marked the dawn of a new age, which was tracked using the "long count," a linear record of days beginning with the Maya fourth creation in 3114 B.C. and predicting the end of the current universe on December 23, 2012 A.D.[98] Within the universe's five-thousand-year life span, many smaller time cycles marked the durations of intermeshing astronomical, natural, and political rhythms.

Tedlock, in translating the Popol Vuh, worked extensively with Andres Xiloj Peruch, a modern Mayan spiritual leader, to interpret the ancient text according to still-existent Mayan beliefs. Present-day Mayan fireplaces include three hearthstones set in a triangle, a representation of a modern Quiché Mayan hearthstone constellation formed by three stars in Orion—Alnitak, Saiph, and Rigel.[99] During the destruction of the third creation, the Popol Vuh states, "The . . . hearthstones were shooting out, coming right out of the fire, going for [the men's] heads."[100] This, according to Xiloj Peruch, is an image of a volcano and an oblique reference to the cosmic hearth. Further evidence comes from ancient Mayan scribes at Palenque and Quirigu, who wrote that at the end of the previous age, three hearthstones ushered in a new age.[101] (Schele and Tedlock disagree over whether the August 3114 date and the ascension of the stars in Orion represent the end of the old age or the beginning of the new age,[102] but it's clear that the hearthstones represent a major turning point.)

Another creation story from the Yucatán Maya reinforces the meshing of cosmic and political cycles. According to Aveni, when Shield Pacal, king of Palenque, died in the mid–eighth century A.D., he had successfully consolidated the power of his city-state against challenges by neighboring cities. Chan Bahlum, Pascal's son, needed a celestial sign to cement his political legitimacy by tying his ancestry to the progenitors of the Palenque royal family, three gods born four thousand years earlier. The divine lineage, carved on a temple, depicts the second-born god as the sun and the firstborn as Venus. The identity of the third god is unknown.[103]

In any case, in A.D. 690, early in Chan Bahlum's reign, a planetary conjunction lined up Saturn, Jupiter, Mars, and the moon, which moved together across the sky to set directly over the old king's temple. Thus Chan Bahlum's royal mandate and divine lineage was affirmed by a clear sign from the gods. That this event involved neither Venus nor the sun was not something that likely troubled the ancient Mayas. Aveni points out that Mayan cosmology does not demand a one-to-one correspondence—only any connection between the astral plane and the human one.[104]

All our cosmologies, from the Sumerian and Mayan to those of modern leather-chaired professors at Caltech and Cambridge, are limited by a fatal lack of vision. Timothy Ferris starts off his book *Coming of Age in the Milky Way* with this remark: "When the ancient Sumerian, Chinese, and Korean astronomers trudged up the steps of their squat stone ziggurats to study the stars, they had reason to assume that they obtained a better view that way . . . because they had got themselves appreciably closer to the stars."[105]

Of what use is climbing a few hundred feet when we know today that the nearest star is six light-years distant? In human terms, we have significantly improved our viewing power with enormous land-based telescopes and more so with the orbiting Hubble telescope, which elevates us beyond the obfuscation of the atmosphere. In cosmic terms, however, a satellite at 675 kilometers brings us barely closer to the most distant skies, several billion light-years away, than standing on ziggurats did. Especially if the other galaxies are zooming away from us every second. As mentioned, the size of our universe is unknown.[106] The visible universe may be only a small part of the unabridged universe, and it's possible that some light may never reach us. We live in what is called the sub-Hubble sphere; it's possible that the unseen part of the universe is a squillion times bigger, in which case what we observe through our telescopes are the random and esoteric motions of local galaxies, not the true flow of space itself.[107]

BIG BANG cosmologists say we can bypass these observational problems by studying the early universe in particle accelerators, "atom smashers." Fermilab's Rocky Kolb says that one thing that separates our cosmology from those of the ancients is that we can replicate ours. He relates a Chinese creation myth from the third century A.D.:

> The world was never finished until P'an Ku died. Only his death could perfect the universe. From his skull was shaped the dome of the sky. . . . His right eye became the moon, his left eye the Sun. From his saliva or sweat came rain. And from the vermin that covered his body came forth mankind.[108]

Kolb notes that the above is not so different in degree from the big bang, but it is a myth because "no one can reproduce the death and decay of P'an Ku." Kolb says, however, that we can reproduce the big bang.

Kolb notes that the mathematical machinery of the big bang model can predict the temperature of the universe at any time during its expansion. This is an audacious statement, given that there's no empirical evidence for these temperatures, but, for the sake of argument, let's assume Kolb is correct. (The average temperature of the universe today is about 3 degrees Kelvin, or 3 degrees above absolute zero, and as we go back in time it gets hotter and hotter.)[109] "We study the very early universe by making a little piece of it in the laboratory," says Kolb.[110] By which he means that at Fermilab, protons are circulated in a four-mile-around tube and collide with antiprotons speeding in the opposite direction. The resultant collisions can achieve temperatures of 3,000,000,000,000,000 (3×10^{15}) degrees Celsius, the temperature of the universe approximately 0.000000000004 (about 10^{-12}) seconds (a millionth of a millionth of a second) after the big bang.[111] Because of this, Kolb claims that we can now re-create for a brief instant "the *conditions* that have not existed in the universe for fourteen billion years." (Italics mine.) It is in vogue today for cosmologists to say that these high-energy particle collisions verify their theories, that these little collisions inside the beryllium tubes of accelerators are eensy-weensy big bangs. "That we struggle so hard to try to show that we are superior to all our predecessors," says Anthony Aveni, "may tell us something about ourselves."[112]

Kolb works at Fermilab, but he admits that, while there's an accelerator within walking distance of his office, he's never conducted an experiment there, has never produced one of these little universes he writes about. Kolb is candid: "I never look at the [particle] events themselves. I don't look at pictures from telescopes, either."

Henry Frisch, a University of Chicago physicist, has conducted many experiments at Fermilab. He calls the particle-collision evidence for the big bang a "pile of baloney." There are a number of problems.

Frisch says that, first of all, cosmologists don't understand how few "events" (as they call them) attain the energy levels cited because they don't understand the physics involved. The protons in the beam at Fermilab do have enough energy to duplicate the temperature of the 10^{-12}-second-old universe, but such events are rare. The proton is not an elementary particle but a conglomerate (a "garbage can," in the words of physicist Leon Lederman) of quarks and gluons. One gets optimum energies only in those unlikely events in which a quark collides directly with another quark. In the year 2000, Frisch says, this happened no more than six hundred times—or one event every 500 *billion* collisions—at Fermilab.[113] Hardly enough to establish what the universe was like at $T = 10^{-12}$ seconds; and, in any case, no one was looking at the collisions for that purpose.[114]

More important, these particle events do not re-create the *conditions* (plural) of the early universe, as cosmologists insist. They replicate only one condition: temperature. Let's say you want to vacation in Akumal, on the Yucatán peninsula, where it's 85 degrees F in January. You live on Baffin Island, where it's -20 degrees. So you turn up your thermostat to 85 degrees. Now in your den you have "re-created the conditions of Akumal." Something's missing, though, no? Even if you wear a bikini and drink a piña colada with one of those little paper umbrellas, it's not the same as vacationing in Mexico.

It's the same deal with particle collisions and the early universe; the only commonality is temperature. Frisch says that the "densities of particles and fields" are entirely different. In other words, the big bang had more stuff in it, or we wouldn't be here today. All the matter you see around you—your MG, the bacon grease in the jar under the sink, various galaxies—was compressed into about the same space as an accelerator particle collision. In the accelerator, however, there are only two quarks in that volume. You won't get much of a universe from that. Also, while such collisions are very hot, the *volume* of heat is umimpressive: equal to a burning match. Again, not much of a bang.

Beyond such obvious details, Frisch says he is uncomfortable with the particle-cosmos connection, because "I am descended from a long line of rabbis." We cannot re-create the initial conditions of the universe, and therefore we can never recreate T = 0. "I'm not comfortable with discussing T equals zero," Frish says. "And if I don't know T equals zero, I'm not comfortable with T equals ten to the minus twelve."

COSMOLOGY REMAINS an interesting discipline, grounded in astronomy and physics. We need to imagine our world, even if that vision is inaccurate or incomplete. The ancient Indians, Babylonians, and Maya combined science with religion and social constructs to complete the picture. That we have done any differently is a delusion. If our cosmology appears free of religion, it's because we've made it into its own secular religion.[115] Unlike physicists or chemists, who welcome threats to their paradigms, modern cosmologists are Lagashians, defending their chosen model against all evidence. As the Russian physicist Lev Landau said, "Cosmologists are often in error, but never in doubt."[116]

The world of orthodox big bang cosmology does not suffer dissidents gladly. The Harvard- and Caltech-educated Halton Arp, a protégé of Edwin Hubble's,[117] never relinquished the intellectual rigor of his mentor, holding that redshifts are not necessarily evidence of an expanding universe. A skilled astronomer, Arp found objects in the sky that made national headlines and challenged big bang orthodoxy. He photographed highly redshifted quasars in the same area as low-redshift galaxies, with some evidence, albeit sketchy, that the quasars and galaxies are linked by hydrogen gas. If redshifts mean what big bangers think, then a high-redshift object cannot be in the same part of the sky as a low-redshift object. How did Arp's employers at the Carnegie Observatories respond to his discoveries? They pulled him off their two-hundred-inch Mount Palomar telescope. He was forced to work "in exile," as the journal *Science* put it, at the Max Planck Institute for Astrophysics in Germany.[118] Carnegie (now part of Caltech) acted in the time-honored tradition of Harlow Shapley, wiping the plates clean.

Even orthodox scientists can't resist invoking the name of God. One of the recent saviors of the big bang is astrophysicist George

Smoot, who in 1992 showed that the early universe, three hundred thousand years after the big bang, was "wrinkled." That is, using the COBE satellite, he found tiny temperature fluctuations in the ancient sky. The differences were only a few millionths of a degree, but the less hot areas were deemed to be the "seeds" for galaxy formation. These "colder" areas were denser, showing that the early universe was not homogeneous but just lumpy enough to result eventually in galaxies, stars, and us. As we have seen, it was more likely that a big bang would result in a cloud of gas, with an "unrelenting black" sky, as Smoot put it. And, as he continued, "We would not be here to observe it."[119] He probably should have left it at that, but there were reporters around, so finally he said, "If you're religious, it's like seeing God."[120] A few weeks later, a reporter traveled to the Lawrence Berkeley Laboratory to interview Smoot, and in the hallway found an interesting graffito: "If you're God, it's like seeing George Smoot."

The so-called wrinkles in the universe have been cited by many cosmologists as "proof" of Guth's inflation. Kolb prefers the phrase "supportive of inflation." What if I believed in unicorns, I asked Kolb, and I found manure in the woods. Is that supportive of unicorns, since they probably defecate? He replied, "Well, if you found a honking big pile of manure, you could at least say that it's not a rabbit." Which is not to say that inflation theory is a honking big pile of anything.

5

PHYSICS
Particles, Voids, and Fields

THE largest particle accelerator in the United States cuts a four-mile-around circle in the northern Illinois prairie, outside the town of Batavia. At Fermilab (Fermi National Accelerator Laboratory), beams of protons and antiprotons circulate in the long stainless steel tube and are squeezed together at two points, inside detectors, so that the particles and antiparticles collide, producing tremendous energies. Physicists examine the aftermaths of these collisions, as new particles—some not seen in this universe since a fraction of a second after the putative big bang—coalesce out of the bursts of energy.

When the accelerator is down for maintenance, Fermilab guides lead tours through the brightly painted accelerator tunnel. The tours start out in the atrium of Wilson Hall, the administration building, and then cross the road to the accelerator. Wilson Hall's atrium also contains the lab cafeteria.

A guide told me that one day she noticed an older man in her tour group who looked familiar. He was fascinated with the accelerator, rubbernecking to see every feature she pointed out. Back at Wilson Hall, he thanked her for the tour, saying he was amazed at what he had seen that day. She said, "You look familiar. Have we met before?" He said yes and gave his name. The man had been employed at Fermilab for more than a decade. He was a theorist, working in the lab's theoretical department in Wilson Hall—as opposed to the experimenters, who work in the detector control rooms over at the accelerator itself. The theorist again profusely thanked the guide, admitting that it was the first time he had actually seen the accelerator. His visit was an accident, of course. The man had thought he was in the lunch line.

This story exemplifies the disconnect between theory and experiment in modern physics. The chairman of Fermilab's theoretical department nodded when I told him the story. "We don't require our theorists to actually visit the accelerator," he said. "But we insist that they all know there's one out there on the grounds somewhere." (This separation of theorists and experimenters is a recent phenomenon.)[1]

Western physics isn't supposed to be this way. Galileo is usually cited as the first real physicist, the one who decided that the laws of nature could not be ascertained through pure reason. And, despite being a mathematician, Galileo made math secondary to experiment. Mathematics was an apt language for describing the results of an experiment, but you had to do the experiment. He dropped objects from a tall, slanted building, and rolled them down inclined planes. He measured and compared their rates of acceleration, and thus destroyed an important piece of Aristotelian theory. This combination of experiment and theory, of action and mathematics, is the key to Western physics. In recent years, the press has fixated on theory alone, but experiment remains the foundation of modern physics.

Physics is often called the queen of the sciences. Ernest Rutherford, the experimenter who discovered the atom's nucleus, said, "All science is either physics or stamp collecting."[2] That's a bit harsh, but what distinguishes physics from the other disciplines is its search for simplicity, for overarching principles. Through the millennia, physicists have asked basic questions. What is matter? What is energy? What is light? (They are also in pursuit of more mundane things, of course, points out Williams College physicist David Park, such as "a high-temperature superconductor with good mechanical properties.")[3]

The quest of fundamental physics is to reduce the laws of nature to a final, simple theory that explains everything. Nobel Prize–winning physicist Steven Weinberg notes that fundamental rules are the most satisfying (at least to him). Isaac Newton's basic laws predicting the behavior of the planets are more satisfying than, say, an almanac showing each planet's position at every point in time. Weinberg cautions that physics can't explain everything, and that it cannot explain events except in terms of other events and rules. For example, the orbits of the planets are the results of rules, but the distances of the planets from the sun are accidents, not a consequence of a fundamental law. Of course, he adds, our laws, too, might be accidents. Weinberg says that physicists are more interested in rules than events, in things that are timeless—the mass of the electron, for instance, as opposed to a tornado near Tulsa.

He set down these thoughts recently in a lecture entitled "Can Science Explain Everything? Can Science Explain *Anything*?" Weinberg demonstrated that there are limits to physics. He was still befuddled about the foreign advances on his recent book *Dreams of a Final Theory.* Why, he pondered, should France pay only 10 percent of what Italy pays? Physicists will never explain the French; the planets are easier to contemplate. Still, Weinberg feels there will be a final theory. "We are moving toward an explanation of the world," he said. "That picture will be a satisfying worldview." And to be satisfying, he added, any final explanation "must be rich enough to include *us.*" Of course, such a final theory won't answer all questions, explains Park: "For example, we have known *all* the fundamental physics of the water molecule for sixty years, but there is still nobody who can explain why water boils at 212 degrees Farhenheit. Why can't we? We're too dumb. I would guess that we will still be too dumb when what Weinberg calls the explanation of the world is in hand."[4]

Physicists' current explanation of the substructure of matter is called "the standard model." It includes the twelve elementary particles and three forces that, when mixed and matched, can build everything in the universe, from soup to galaxies, and can explain all actions. The particles include the six famous quarks (up, down, strange, charm, bottom, top—we don't think there are any more) and six leptons (the electron and its two heavier cousins, the muon and the tau, and their three associated neutrions). The three forces are electromagnetism, the strong nuclear force (which holds quarks together), and the weak nuclear force (responsible for radioactivity). There's a fourth force—gravitation. It's important, but nobody knows how it fits into the standard model. All the particles and forces in the model are quantized; that is, they follow the rules of quantum theory. There is yet no theory of quantum gravitation.

The standard model is less than satisfying. Scientists think that besides being incomplete, it's too complicated. There must be a simpler plan. Nobel Prize–winning physicist Leon Lederman says that a good final theory should be concise enough to fit on the front of a T-shirt. The present model requires two people walking side by side, one with the particles, the other with the force picture.

Another problem is mass. All the particles have different masses, and no one knows where these come from. There's no formula, for instance, that says the strange quark should weigh twice (or whatever) as much as the up quark, or that the electron should have 1⁄200 (or whatever) the mass of the muon. The masses are all over the board; they have to be "put in by hand," as the expression goes—each measured individually by experiment. Why, in fact, should the particles have any mass at all? Where does it come from?

To solve this problem, many particle physicists today believe in something called the Higgs field. This is a mysterious, invisible, etherlike field that pervades all of space. It makes matter *seem* heavy, like a man running in invisible oil. If we could find this field or, rather, the particle that is the manifestation of the field—called the Higgs boson—we would go a long way toward understanding the universe. It was the Higgs boson that the Superconducting Super Collider (SSC), the fifty-four-mile-around accelerator proposed to be built in Texas, was supposed to find. Congress voted down funds for the SSC back in 1993.

As we shall see, the Higgs field showed up many centuries ago in ancient India, under the name *maya,* which describes a veil of illusion that gives weight to objects in the material world.

THERE HAVE BEEN two great shifts in Western physics. The first came with Galileo and Newton, who moved science away from ancient

Greek ideals of pure reason, making it hard-core and dependent on experimental data and causation—rejecting notions, for example, that light is a "quality" and attempting to quantify things such as light, force, and matter. Weinberg still sees Newton as the most important scientist: "He transformed the intellectual world put in place by Aristotle." In terms of methodology and worldview, Weinberg says we still live in Isaac's world. (Contemporary physicists treat Aristotle with some contempt. Alternate views exist.)[5]

Despite this, another great shift came in the twentieth century with the advent of quantum theory. Galileo, Newton, Michael Faraday, the nineteenth-century English experimenter, and his theorist counterpart James Clerk Maxwell, among others, had put in place the magnificent edifice of classical physics. We knew how things moved mechanically, how light bounced off objects, how electromagnetic radiation propagated through the universe—a host of knowledge about the physical world. Newton's second law, F = ma (force equals mass times acceleration), for instance, is one of the mantras of classical physics. Then quantum physicists went down into the atom and discovered a new world.

Well, actually, not a new world. It is often said that Newtonian and Maxwellian laws rule the macroworld, and that quantum theory holds in the microworld. This implies that there are two universes with their own separate laws. Not true. This is just a convenient, if sloppy, way of describing the situation. There is only one world, and the real world is the quantum world. However, our classical laws are *good enough* to work in the bigger macroworld.

Newton and others thought of particles as hard little balls, following classical laws. In reality, twentieth-century European physicists—Max Planck, Rutherford, Niels Bohr, Werner Heisenberg, Max Born, Erwin Schrödinger, Paul Dirac, Wolfgang Pauli, and many others—discovered that subatomic particles such as electrons are really squishy, indecisive things. We cannot predict precisely where they are at any moment in time. We can only determine percentages—there is a 70 percent chance the electron will be *here,* a 30 percent chance that it will be *there.* And it gets worse. Sometimes light behaves like radiation (it's continuous, a wave) and sometimes like matter (it's a particle—in this case, the photon). Conversely, matter particles can act like waves. It's an uncertain world down there.

The reason that it seems like a whole other world up here on the surface is that all this uncertainty averages out when you gather squillions of particles together. Newtonian laws such as F = ma are good averages, sort of like insurance company actuarial tables. A baseball is chock full of particles, all of which are allowed the luxury of chance and uncertainty. But lumped together, the uncertainty tends to wash out. Your insurance company doesn't know when you're going to die. But it knows how a couple of million people just like you are going to behave, and how many of you will die in any particular year. Roger Maris knew how much force it would take to power a baseball over the right field fence. A few quarks or electrons might rebel; as a group, though, they obey Newton.

MANY ANCIENT CULTURES had inklings of quantum theory. Where did this come from? Rutherford had a radioactive source of alpha particles to probe the nucleus. J. J. Thomson had cathode-ray tubes for discovering the electron. The ancients had no such equipment. Where did the concept of field come from, so new to us and yet prevalent in past cultures? Michael Faraday had to build a dynamo (electrical generator) to fabricate his field theory.

Democritus of Abdera, the fifth-century B.C. Greek philosopher known as the "laughing philosopher" because he was amused by the foibles of men, is sometimes also called the father of particle physics. He had some prescient ideas. At one point in his life, he sank into a deep depression, determined to kill himself by starvation. His sisters tricked him. Preparing food to celebrate the feast of Demeter, they baked bread. The smell wafted up to his room, where he was revived—not only physically but intellectually. He asked himself, How does the essence of bread travel from his kitchen below to his bedroom above? His solution was the atom, literally "that which cannot be cut." He figured that the loaves of bread sloughed off atoms that traveled to his nose. Democritus proposed that all matter is composed of finite, invisible, and indivisible particles that combine in various ways to produce all the objects we see around us.

Today we use the word *atom* to refer to the individual chunks of elements in the periodic table: hydrogen, oxygen, lead, uranium, and so on. Those are hardly a-toms (uncuttable) in the Democritan sense. They are complicated and eminently cuttable into smaller parts. Our quarks and leptons are Democritus's atoms. (Though there is always the possibility that quarks might turn out to be cuttable.)

Despite the fact that we call Democritus the first particle physicist, his was not a new idea. Steven Weinberg says that Indian "metaphysicists" came upon the idea of atoms centuries before Democritus,[6] who came up with many ideas embraced today by physicists. Democritus's statement "Everything is the fruit of chance or necessity" could be the defining statement of quantum theory. That is, randomness and causality work hand in hand. We cannot predict, say, when any particular pion will decay, but we can predict when half of a large group of them will decay (hence the term *half-life*). Democritus also believed in the void, in the vacuum, in nothingness. Once he had thought of atoms, he needed someplace to put them. "Nothing exists except atoms and empty space," he wrote. "Everything else is opinion."[7]

We see these ideas in non-Western cultures as well. Where we don't see them is in ancient Greece or in Europe, until very recent centuries. Democritus was a Greek, and some of his ideas were shared by other pre-Socratic philosophers. In general, though, his scientific theories were rejected by later mainstream Greeks—Aristotle and Plato, to name two. In fact, Plato wanted to burn all of Democritus's books.[8] Leucippus, Democritus, and the other pre-Socratic Greek philosophers sought quantitative rather than qualitative explanations for the world and asked the question "how?" rather than "why?" in distinction to the more abstract, teleological approach of the later Greeks.

The one concept of Democritus's that was embraced by his fellow Greeks, and that endured into the Renaissance, was his concept of the eidolon. It was not one of his better ideas. Light fascinated all ancient and medieval cultures, and much of their physics was focused on it. Two concepts hatched in ancient Greece that entranced the West for centuries were those of the ray and the eidolon.

In the fifth century B.C., Empedocles (best known for positing that all matter is constructed from earth, air, fire, and water) suggested that vision occurs because a visual ray reaches out of the eye and feels what is in front of it.[9] The eye is an active participant in vision, sending out rays as probes to gather visual information. A few decades later, Democritus came up with the eidolon. Just as objects slough off atoms, he said, they also slough off a thin visual layer of themselves, perhaps one atom deep. This is the eidolon, a physical shell of the object, floating through space to the eye of the beholder.[10] We know now that rays do not emanate from the eyes. Arab scientists dismissed these ideas.

No ancient or medieval culture practiced physics at the level we've witnessed in the West during the past four hundred years. What has marked Western physics is the intertwining of experiment and theory. Some of the cultures we are about to examine were strong in theory but weak in experiment, or vice versa. We will start with a civilization that valued data over hypothesis.

China

Ancient and medieval Chinese physicists did not envision aspects of quantum theory. They did, however, conduct experiments; their knowledge was empirical rather than intuitive. Perhaps because of this, Chinese physics of days gone by mirror the classical physics of the West from the era of Galileo to the beginning of the twentieth century—before the quantum age began. Experimental techniques of the ancient and medieval world would, quite naturally, yield classical results. Let us hasten to add that the Chinese, says David Park, never put together an all-encompassing dynamic theory as Isaac Newton did.[11]

ACCORDING TO BRITISH sinologist Joseph Needham, the ancient Chinese, like Aristotle,[12] viewed the universe as a continuity, rather than as a collection of atoms. The yin-yang duality dominating nature was viewed as a rising and falling, as a wave crests and troughs, and loosely connected with the tides. A first-century A.D. Chinese writer says, "The Yang having reached its climax retreats in favour of Yin; the Yin having reached its climax retreats in favour of Yang."[13] As basic forces oscillate, individual objects also oscillate in a web of "mutual influences," reflecting the Chinese belief in the inherent rhythms in all matter. *Ch'i,* the Chinese concept of energy, soul, ether, was not made up of particles but acted on objects and connected them. Needham says these influences worked over large distances, vibrating according to the specific rhythms of tangible matter and with the cosmic oscillation of the yin-yang cycle.[14]

A phrase translated as "bright window dust"—referring to the dust notes caught by sunlight—was used by Chinese alchemists in the second century A.D. as a metaphor for potable gold, and also reflected their view of light as emanation. (Some believed that gold was a solid form of sunlight.) In the twelfth century A.D., Wu Tsheng commented, "If the elixir [of life] succeeds it will appear as an impalpable powder like bright window dust. If such an elixir (so full of motion, energy, and vitality) is ingested, it will irrigate . . . the body of man (with a life-giving water.)"[15] Needham writes:

> The idea of a solid substance so finely comminuted as to become an impalpable dust able to penetrate everywhere, even through apparently impenetrable solids, caught [the Chinese] imagination strongly. Hence the expression "bright window dust." . . . It was perhaps rather characteristically Chinese that these observations did not arouse . . . any ideas of an atomist nature. On the contrary, the poets laid their emphasis on permeation, penetration, and rest as opposed to the ceaseless motion. They felt that the elixirs, if made correctly . . . must consist of such subtle matter, able to pass like incense smoke. . . . Here we touch upon something very deep-seated in Chinese medieval natural philosophy . . . the assimilation of matter, almost infinitely divided, to *chii, pneuma,* vapour or emanation.[16]

Luminescent emanations caught the Chinese imagination. The ancient and medieval Chinese described static electricity, phosphorescent organisms, marsh lights, and fluorspar (which glowed when under friction). Needham suggests that artificial phosphors were manufactured by the Sung Chinese. An eleventh-century manuscript describes a painting of an ox "which during the day appeared to be eating grass outside a pen, but at night seemed to be lying down inside it." The Sung alchemist Lu Tsan-Ning explained that secretions from a certain oyster could be mixed with paint to create colors that only appeared in darkness. This story would seem fantastic except that in 1768 John Canton described producing a calcium sulfide phosphor from oyster shells ("Canton's phosphorus"). When mixed with other chemicals, the phosphor could create luminance of different colors.[17]

Marsh lights (ignes fatuis), the glowing lights seen over swamps and decayed matter, were associated by the Chinese with blood and death. (Perhaps the notion of *Ch'i* as the life emanation in human blood and also as vapor informed this association. The Aztecs and Indians had a similar connection of blood with energy.) The second-century A.D. *Po Wu Chih* (Record of the Investigation of Things) describes marsh lights and suggests a connection with electricity:

> These lights stick to the ground and to shrubs and trees like dew . . . wayfarers come into contact with them sometimes; then they cling to their bodies and become luminous. When wiped away with the hand, they divide into innumerable other lights, giving out a soft crackling noise, as of peas being roasted. . . .
>
> Nowadays it happens that when people are combing their hair, or when dressing and undressing, such lights follow the comb, or appear at the buttons when they are done up or undone, accompanied likewise by a crackling sound.[18]

Ideas around sound were also based in wave concepts. During the first and second centuries A.D., Wang Chong, in *Discourses Weighed in the Balance,* compared the propagation of sound to water waves:

> A fish one *chi* [24 centimeters] in length moving in water will cause the water on either side to vibrate. The central area of vibration would be only a few *chi* in diameter. . . . The extent of the vibration would reach no farther than a hundred steps, and at a distance of one *li* [1,800 *chi*] all would be quiet . . . because the distance is too great. A man producing sound by manipulating air is like a fish, the change of air is like that of water.[19]

Wang Chong does not specifically state above that sound is a wave. That can be inferred or not, depending on the reader.

Much later, Ming dynasty (1368–1644) scholar Song Yingxing asserted: "Air has substance. . . . When an arrow flies through it, sound is produced by striking it; when the string of a musical instrument is plucked, sound is produced by vibration. . . . When one throws a stone into water . . . where the stone drops is no larger than a fist, but waves will spread outwards circularly. The vibration of air is the same."[20]

Chinese application and understanding of acoustics is also associated with vibration and wave motion. A set of sixty-four bronze bells from the fifth century B.C. illustrates Chinese technology related to acoustics.[21] From a physics standpoint, most interesting is that each bell had two "strike points" voicing two notes, which required an asymmetrical mass. According to historian Cheng-Yih Chen:

> The use of asymmetry in mass distribution to obtain an extra mode of vibration . . . requires rather advanced acoustic analysis so that each mode can be individually excited without appreciable interference. . . . Only when the nodal lines of one of the vibrational patterns fall along the antinodal lines of the other can the two modes of vibration . . . be individually excited to produce their corresponding resonance frequencies without interference.[22]

Thus the "front strike point" is situated exactly where the lower vibrational antinode meets the higher vibrational mode, while the side notes are the reverse.[23]

Theoretical explorations came later. (Perhaps listening to the beats caused by notes not exactly in tune led to the concept of vibration.) In any case, the Chinese recognized that slow beats of vibration were related to low notes and fast beats to higher notes.[24] Resonance is described as early as the fourth or third century B.C., when musicians noted how if one string of a zither was struck, other strings of the same notes would also vibrate. In the Tang dynasty (618–907) Nianzu related a story of a monk whose chime hanging in his room sounded without any perceptible cause. The monk became ill from this; a visiting friend, noting that the chime sounded when the central monastery bell rang, cured him by filing down part of the chime. The friend's reasoning? The monastery bell was the cause of the chime's sounding, and by filing off part of the chime, the bell and chime no longer sounded at the same frequency.[25]

In contrast to light and sound, Chinese advances in both optics and mechanics were largely based on logic and deduction, rather than on harmonic theory. Mo Zi (circa 450 B.C.) is credited with founding the Mohist school, a logical and philosophical system with communities in ancient China from the fourth through the second centuries B.C.[26] The Mohists compiled the *Mo Jing,* a work of canons and explanations, which covered topics from mechanics and optics to logic.[27]

Mo Zi, or the people who followed him, experimented on light and concluded that it traveled in straight lines. The Mohists created a pinhole upside-down image using a wall with a small hole in it. The interior room was dark; the outside was in sunlight, with the hole in line with the sun. The Mohists found that a person standing between the hole and the sun cast an inverted shadow on the back wall of the interior room, thus predating by sixteen hundred years the camera obscura ("dark room") of thirteenth-century Europe. Mohists analyzed the phenomenon this way: because the person's head blocked sunlight coming from above, the shadow of the head appeared below, and because the foot blocked sun coming from a low angle, the foot's shadow appeared above.[28]

Mohists also observed the shadow of flying birds and applied the idea of straight lines to this. In any instant, the bird's shadow is not moving, because the bird's body blocks the rays of light. Therefore the "moving shadow" was really a succession of still shadows.[29] In a precursor to understanding diffuse and direct sunlight, they suggested that shadows resulted from the "absence of light." Partial shade resulted from several sources of light, where light from one source strikes an object and is blocked, while light from another source passes the object and partially illuminates the shade.[30]

The Mohists explored how shadows change in size, how images are formed on a flat mirror, and how a concave mirror creates both inverted and upright images, while a convex mirror creates only an upright image. Using spherical mirrors, they discovered that an object placed at the center of the sphere will merge with its image. They thus understood the difference between a mirror's center and its focal point (called "central fire").[31] Fifteen hundred years later, Jin Dynasty writer Zhang Hua used a piece of spherical ice to focus sunlight and set fire to dried leaves.[32] The Mohist canon vanished in the fourth century A.D. and was not in wide circulation again until the eighteenth century.[33]

The Sung dynasty scientist Shen Gua (A.D. 1033–1097) studied images in relation to concave mirrors. Whether he knew of the earlier Mohist work is not clear. "The burning mirror," he wrote, "reflects light so as to form inverted images. This is due to a focal point being between the object and the mirror. . . . It is analogous to rowing where an oar moves against the oarlock."[34] Later Shen Gua noted, "The oarlock constitutes a kind of 'pivot point' (. . . literally 'waist'). Such opposite motion can also be observed as follows: when one's hand moves upward, the pinhole image moves downward and vice versa."[35]

In the thirteenth century, the Taoist Zhao Youqin pursued the pinhole experiments. His experiment was done in a room with two circular wells, one four feet deep by four feet in diameter, the other eight feet deep by four feet in diameter. A four-foot-high table was placed in the deeper well, bringing the well's effective bottom to 4 feet.

A thousand candles were placed on each surface and the top of each well was covered, except for a single centered hole. Suspended from the room's ceiling was a movable screen, on which the light from the candles was projected. The separate wells allowed a number of variables to be studied—such as the distance between the light and the screen, or between the light and the object—while the stability of the candles in the well and the source of egress remained steady. Different sized tables could be placed in the wells to vary the distance from the cover.[36]

Zhao discovered that a small pinhole resulted in an inverse image shaped like the light source, regardless of the pinhole's actual shape, while a sufficiently large hole produced an image that was not inverse and also followed the shape of the hole. He also found that the brightness of the projected candles on the screen decreased as the size of the hole decreased; the brightness also decreased as the distance between the candles and the screen increased.[37]

A twentieth-century scholar, Jing-Guang Wang, wrote that Zhao's "basic idea was: 1) there exists a light spot on the screen corresponding to a single candle; 2) if one thousand candles are burning there should exist one thousand images. These images may overlap. The whole image changes as the spacing of the candles change. It is evident that Zhao understood the principle of the rectilinear propagation and superposition of light."[38]

ANCIENT CHINESE technology used the concept of the center of gravity, as reflected in the Chin dynasty (221–207 B.C.). Water pots were weighted to stand upright when full of water but to fall over when empty.

Force is suggested as a concept in the *Mo Jing,* coming from people's experience with work, though the sinologist A. C. Graham claims the Mohists thought only in terms of "weights and pulls," not in forces.[39] The *Mo Jing* ties mechanical force in with human strength, calling the body *xing,* or "shape," while action done by the body, such as lifting, is called *fen,* or "exertion." "Force," according to the *Mo Jing,* "is that which causes the 'shape' to 'exert.' "[40]

The Chinese saw physics in terms of balance. What Dai Nianzu, a modern scholar of Chinese technology, calls the "moment of force" is also discussed in relation to weights on a balance beam. As in optics, the Mohists seem interested in the central point, where an object would be in balance with a weight. Hundreds of years before Archimedes, they realized that the distances between the fulcrum and both the object and the weight were critical in maintaining balance. They called the distance between the fulcrum and the object *ben,* and the distance between the fulcrum and the sliding weight *biao,* corresponding to current concepts of the arm of load and the arm of effort.[41]

"If the mass is heavier than the sliding weight yet the level is horizontally balanced, this is because *ben* is shorter than *biao.* If now at both points of suspension the same weight is added, the *biao* side must go down."[42] When one side went down, it was because of both the weight and the *ch'uan,* a term roughly correlating with the "power, leverage, [and] positional advantage of a [human] ruler."[43]

The *Mo Jing* makes tentative analyses into stress and deformation of materials. Mohists noted that a wooden beam that did not bend under a load was strong enough to bear its load, and compared that to a horizontal rope bending under its own weight: "Ropes in that position are very poor . . . in withstanding a perpendicular load," they concluded. Mohists explored the reasons behind this in analyzing hair. How easily a hair breaks depends, says Nianzu, "on whether the cohesive substance in a hair is homogeneously distributed along the whole length, and whether the load is evenly born . . . without a weak link when the hair is taut."[44] Graham says that the bending of a horizontal object depends both on weight and on its *ch'i,* here meaning the "full extension of the weight-bearer."[45]

According to Mohist thought, a vertical object is supported either by suspension or by resting on something below: "Pillaring, supporting from below, is explained by the principle . . . that all weight tends vertically downward. 'Pillaring' in statics is the counterpart of *ch'ieh* 'pulling up, suspending,' just as the counterpart in dynamics is *shou* 'receiving from below,' " says Graham.[46] The canon illustrates this with:

> Let a square stone be one foot from the ground, put stones underneath it, hang a thread above it. . . . That the square stone does not fall is because it is supported from below. Attach the

> thread, get rid of the stones: that it does not fall is because it is suspended from above. When the thread snaps it is because of the pull of the square stone. Without any alteration except the substitution of a name, it is a case of "receiving from below."[47]

Modern physicists also think in such terms. For example, contemporary physicists have asked the question, If gravity pulls us toward the center of the earth, what keeps us from going there? The answer is electromagnetism, the primary force that holds matter together and resists gravitation. In fact, physicists have calculated how high mountains can be on a planet with the earth's gravity. (The not too surprising answer: about as high as the Himalayas, the highest range on earth.)[48]

Early concepts of floating predate Archimedes. The Mohists state, "When a very large body floats on water with only a very small part of it submerged, that means the constant equilibrium between the submerged part and the whole body has already been established." They do not take this idea further to the displacement of water, however,[49] a related concept developed by Archimedes (Archimedes' principle, third century B.C.), supposedly while he was immersed in a bath, and leading to his running naked through the streets of ancient Greece screaming, "Eureka!"

More than two thousand years before Newton, the Mohists tackled the laws of motion. They noticed that "when a cart is moving forward drawn by a horse but the horse is suddenly halted, there is a tendency for the cart itself to keep moving forward."[50] They took this further: "The cessation of motion is due to the opposing force. . . . If there is no opposing force . . . the motion will never stop. This is as true as that an ox is not a horse."[51] Today, Newton's first law of motion is generally stated, "Every body continues in its state of rest, or of uniform motion in a straight line, unless it is compelled to change that state by forces impressed upon it."[52]

The difference in Newton's statement is that he begins with a "state of rest" as a default before going on to motion. In a sense, we can infer that the Mohists are a bit more modern, beginning with motion as a default. Particle physicists today see the universe as roiling in motion, not static. In any case, it is not obvious to the earthbound observer that a object in motion will remain in motion unless acted upon. It is not our experience with oxcarts or automobiles, in which one must keep whipping the animals or pressing down on the accelerator to maintain speed. The Mohists had to imagine the universe unobfuscated by air resistance and mechanical friction, as Galileo, Newton, and Descartes did. How they envisioned this from their earthbound experiments is not recorded. (As Park points out, however, neither the Mohists nor Galileo nor Descartes came up with Newton's second law of motion: force equals mass times acceleration, or F = ma.)[53]

The Mohists also bent their minds toward analyzing time and space. Without much elaboration, the *Mo Jing* sets down theories of space, infinity, motion, time, duration, and relativity. Here are some short excerpts:

> CANON: Space includes all the different places.
> EXPOSITION: East, west, south, and north are all enclosed in space. . . .
>
> CANON: Outside bounded space no line can be included.
> EXPOSITION: A plane area cannot include every line since it has a limit. But there is no line that could not be included if the area were unbounded. . . .
>
> CANON: Finiteness is possible for a limited area within an unbounded area of space.
> EXPOSITION: Finiteness signifies that the motion of the body is restricted to a limited area of space. . . .
>
> CANON: The boundaries of space . . . are constantly shifting. The reason for this refers to extension-like concepts, such as length and duration that are measurable.
> EXPOSITION: Extension: the body in motion that goes through definite length, occupies a position in the spatial universe. . . . Space: length: that the south is opposite the north is equivalent

> to the opposition between east and west. The motion of any body, in spite of the sun may still be measured in space (length) and time. . . .
>
> CANON: Spatial positions are names for that which is already past. The reason for this refers to reality.
> EXPOSITION: Knowing that "this" is no longer "this," and that "this" is no longer "here," we still call it *north* and *south*. That is, what is already past is regarded as if it were still present. . . .
>
> CANON: Duration includes all particular (different) times.
> EXPOSITION: Former times, the present time, the morning, and the evening are combined together to form duration. . . .[54]

Mohist scholar Zhang Yinzhi seems to connect with Newton's statement that "absolute, true and mathematical time . . . of itself, and from its own nature flows equably without regard to anything external."[55]

India

The Indians came closest to modern ideas of atomism, quantum physics, and other current theories. India developed very early, enduring atomist theories of matter. Possibly Greek atomistic thought was influenced by India, via the Persian civilizations.[56] The Indians lacked, however, the experimental sophistication of the ancient Chinese, medieval Arabs, or Europeans since the Enlightenment. The Rig-Veda, dating somewhere between 2000 and 1500 B.C., is the first Indian literature to set down ideas resembling universal natural laws.[57] Cosmic law is connected with cosmic light, with gods, and, later, specifically with Brahman.

Around the time of Buddha (500 B.C.), the Upanishads, written over a period of several centuries, mentioned the concept of *svabhava,* defined as "the inherent nature of the respective material objects,"—that is, their unique causal efficacy, such as burning in the case of fire and flowing downward in the case of water.[58] The Jainist thinker Bunaratna says, "Everything that exists comes into being because of the operation of *svabhava.* Thus . . . earth is transformed into a pot and not into cloth. . . . From the threads is produced the cloth and not the pot."

In contrast, the concept of *yadrccha,* or chance, had also existed from very ancient times, although it was not widely accepted. *Yadrccha* implied the lack of order and the randomness of causality.[59] Both concepts add up to the assertion of the Greek Democritus, uttered half a century later: "Everything in the universe is the fruit of chance and necessity."[60]

The example Democritus gave—similar to the threads of the cloth—was the poppy. Whether the poppy seed takes root or dies is a matter of chance, depending on whether it lands on fertile soil or a barren rock. But that it grows into a poppy and not into an olive tree is a matter of causality. The importance of chance, or *yadrccha,* was rejected by Aristotle and other ancient Greeks who came after Democritus.

The traditional Western argument is that Democritus was writing about physics and what the Upanishads describe is metaphysics, though the words are similar. Park sums up: "The *Upanishads* refer to an imaginary symbolic cosmos. Democritus was talking about the way things really are or (better) might be. These are different worlds of discourse. They cannot be compared."[61] On the other hand, neither the ancient Indians nor Democritus derived their ideas from experiments, and in that sense, we could dismiss both as philosophy rather than science. Or we could be more catholic and accept that two different ancient cultures came to similar conclusions about the world. Both the ancient Indians and the pre-Socratic Greeks came to their belief in atoms through logic, though using different logical pathways. Democritus simply assumed that there had to be atoms—uncuttable pieces of matter. Imagine, he said, a magic knife, with which one can cut a wedge of cheese into smaller and smaller bits. Can you cut forever? No, he concluded. Eventually you come to the atom. But this is just an assumption, a good guess. Why *not* cut forever? The Indians came to the same conclusion via a different route. Take a mountain and a molehill, they said. Which has more particles? The mountain, obviously. That means you cannot cut forever, that there is a finite, uncuttable particle. If the particles were infinitesimal, the mountain and molehill would have equal numbers of particles, and they would lose any real meaning—again, an assumption but, in a way, more hard-minded than Democritus's guess. And the Indians, unlike Democritus, displayed a rudimentary understanding of infinite sets.

Out of the Brahman's unifying law and light, the Upanishads of the seventh century B.C. developed the first early classifications of matter: "Brahman, desiring to be many, created *tejas* (fire), *ap* (water) and *ksiti* (earth), and entered into these three," states the Chandogya Upanishad. Later, this evolved into five elements, adding air and *akasa* (roughly translatable as "space," "ether," or another pervasive nonmaterial entity) to the original three.[62] This predates the Greek Empedocles and his four elements: earth, air, fire, and water (circa 460 B.C.).

The Samkhya philosophical system (sixth to fifth centuries B.C.) related each of these elements to the five senses and the qualities perceivable by these senses: touch, sight, sound, taste, and odor. In this way, the intangible and eternal universe seen through its laws became tangible in "gross" matter. Different materials were formed by different configurations of the five elements. Very modern and Western in its approach, the Samkhya asserted that matter could not come out of nonbeing but resulted from what was "potentially present," that is "the unmanifest becomes manifest." Thus it posited a perpetual transformation of matter from potential matter:[63] "the material universe emanates out of *prakrti* . . . the 'rootless root of the universe.' "[64] As Park quips, "Sounds quite Aristotelian."

Samkhya philosophers saw both the external world and the internal phenomena of the self—the source of all experience—as belonging to "the sphere of mutation or change," that is, the world of reality. Yet they also developed the concept of "nonbeing," what the Buddhists called *maya:* illusion. According to S. N. Dasgupta, an Indian natural scientist and philosopher, "Only the inmost sheath of the self as the 'supremely blissful' began to be held as the one unchanging reality."[65]

The *maya,* under Buddhist philosophy, gives illusory weight to the universe. In this and other ways, the *maya* is similar to the Higgs field, an all-pervasive, invisible field that fills the universe like an ether, or so some contemporary particle physicists propose. They are confused by the various weights of particles in the standard model. There seems to be no formula that generates these masses. In fact, why should mass exist at all? A possible solution is the Higgs field, which imparts more weight to some particles than others, making them appear heavy. The particle responsible for this field, the Higgs boson, is still being sought in accelerator laboratories today.

THE INDIANS EXPLAINED the visible universe in terms of atoms, the smallest unit of matter that could not be created or destroyed. Three fundamental philosophical systems are important in Indian atomism: the Nyaya-Vaisesika, the Jainist, and the Buddhist schools. Although Indian atomism (developed around 600 B.C.) seems to have evolved about the same time as Greek atomism (roughly 430 B.C.), Indian atomism endured as an accepted concept straight through the Middle Ages. Whether Indian culture influenced Greek or vice versa, or whether both evolved independently is a matter of dispute.[66]

The orthodox Nyaya-Vaisesika school developed India's most enduring and established atomic theory. In (600 B.C.) Kanada founded the Vaisesika school; he was one of the (or perhaps *the*) earliest exponents of atomism. (His name, which means "one who eats grain," is apparently a reference to his atomic theory.) The earliest actual text appears in the first century A.D., called the *Vaisesika-sutra* of Kanada.[67] It describes a universe in continual change and posits a theory of causation that asserts that cause and effect are different, yet connected. A connection is made between the whole and its parts: on one hand the Nyaya-Vaisesika school said that the whole has an existence of its own and does not exist as separate parts—almost a force or wave concept of matter—and on the other hand it said that when the whole disintegrates, the parts continue with their own discrete existence.[68] This reasoning is counterintuitive, but then so is much of quantum theory today.

The ancient Indians reasoned that moving atoms had the potential to combine with other atoms of the same class, forming a dyad, also believed to be sizeless. Yet the concept of "two" itself gave the dyad a magnitude, albeit an invisible one. To be perceptible, a triad had to form, composed of three dyads.[69] There are vague parallels here to quark theory, in which three quarks combine to form protons, neutrons, and other hadrons.

The Jainist system reflects some modern atomistic thinking. (Jainism is an offshoot of Hinduism resembling Buddhism.) Jainist sutras from 100 to 200 B.C. discuss the nature of matter and how it combines. (Much of what the English chemist John Dalton stated in his atomic theory in 1803 is a reiteration of basic Jainist ideas.)[70]

Jainist thinkers rejected the notion of a whole-part duality and asserted that the atom is both the cause and the effect of matter, rather than being the effect, as believed by the Nyaya-Vaisesika school. To Jainists, the universe had no beginning or end,[71] and substances were eternal and unchanging.[72] The Tattvarthadhigama Sutra (c. 150 B.C.–A.D. 100), describes substances that undergo modifications "while maintaining their primary nature. An ingot of gold can be made into a ring or necklace without losing its goldness. . . . The possession of attributes [unchangeable] and modifications [changeable] is characteristic for being called a substance."[73]

Formless substances include *dharma* (medium of motion, as opposed to the more modern definition of "righteousness"), *adharm* (medium of rest), *akasa* (space), *kala* (time), and *jiva* (soul). But substances with form, *pudgala,* dealt with the world of matter and energy and included *anu* (atom) and *skandha* (molecule).[74] Substances with form took up space and, as in other Indian philosophies, had the attributes of "taste, color, touch, and smell." According to Jain, "the touch can be experienced by hardness, density, temperature, and crystalline or electrical characters."[75]

The Jainist atom, the smallest indivisible unit of matter, is discrete and porous, thus having a capacity of extension and condensation."[76] (Dalton later claimed that the atom was "hard and indivisible.")[77] The atom was a point in space (or "field") and ephemeral in relation to time;[78] or, as Mrinal Kanti Gangopadhya, professor of Sanskrit at Calcutta University, puts it, each atom occupied one point (*pradesa*) of space.[79] This theory appears inconsistent. How can something be both a "point" and "porous?" Where do you put the holes in a point?

Nevertheless, there are similarities between Jainist atoms and today's elementary particles, which are pointlike particles, with zero radius, that create fields. "Points with a pull" is another way of thinking of quarks, electrons, and the like. In this way, the Jainist atomists were echoed by Roger Joseph Boscovich. Boscovich, a Dalmatian geometer, put forth the proposition in 1760 that particles had no size; they are geometrical "points of force" that in turn create fields of force. This got rid of the old European concept of "spooky action at a distance" that allowed one particle to affect another.[80] Boscovich was pretty much ignored. In the nineteenth century, the English experimenter Michael Faraday, building on Boscovich, elaborated on the whole concept of "field," an idea that pervades modern physics.

The idea of particles being geometrical points with no dimension is very Indian, and is still counterintuitive to us today. How can you have a building block of matter that has no radius? How can something be nothing? The proof is in the experiments. You can't measure a zero radius and actually get it down to zero, but as equipment has improved, the measurement of the electron has shrunk and shrunk. In 1990, the electron was measured at less than .000000000000000001 inches. That's as good a zero as physics can supply. Despite this, the electron has mass, electric charge, something called spin. The physicist Leon Lederman compares the electron to the Cheshire Cat. Slowly it disappears until all that's left is its smile: spin, charge, and mass.[81]

Indian theorists posited that atoms combine to form aggregates, which then make up all manifestations of physical matter.[82] The Jainist atom came in two opposing kinds—"*snighda,* positive or soft, and *ruksha,* negative or rough"—which combined, an idea foreshadowing the modern idea of ionic bonding.[83]

Molecules are defined as "aggregates of atoms capable of existing in gross form and undergoing the process of association and dissociation." As the Jainist atoms vibrated, different kinds of molecules took up the vibration in differing intensities.[84] The concept of vibration is modern as well. From each molecule emanates one or more distinctive wavelengths of radiation, forms of vibration. Spectroscopy can be used today to identify chemicals by these vibrations.

There were various classes of molecules in ancient India, the simplest carrying the quaint description "finest molecule formed of two *anus.*" The molecules formed of two *anus* were the Jainist concept of the simplest chemical bonding of two atoms, and reflect how atoms of oxygen, nitrogen, and many other elements combine in pairs.[85]

Of the four Buddhist philosophies, only two, the Vaibhasika and Sautrantika schools, embraced the reality of the external world. (The later Yoga and Madhyamika schools, collectively called Mahayana, taught that all reality is illusion.) Vaibhasika and Sautrantika both defined an atom as the smallest unit perceptible to the senses. The Buddhists described atoms as "indivisible, unanalyzable, invisible, inaudible, untestable, and intangible." The one word inconsistent with modern scientific thought is "untestable." As Park comments, "This word gives the whole show away. It is all talk, no experience." On the other hand, today we consider superstring theory, a so-called Theory of Everything promoted by some theoretical physicists, to be scientific despite the fact it is presently untestable." (String theory would require a particle accelerator ten light-years in diameter to verify it.)

However, and perhaps more relevant to modern physics, the Buddhist atom was seen as transitory, continually going through phase changes.[86] Says science historian D. M. Bose, the Buddhist atom was more of a force or energy present in all matter—that is, an "earthly atom-force of repulsion." This tied in with Buddhist beliefs that all existence is momentary, and stable matter an illusion (I quote here a modern source, B. V. Subbarayappa, not an ancient source):

> What is ultimately real is instantaneous being. As things have momentary existence, i.e., they disappear as soon as they appear, the Buddhists do not consider motion with reference to matter

> at all. But as Santiraksita says, "The essence of reality is motion. Reality is indeed kinetic . . . the interdependence of the moments following one another, evokes the illusion of stability of duration, but they are forces . . . flashing into existence without any real enduring substance."[87]

The Mahayana doctrine of *sunyata* (emptiness) also tied in with this, conceiving of a true void as the foundation of all existence. (Mahayana is one of the two great schools of Buddhism, the other being Hinayana. The Buddhist sage Nagarjuna suggested that the void lies at the end and beginning of all physical matter, which arises from emptiness and returns to it.[88] The Madhyamika compared reality to a swatch of cloth, which from a distance appears solid but on close inspection reveals only a loose assemblage of threads. In the twentieth century Western physicists discovered quantum reality. The "solid" table that supports our dinner plates is not solid at all, atoms being composed mostly of empty space, the interaction of particles in constant violent motion providing the illusion of stability and solidity.

In contrast to atomic theory, the Indians conceived of an etheric field permeating the known universe. John Maxson Stillman, a scholar of Indian alchemy, describes it as "infinite in extent, continuous, and eternal. It cannot be apprehended by the senses. . . . It is also described by certain authorities as . . . occupying the same space that is occupied by the various forms of matter."[89] To make the physics of their eras to work out properly, Newton and James Clerk Maxwell also posited a mysterious, invisible ether that pervaded all of space. It turned out not to exist.

In 600 B.C. Kanada applied his logic to light and concluded that light and heat are two forms of the same substance: "Light is coloured, and illumines other substances; and to the feel is hot: which is its distinguishing quality. It is defined as a substance hot to the feel." Like other substances, light existed both in an actual and a potential state, proved to Kanada by its associated sensory qualities. He made observations such as this: "The heat of hot water is felt but not seen; moonshine is seen, but not felt."[90]

Kanada also asserted that light had both a holistic and a particulate nature: "[Light] is eternal, as atoms; not so, as aggregates." This statement reflected the whole-part duality of the Nyaya-Vaisesika school: that in atomic "subtle" form, light would be sizeless and eternal, whereas in molecular "gross" form, it would be temporary.[91] He went on to say, "The mote, which is seen in a sunbeam, is the smallest perceptible quantity. Being a substance and an effect, it must be composed of what is less than itself; and this likewise is a substance and an effect; for the component part of a substance that has magnitude must be an effect. This again must be composed of what is smaller; and that smaller thing is an atom." Today, of course, we refer routinely to the wave-particle duality of light. It behaves both "holistically," continuously, as radiation (waves) and in quantized form, as photons.

The Middle East: Persia

The traditions of ancient Zoroastrianism and, later, Manichaeanism provide an important link between concepts of light and order from India on one side and those of ancient Egypt and Greece on the other. According to the historian Jacques Duchesne-Guillemin, writing in the 1960s, early Zoroastrianism was rooted in the philosophy of the sixteenth-century B.C. Aryans, who at various times occupied modern-day Iran, Pakistan, and Turkestan and who swept into India as the Vedic Aryans. More recent scholars think this connection is unclear but do not dispute that both cultures arose from Indo-Europeans originating in central Asia. Zoroastrianism became the official religion of Persia in the third century A.D.[92] Manichaeanism arose from the teachings of Manes, a Persian in third-century Baghdad, at a time when Christianity was on the rise.

Zoroastrianism focused on the duality of light and dark, pitching Ahura Mazda (Mithra), the source of light, truth, and goodness, against his twin, Angra Mainyu (Ahriman), the source of darkness and evil. Both were accompanied by a pantheon of deities and angelic forces.

At first, Ahura Mazda was a creator god incorporating both dualities. All powers of order and creation, and of chaos and destruction, emanated from him. In this way, he was the basic physical forces of the universe personified. One very ancient manifestation of Ahura Mazda is Asha (Arta), a word derived from the Indian *rta,* or cosmic order. In Zoroastrianism, Asha is both truth and natural order.[93] Yet just as the Indian *rta* was manifested by light, so Zoroastrian texts state that "Asha filled the space with lights."[94] Asha was also associated with fire.

Later, Ahura Mazda was associated solely with the powers of light. Inscriptions carved in rock during the time of the Persian king Darius (522–486 B.C.) show Ahura Mazda arising from the Egyptian winged solar disk, along with symbols for the moon. The Egyptian influence apparently came via the nearby Hittite and Assyrian civilizations. Hence the solar and lunar Eye of Horus influenced Zoroastrian theology.[95]

In Zoroastrianism's offshoot Manichaeanism, light was a transcendental quality, a concept echoing the ideas of Plato and the Neoplatonists. Plotinus in Egypt (third century A.D.) believed that goodness was like a field, suffusing all of the universe from the supreme deity in "a sort of radiation like the sun's," says David Park, author of *The Fire Within the Eye,* a definitive history of light.[96] Plato's "good," in *The Republic,* is more than just goodness. It is the Idea that actualizes all the other Ideas. Manes (A.D. 215–275) probably read Plotinus (there were plenty of Neoplatonists around),[97] and his theology was similar. God was goodness. God *was* light, literally, just as the Devil *was* darkness. Personified deities and physical forces seemed to fuse in Manes's thinking. Frankly, God's force had no effect on the Devil's because his powers were "contemplative" (thought, intelligence, etc.), whereas the Devil's were energetic (fire, wind, etc.). But God made agents to carry out his work: Buddha, Zoroaster, Jesus, to name a few. Jesus was made of light except for a small bit of earth that allowed him to suffer,[98] thus linking light to matter.

The essential struggle between Ahura Mazda and his evil twin, Angra Mainyu, as well as Manes's spin on this story, suggests a theory of matter and energy. Ahura Mazda first made the world to be pure light; later a second world came into being that was material in nature. The material world was conquered by Angra Mainyu, which caused the pure light of existence to be mixed with darkness. Thus, according to Amherst College physicist Arthur Zajonc, "All physical existence became a mixture of good and evil, light and darkness."[99] For example, wood was composed of light and dark, since one could release the light from its physical bonds by burning.[100] In addition, rubbing two sticks together would yield a spark.[101]

Matter was composed following classic schemes of Chinese, Indian, and Greek thinking, including fire, earth, water,[102] and metal (power).[103] But of these, fire was paramount. Zoroastrian priests made offerings of fuel, incense, and fat to their holy fires.[104] The fires represented purifying agents, by which matter could return to its spiritual (light) form, as animal fat is rendered pure by melting.[105]

An interesting aspect of Manichaeanism is that it is diametrically opposed to Aristotelian ideas of light and matter. Light has more of a corpuscular nature in Manichaeanism and is more interchangeable with matter. For example, in Manichaeanism, all matter once again belonged to the "Earth of Darkness." Matter also contained tiny bits of divine light because First Man, made of pure light, was overcome by the Devil, thereby losing most of his light to physical matter. Manes's light seems more atomized, and less of a field concept. To ensure that light remained trapped on earth, the Devil created Adam and Eve, each with their bit of holy light. When they procreated, holy light was dispersed in the corrupted matter of their offspring. According to Park, in trying to recover this light, God created the current world:

> containing the sun and moon as well as the Earth of Darkness with its small quantity of stolen light. Matter craves soul to make it alive. . . . This soul is not like First Man, since its elements are corrupted by their connection with matter, and the bonding of light into matter makes its recovery and purification more difficult.[106]

The sun and moon were bits of pure light; next came the stars, made of partially corrupted light; the rest of First Man's light was completely imprisoned in earthly matter.[107] Mani taught that priests could free this light by their actions. The light in food they ate could be released into their bodies, which they then returned to heaven after death. Specifically, the priests' spirits carried the stored light to the moon, which gradually grew larger until, when full, it beamed all the saved light back to the sun.[108] Okay, the astronomy here is all wrong, but the concept of transferring matter to energy in the body was right on.

This was an elaborate scheme, but in the thirteenth century, Robert Grosseteste wrote in *De luce,* "Light . . . was the first form of corporeity, and from it all else followed. Multiplying itself from a single point infinitely and equally . . . light formed a sphere and together with this action arose matter."[109] Light, to Grosseteste, was matter condensed, and light could multiply.

Park feels that one shouldn't compare Manichaeanism and Aristotelian concepts of light because the first was meant to be allegorical and the latter was meant to be taken literally.[110] We have seen similar developments in modern physics. The contemporary theorist Murray Gell-Mann posited the concept of the quark to make sense of all the hadrons (protonlike particles) that had been discovered in the 1950s. Gell-Mann figured out that they could all be built up from basic particles which he called quarks. He did not think these quarks were real, however, but rather bookkeeping devices. Quarks were soon found in accelerator experiments, and few doubt their existence today, despite their having been conceived as incorporeal mathematical concepts.

Islam

In A.D. 786, Haroun al-Rashid, the caliph of Baghdad, set the stage for Arabic transmission of classical works. Highly educated, he scoured the known world for Greek and Syriac texts to translate into Arabic, at a time, says David Park, when "Charlemagne and his peers were trying to learn to write their names."[111] In a sense, the Arabs' fascination with the

Greeks may have been their undoing, for they partially absorbed and transmitted ideas that were to be abandoned by Europeans during the Enlightenment. (Which is not to say that the Europeans didn't first absorb ancient Greek ideas before moving on.)[112]

In Baghdad, Yaqub ibn Ishaq al-Kindi (A.D. 801–866), perhaps Islam's first philosopher, benefited from al-Rashid's efforts. In *On First Philosophy,* al-Kindi drew on both Plato and Aristotle to address cause and effect, matter, motion, and time. According to Alfred L. Ivry,[113] who has compared the philosophies of al-Kindi and Aristotle, al-Kindi believed that the qualities of "unity" and "plurality" existed in all matter, though unity was not essential but "accidental," that is, was subject to chance and unpredictability. He therefore posited a "Unique True One" who caused all matter and the earth to be created out of nothing through a process of emanation. Ivry says the concept of emanation was rooted in Plato, while the "True One" was rooted in Aristotle's "true principles, though clearly also connected with Allah."[114] In fact, both concepts were associated more with Plotinus.[115]

Al-Kindi was fascinated with emanation, from which he developed a theory of rays. Borrowing Aristotle's belief that all movement is generated by the motion of the heavenly spheres, he suggested that the force behind the spheres came from rays. He explored both light and "visual" rays, an explanation of vision coming from Empedocles, Euclid, and Ptolemy. Like Ptolemy, he believed that visual rays emanated from the eyes in the shape of a cone and had physical dimensions and warmth.[116] Both Ptolemy and later Theon of Alexandria (fourth century A.D.) had used Euclid's optics and Aristotle's physics to imply a connection between visual rays and light rays. But al-Kindi decided that the two kinds of rays were one and the same, and that therefore the eye radiated like any other light source.[117]

Hence, in Park's words, al-Kindi concluded that "everything that actually exists emits rays in every direction, so that the whole universe is causally bound together by a web of radiation. . . . Rays originate in substance [which is made by the four elements] and act on form."[118]

Later, in 984, mathematician Ibn Sahl wrote a text on lenses. His genius led him to discover Snell's law,[119] which states that when light hits the boundary of two materials, the angle of incidence is equal to the angle of refraction. (Willebrord Snell Von Roijen, a seventeenth-century Dutch mathematician, is usually credited with this discovery.)[120] Park says, "[Ibn Sahl] had everything that was needed to create a theory of optical instruments more than 725 years before Kepler—except, apparently, the concept of an optical instrument."[121]

From the standpoint of modern optics, Abu Ali al-Hasan ibn al-Haytham (known as Alhazen in the West) was one of Islam's most influential thinkers. Born south of Baghdad in A.D. 965, he lived most of his life in Cairo at a theological university,[122] where he drew from the ancient Greeks but then applied careful mathematics to physical paradigms and tested them through experimentation. That the sun's rays moved in a straight line had been clear to many before Alhazen, but he analyzed this geometrically: "The sun's rays proceed from the sun along straight lines and are reflected from every polished object at equal angles, i.e., the reflected ray subtends, together with the line tangential to the polished object which is in the plane of the reflected ray, subtends two equal angles."[123]

He also demonstrated through experiments that the sun radiated light in all directions, thus proving what al-Kindi had theorized a century earlier. Alhazen writes in his *Kitab al-Manazir:* "1. All lights, no matter what their source of emission, propagate rectilinearly. 2. Every point of a luminous object, whether it is self-luminous or accidentally so, radiates light along every straight line that can be imagined to extend from it. . . . I mean in all directions."[124]

From these conclusions he discovered that the sun's rays are focused at one point in a parabolic mirror, which explained why these mirrors could be used to set fire to objects. He also manufactured actual lenses and mirrors to use in demonstrating his theories.[125]

Alhazen experimented with refraction of objects in water and used the "rectangle of velocities" to measure refraction. He also discovered Snell's law applied to small angles, but was apparently unaware of the work of his contemporary, Ibn Sahl.[126] For Alhazen, color and form were strictly visual phenomena, not, as Aristotle postulated, an object's actual form entering our mind.

The long-cherished Greek idea of rays emanating from the eye to the object was also rejected by Alhazen, through several observations: 1) Looking at the sun hurt one's eyes, thus it seemed that the eye was struck by something from the sun—the rays are *coming into* the eye, not going out; 2) the eye registers detail only under specific light conditions; if sight depended on a ray from the eye, the conditions outside of the eye would not matter; 3) if the ray emanating from the eye was material substance, then the whole area between the eye and the observed object would be filled with this material, even if one were observing a distant mountain. To Alhazen this was logically absurd. (To Park, it is even more absurd. Look at a star, he suggests, and you see it immediately. If the eye required a ray to contact the stars, it would take more than a lifetime to see most of them.) Yet if the ray was not material, how then could it transmit information about the perceived object? For, in Alhazen's terms, "sensation belongs only to animate bodies."[127]

Alhazen made important strides in explaining how the human eye works. What we call the crystalline lens, he called the "crystalline humor," which he placed roughly in the center of the eye's sphere, whereas it should be near the front (under Islamic law, he would not have been permitted to dissect an eye to verify this). Alhazen then attacked the problem of perceiving whole images. Says Park, "If each point of the object corresponds to one point of the image, then the whole image will be a faithful small-scale representation of the object," which then is projected onto the crystalline humor.[128] There will be a tiny replica of the object in the eye.

How each point of a much larger object, such as a mountain, fits on the crystalline humor posed a mathematical problem. Alhazen solved it experimentally by using the camera obscura. This precursor to the modern camera showed that many images could be clearly projected onto a screen as points without interfering with one another.[129]

Theoretically, there remained a problem. What Alhazen proposed was that objects we see in our minds are reproductions of real objects. A mountain is reproduced point for point on the eye's lens, only tiny. He thus fomented a mathematical crisis. How can the miniature mountain in the eye have the all the points of the real mountain? The answer is that both have an infinite number of points, points being dimensionless. It wasn't until the nineteenth century that Georg Cantor made infinite sets a part of respectable mathematics.[130] You can demonstrate this to yourself by drawing two inequal lines, like so:

———————————

————————

Draw straight lines from every point on the long line, starting at the ends, to every point on the shorter line. Of course, you can't physically complete the project, there being an infinite number of points on the top line. But you'll get the idea of how a large image translates to a small image that is a point-for-point copy. Because points are dimensionless, there are infinite points on both lines. These two infinities are equal.[131]

In the tenth century, burgeoning Islamic philosophy moved away from these experimental traditions. Al-Suhrawardi (1153–91) focused on intuitive knowledge that drew from Platonic, Aristotelian, Zoroastrian, and alchemical ideas and combined them with Sufism, an Islamic mystical tradition. To Al-Suhrawardi, light was essence and "being," as its absence was "non-being."[132] Yet Suhrawardi's work brought a renewed interest in practical optics; a later thirteenth-century commentator on his work, Qutb al-Din al-Shirazi, discovered that rainbows resulted from both reflection and refraction. Then writings of a student of al-Shirazi's returned Alhazen's discoveries to the forefront of Arabic thinking. From there Alhazen's masterpiece, *Kitab al-Manazir,* entered western Europe and remained the primary work on optics until Kepler arrived on the scene in 1610.[133]

TWO DIRECTIONS of Islamic philosophy informed a debate over the nature of matter and space. One grew out of the Kalam, a system of logical argument[134] somewhat analogous to the European scholastic movement. In the Kalam, orthodox theologians sought to disprove Aristotle's theory that the universe was governed by cause and effect rather than through the will of Allah.[135] The other movement, represented by al-Kindi, al-Farabi, al-Razi, Avicenna, and by the Ikhwan al-Safa, a philosophical scientific group (a.k.a. the Brothers of Purity), espoused Aristotle's empirical methods while developing their own theories of physical phenomena.[136]

In the tenth century the alchemist-physician-philosopher Abu Bakr al-Razi (ca. 923) posited a three-dimensional, absolute space independent of material matter. According to Arabic scholar Shlomo Pines (and in violent opposition to Islamic orthodoxy) al-Razi believed that "absolute space and time were more fundamental than the cosmos; and this held also for matter, which was eternal, had an atomic structure and subsisted before bodies were formed in a state of dispersion."[137] In the West, the idea of absolute space independent of matter was alien until Isaac Newton embraced the concept. Europeans generally believed that space was defined by the object that occupied it and did not have a life of its own. (In all fairness, Democritus and some of his pre-Socratic colleagues held ideas of space closer to Newton's, but their beliefs were rejected by the later Greek philosophical establishment.)

Meanwhile, adherents of the Kalam rejected Aristotle's effort to understand matter in terms of substance and properties.[138] Qadi Abu Bakr Al-Baquillani, an eleventh-century philosopher, posited that existence was not inherent in things, which meant that matter itself was contingent on Allah, and not on chains of cause and effect. Therefore, matter was inert. This conclusion opened the door to general Islamic acceptance of atoms, which appeared from nothing and returned to nothing at the will of Allah.[139]

Orthodox Arabic thinkers believed that atoms were identical in substance but not in qualities and without size. A spatial content only occurred in context of atomic combinations to form matter. According to the physicist Max Jammer, in his classic book *Concepts of Space,* "Although a definite position . . . belongs to each individual atom, it does not occupy space. It is rather the set of these positions—one is almost tempted to say, the system of relations—that constitutes spatial extension."[140] Length occurred when two atoms connected; area when four atoms connected; and volume when at least two four-atom, two-dimensional complexes piled up.[141]

These developments presaged later theories of Gottfried Leibniz (1646–1716), the German mathematician who theorized that the universe was made of an infinite number of harmonious "energy centers" called monads. His ideas mirrored the Kalam both in the nature of space as extension and in the method of atomic combination. Jammer suggests that Leibniz was influenced by these theories, which were transmitted to Europe by the Jewish philosopher Maimonides in his *Guide for the Perplexed*.[142] Unlike Leibniz, however, the Arabs did not need to explain why atoms had order and harmony, since both emanated from Allah.[143]

Classic Greek thinking denied the possibility of the void. The Islamic Ikhwan al-Safa echoed this belief. For Muslims, physical objects could not exist without space; the concept of the void was irrelevant since, according to Islamic scholar Seyyed Hossein Nasr, "There is no space outside the cosmos and the Universe cannot be said to be in space." Therefore space was always occupied, even if it appeared empty, since the Ikhwan believed that spirits as well as physical matter could fill space.[144] Modern particle physicists only partially believe in the void. The vacuum is necessary to particle physics, but the theorists have quickly filled it up with "virtual particles" that momentarily pop in and out of existence. It seems very Ikhwan in spirit.

Yet al-Razi and Avicenna (980–1037) contested Aristotle's conclusions that a void was impossible, which he had deduced from water rising in the "exhausted tubes" of the clepsydra (water clock). Specifically, al-Razi interpreted such experiments as showing the presence of an attractive force occurring in empty space. He posited two voids, one existing within and one existing outside of the material world. The attractive force resulted either from a tendency of the first void to pull together or from the first void's attraction to the otherworldly second void.[145]

Under the Kalam, atomism inevitably led people to believe that empty space existed, since atoms were separate entities that combined and separated. Once they separated, what was between them except emptiness? Following logically, the notion of discrete time, and then discrete motion, emerged. According to Jammer, motion became "a sequence of momentary leaps: the atom occupies in succession different individual space-elements."[146]

Aristotle had claimed that an object moved only if an outside "mover" exerted force on it throughout the course of movement. In the eleventh century, Avicenna, in agreement with the Mohists and in opposition to Aristotle, foreshadowed Newton's law of inertia by claiming that a projectile would never stop in a vacuum because no outside factor was there to interfere.[147] Commenting on Aristotle, he said, "We found the most valid opinion to be that of those who hold that the moved [object] receives an inclination from the mover [source of force]. The inclination is that which is . . . resisting a forceful effort to bring natural motion to rest."[148] Like Newton, Avicenna came to the conclusion that one need only apply force momentarily to propel an object forever in a vacuum.

Finally, Abul-Barakat al-Baghdadi (c. A.D. 1080–1164) studied objects in free fall. He applied the notion of "violent inclination" to conclude that falling objects were subject to a steady, accelerated motion as a result of the force on them, an idea foreshadowing Newton's. That is, applying a force constantly will cause an object to accelerate continuously. According to Pines, these conclusions influenced fourteenth-century European thought, especially the decline of Aristotelianism.[149]

It is difficult to compare any ancient or medieval culture favorably to modern physics. And by "modern," in this case, I mean the period from

Galileo to the present. In addition to the great equipment, the interplay of theory and experiment in modern natural philosophy—as it is sometimes called—has never been duplicated.

That might be changing. I hope not, but there are signs of decay. Our society's, and the media's, obsession with theory is an ominous sign. On the floor of my office is a spiffy new book on time travel I'm supposed to review. It was written by a professor of physics at a famous university. Really. Time travel presented as physics. Browse the physics section of a bookstore and you will find books on string theory, more books on string theory, and even more books on string theory. Other authors promise to reveal the "Theory of Everything." It is true that Steven Weinberg, one of most important theorists of the twentieth century, called one of his books *Dreams of a Final Theory,* but the operative word was *dreams.* These new physicists display no such modesty. And the media eat it up.

Meanwhile, out of the public eye, experimenters are doing some exciting and important work. Current experiments involving a principle called CP violation may undermine the standard model. It is always exciting when a respected theory gets overturned. At Fermilab and CERN (the European accelerator laboratory in Geneva, Switzerland), experimenters have violated time symmetry. That is, they have demonstrated that the universe prefers going forward in time rather than backward.

This may seem obvious to us, living in the macroscopic world, where time always goes forward. You can't unring a bell, unscramble an egg, or grow younger and return to the womb. But at more basic levels, it was always thought that time was symmetric. Newton and Galileo, for example, wrote laws that are time symmetric; they go backward and forward equally well. In the following chapter you'll see that Lavoisier turned hydrogen and oxygen into water, and then back again. The subatomic world was believed for many years to be even more symmetric. A movie of particle reactions should look pretty much the same run backward or forward. The CERN and Fermilab scientists found out this isn't true. Experimenting with particles called kaons, they found the particles more likely to go forward in time than backward. This has been suspected since the 1960s, but was never shown before in the laboratory.

Worldwide, experimenters continue to look for the Higgs boson, which if found would prove that the ancient Indians were correct: there is *maya,* or a Higgs field, an all-pervasive ether that gives substance to illusion. And as I write this, at Fermilab experimenters are searching for extra dimensions. They're not dreaming about them; they're actually trying to pin down dimensions beyond the normal four. I asked one of the experimenters, Henry Frisch, whether string theory was driving the search for extra dimensions. "No," he said. "We expect to find all those lost socks."

Which is not to say that theory isn't important. It's just that the kind of theory popular today focuses on cosmology, "theories of everything," and the like, things remote from experiment. This is often called "deep theory" or, less euphemistically, "recreational mathematical theology."

In the early 1990s in Chicago, I heard Murray Gell-Mann give a lecture on deep theory to a group of scientists. Gell-Mann has been one of the most important—and productive—theoretical physicists of the past fifty years. Gell-Mann explained the importance of deep theory to the audience. "If we doubt the importance of such theory," he said, "then we must doubt the importance of the theoretical physicist. And if we doubt the importance of the theoretical physicist, we must question the salary of the theoretical physicist." At this point Gell-Mann paused, presumably for laughter. But his colleagues were too busy taking notes.

6

GEOLOGY
Stories of Earth Itself

In some ways, geology is the most evident of sciences, with the processes of sedimentation and erosion etched in the land around us. Modern field geologists still rely heavily on a rock hammer and their eyes for observation, both of which were available to early peoples. Geology, like other sciences, has its roots in a slow compilation of observations and practical knowledge, which eventually was fused with philosophical ideas to yield theories.

Prehistoric peoples must have had intimate knowledge of the qualities of the stones they depended on in order to live. Neanderthal humans in the Middle Pleistocene crafted stone tools of a specific form known as Mousterian.[1] They used two methods: by chipping at the stone core to create the tool, and by using the chips themselves as the tools.[2] Geologist Gordon Childe says that "both procedures demand both great dexterity and considerable familiarity with the properties of the stone utilized. Just bashing two stones together is not likely to yield a useable flake or core tool. To produce either the blow must be struck with precisely the right force and at the correct angle on a flat surface."[3] Modern geology students who have attempted to make their own tools in this fashion can vouch for the difficulty involved. One student told me she spent a full morning trying to make a stone cutting tool from two pieces of flint she found on the beach.

A distinctive side-scraper tool was used for half a million years, and the struck-flake tool was characteristic from southern Africa to Europe.[4] Childe asserts that such skill was passed down through thousands of generations. Museums today contain tens of thousands of hand axes, all of the same pattern and hewn from the same materials. Childe finds it "highly improbable that so many hominids in the long Lower Paleolithic Age should each by individual trial and error [have] selected flint of the nearest available microcrystalline stone and independently hit upon the same method for shaping it."[5]

Early peoples also used the sedimentary nature of rock outcrops to their advantage. Even if they didn't speculate on the origins of this layering of stone, they surely knew that the sediments predictably repeated themselves from outcrop to outcrop. Childe writes that Neolithic miners sank shafts through ten feet or more of solid chalk in order to reach layers of flint. There would have been no surface indications of the flint. "They must have observed an exposure of the flint-bearing layer in a ravine or on a scarp," says Childe, "and correctly inferred that it continued below the surface into the hillside."[6]

Ancient World: The Middle East, Western Asia

Much has been made of the influence of ancient Greece on contemporary and later civilizations, but in geology it is clear that the Greeks learned from earlier civilizations: Sumerian, Babylonian, Assyrian, Egyptian, Indian, and Chinese.

Mining activities demonstrate ancient peoples' understanding of stones and minerals. "Even in ancient times," asserts a geologist writing about the area formerly called the USSR, "the population included competent men who knew certain properties of minerals—hardness, friability, malleability, etc., . . . —that were taken into account in processing and using these minerals."[7] The evolution of metallurgy from the casting of metals to extracting them from ores reflects an even greater geological sophistication. Searching for copper, for example, is not so easy. The ancient metallurgist must tell from superficial clues which rock contains ore, and which minerals in the ore yield copper when heated with charcoal. The classification of various copper ores, says Childe, is highly abstract.[8]

The mountains in Turkey, Armenia, and Afghanistan yield evidence of some of the earliest metalworking knowledge. Smelting equipment from around 4000 B.C. has been unearthed in the Kerman range in Iran. Smelting—the heating of ore to separate impurities from the desired metal—requires getting the metal out of the natural ores and using (after first making) charcoal.[9] In the Urals, Kazakhstan, and Central Asia, we find remains of old mines from as early as the Stone Age. Smelting copper from malachite occurred around 1500 B.C. in the Urals, while gold and copper mining is evident from the same period in Kazakhstan. Silver mining appeared in Central Asia during the Bronze Age, while iron-smelting furnaces have been found in ancient Armenia, along with salt mines. There is also evidence that Iron Age people in central and northern Russia used surface iron ores extracted from bogs.[10]

To the south, the burgeoning Sumerian civilization first worked copper. Bronze appeared by the end of the fourth millennium, and the Sumerians also worked gold and silver. Iron was not in common use in 2700 B.C.[11] Given the extreme rarity of stone of any sort in the lower Tigris-Euphrates river plain, Sumerian skill with copper (including annealing, filigree, and casting)[12] is all the more notable.

The Sumerian penchant for making lists of plants, animals, and stones laid the foundations for data classification.[13] These were simple noun lists, such as the following bird description: "The shepherd-bird says ri-di-ik, ri-di-ik. The shepherd-bird [has] a variegated neck like the *dar*-bird. He has a crest upon his head."[14] Sumerian lists were gradually translated by the Akkadians to the north, who conquered Sumer at the end of the third millennium B.C. This led to the development of a bilingual set of twenty-two tablets in Babylonian times, listing among other things "metals, metal objects, stone, and stone objects."[15]

The later Assyrian civilizations had quite sophisticated knowledge of minerals and rocks. For example, they knew that copper had black and red oxides that gave color. They leached soil to extract the salts, and they knew about acids. They experimented widely with fire and its effects on minerals, and used both flint and pyrite (or pyrite crystals) for starting fires. They knew how to get ammonium chloride from burning dung, which led to the discovery of mercury.[16]

Building on Babylonian and Sumerian word lists, the Assyrians compiled a comprehensive lapidary. In a work translated by R. Campbell Thompson, dozens of minerals and stones are classified by description or characteristics. Here is a short sampling: "fire stone" (pyrite), hematite, blue vitriol, iron sulphide, arsenic, amethyst, corundum, diorite, basalt or other volcanic rock, "bird's eye stone," sulphate of copper or iron ("pollution of the penis of a man"), "pregnant stone" (aeites), and white limestone.[17] Though objective observation is here liberally mixed with imagination—but how much more interesting!—this is the rough beginning of a classification system.

The Assyrian word for *stone* in this lapidary is attached as a suffix to most minerals, as well as to the names for hailstones, coal, chemicals, metals, and glass in formation. Assyrian names could also reflect the stone's properties or origin: "stone of mountains," "heavy stone," "stone of wearing out," "stone of moistening" (cinnabar, which Assyrians knew could yield mercury).[18]

Having studied cuneiform tablets, Thompson also says that the Assyrians had a primitive system of classifying stones into two hardnesses, comparing them to the soft lapis lazuli or to the hard sapphire.[19] If he is correct in his assessment, this may be one of the earliest beginnings of a classification system used by modern field geologists, who test the hardness of minerals by scratching one against the other. This is the Mohs hardness test, with talc as the softest mineral and diamond as the hardest.

We also see in the ancient world a nascent understanding of a major geologic discipline, hydrology. The peoples of Mesopotamia (the Sumerians, Babylonians, and Assyrians) had to live with the flood patterns of the two great rivers around them: the Tigris and the Euphrates. Unlike the Nile, these rivers flooded late in the planting season, so the growing plants had to be shielded from the floods. The Mesopotamians developed elaborate systems of dikes and canals to control the water and to use it for irrigation when needed. Additionally, the soil grew increasingly saline from salts dissolved in this stored water, which evaporated in the summer heat. Although these people may not have understood the role of salt in decreasing the fertility of their soil, surely they must have understood the deposition of silt by rivers, since they continually had to dredge their canals of silt or dig new ones. The monitoring and maintenance of the earth and irrigation works were overseen by the kings and priests, and were essential to the survival of the society. Whole settlements were moved on regular bases in order to find fertile soil.[20]

Ancient World: Egypt and Africa

Mineralogy was a major interest in ancient Africa. In Egypt, copper was in use between 5000 and 4000 B.C. and had been smelted by 3000 B.C.[21] The Egyptians had invented a predecessor of the beam balance to weigh metals by 2500 B.C. and were able to work gold and silver and cast them, and they discovered bronze around 2000 B.C.[22] Iron was in evidence by 1570 B.C. and iron smelting by 800 or 700 B.C. Ancient Egyptian practices influenced peoples in Nubia and Cush (modern Sudan) but did not easily cross the desert into sub-Saharan Africa.[23]

Beginning in 3000 B.C., the Sahara dried up into true desert, a major barrier to communication between the high civilizations of Egypt and the Middle East and the rest of Africa. Historian J. F. A. Ajayi says this is why knowledge of copper and bronze working made its way into Crete and Greece, and from there into the rest of Europe, while sub-Saharan Africa missed the Bronze Age.

The Nile and its annual flooding were of critical importance to ancient Egyptians; they carefully noted its flood crest each year in documents primarily recording political events. One year was named for a vast flood: "Year of the Filling of all the Lakes of the Rekhyt-Folk in the West and the East of Lower Egypt."[24] Otherwise one sees orderly annual measurements: "6 cubits," "4 cubits, 1 palm," "5 cubits, 5 palms, 1 finger,"[25] where one cubit, the equivalent of about 523 millimeters, equaled 7 palms, which equaled 28 fingers.[26] These measurements were made with gauges dubbed Nilometers. The Roman philosopher Seneca in the first century A.D. describes the Nilometer at Aswan:

> The Nilometer is well built of regular hewn stones, on the bank of the Nile, in which is recorded the rise of the stream, not only the maximum but also the minimum and average rise, for the water in the well rises and falls with the stream. On the side of the well are marks, measuring the height sufficient for the irrigation and the other water levels. These are observed and published.

The calibration of these devices was often not perfect. The equivalent of the zero mark seems to have varied. (It wasn't technically a zero line, since the Egyptians didn't have zero, but a baseline to which other levels were compared.) There was some fudging, too. Since, among other factors, the higher the Nile, the higher the taxes, the figures did not always represent true readings.[27]

The Egyptians were also big on lists. They evidently believed that naming items brought them into being, and as such these lists represent mythological texts.[28] Nonetheless, one can see a slow categorization that precedes a scientific inquiry. In compiling original Egyptian texts, Marshall Clagett, a historian of math and science, reflects this by dividing his study into "knowledge" (Egyptian *rekh*) and "order" (Egyptian *maat*). The first section pulls together the works of scribes who were "able to measure, count, and record." The second section reflects "the concept of cosmic rightness or order, one of the meanings of *maat.*" Both concepts, believes Clagett, who taught at the Institute for Advanced Study, in Princeton, New Jersey, were essential to the development of science in Egypt.[29] The *Onomasticon of Amenope* lists, for example, oils, plants, animals, cities, and natural features. A sampling: "flood, river, sea, wave, swampy lake, pond, well in desert, pool, river bank, watercourse, runnel, current, water-hole, shores, island, fresh land, tired land, mud, clay, low-laying shoal, woodland, sand."[30] The *Onomasticon* is part of the *rekh* section. Amenope was a scribe during the time of Ramses IV and was called "scribe of the sacred books in the House of Life." An *onomasticon* is, according to Clagett, a "great list of names arranged under the various categories in which the author or general Egyptian tradition saw the world."

India-Sumeria

Roughly contemporaneous with the Sumerian civilization were the Harappans in the Punjab area of northwest India. Their cities appeared as early as 2300 B.C. and disappeared by 1750 B.C. Shortly thereafter the Vedic Aryans invaded from central Asia (though today, some Indians call the Aryan invasion a myth, promoted by the conquering British to denigrate the cultural accomplishments of the indigenous peoples). Cities equal to the Harappan centers did not reappear for more than a thousand years, around the sixth century B.C.[31] Evidence suggests that the Harappans and Sumerian civilizations were in contact.[32] Harappan seals have been unearthed in Mesopotamia, and Harappan amulets and ornaments resemble Sumerian ones. Motifs of some Harappan seals are similar to Syrian and Cretan designs. Historian D. M. Bose writes: "There seemed to be brisk commercial relations between the Harappan culture and that of Mesopotamia at the time when the former was in a flourishing state. . . . There was a movement of ideas and techniques among the diverse culture-areas in the third and second millennium B.C."[33]

Again, stone and metal technology gives us early indications of what the Harappans knew about minerals. The Harappans worked copper and bronze, and shaped chert and flint for weapons, using techniques from prehistory. They also used rought rocks like basalt, granite, and sandstone for grinding, and soft alabaster for making pots and bowls.[34] They were familiar with gold, silver, and lead.[35] Even more technically difficult, they alloyed tin and arsenic with copper, and extracted copper from sulphide ores. It is difficult to tell, however, if such alloying was done intentionally.[36] According to Bose et al., only 14 percent of copper-bronze tools were alloyed correctly; the Harappans were not consistent in doing large-scale production of metallic implements.[37] Nonetheless, the common use of these materials suggests a knowledge of both their qualities and a tried-and-true method of finding and extracting them.

As early as the fifth century B.C., the fame of Indian steel and iron had made its way to Persia and to Rome. Pliny refers to "swords of good quality made of Indian steel."[38] Even later (300–400 A.D.), the Hindus

forged a huge iron pillar at Meharauli (Delhi), twenty-four feet high and weighing 6.5 tons, a metallurgy feat not equaled elsewhere for centuries.[39] The pillar was made of almost pure iron: 99.72 percent. The pillar has yet to rust, perhaps, Bose speculates, because of a magnetic oxide coating or from the high phosphorous and low sulfur and manganese content of the iron.[40]

It was the Vedic Aryans, who came into the Punjab around 1500 B.C., who gave the world some of the earliest philosophical texts on the nature of matter and the theoretical underpinnings for the chemical makeup of minerals. Sanskrit Vedas from thousands of years before Christ implied that matter could not be created, and that the universe had created itself.[41] Reflecting this, in his Vaiseshika philosophy, Kanada (600 B.C.) claimed that elements could not be destroyed.[42] Kanada's life is somewhat mysterious, but his name is said to mean "one who eats particle or grain," likely referring to his theory that basic particles mix together as the building blocks for all matter.[43] Two, three, four, or more of these elements would combine, just as we conceive of atoms doing.[44] The Greeks would not stumble on this concept for another century.

As for large-scale geologic features, the erosion and deposition of sediment by rivers must have been a fact of life for the Harappans. For pottery they used clay taken from alluvial deposits in rivers. The clay was then mixed with sand and lime.[45] In addition, the Harappans selected high-quality stone when making flint weapons. Their craftsmen recognized eroded flint nodules, which they stripped to get down to fresh exposures that would yield the best material.[46] Though primitive, this practice reflects the essence of field geology: using surface features to extrapolate what may lie beneath.

In India, we see the beginning of theoretical speculation of the size and nature of the whole earth. India's Mesopotamian, Egyptian, and early Greek contemporaries believed the earth was flat.[47] Some one thousand years before Aristotle, the Vedic Aryans asserted that the earth was round and circled the sun.[48] A translation of the Rig-Veda by J. Arunachalan goes: "In the prescribed daily prayers to the Sun we find . . . the Sun is at the center of the solar system. . . . The students ask,

'What is the nature of the entity that holds the Earth?' The teacher answers, 'Risha Vatsa holds the view that the Earth is held in space by the Sun.' "[49] This passage also hints at modern ideas of gravitation. (However, by A.D. 550. Indians had adopted a Greek model that claimed that the solar system circled around a stationary earth.)[50]

Indians also developed a view of their physical earth, which they divided into seven areas or islands called *dvipas.* These have been identified as East Africa, the Middle East, the Mediterranean region, Europe, eastern Asia, Southeast Asia, and much of the rest of Asia, called Jambu. Jambu was surrounded by an "ocean," meaning a physical barrier. Yet another Puranic conception compares the earth with a lotus, each petal representing a continent, each continent situated equidistant from its neighbors and surrounded by oceans. Although the geographical areas named as various *dvipas* are not today called continents, the idea of seven discrete land masses on the earth certainly suggests today's seven continents.[51]

Ancient China

The Yellow River civilizations in China, spanning three millennia (2200 B.C. to ca. A.D. 1300), can lay claim to many significant advances in geology, among which are the first magnetic compasses, the first seismographs, and detailed record keeping.

Like other ancient cultures, the Chinese had their successes, including mining and casting bronze, brass, copper, silver, gold, tin, and zinc. Iron mining and steel working, beginning in the ninth century B.C., led to the discovery of magnetite and magnetism. The Chinese compared iron's attraction to magnetite to a child's attraction to its mother.[52]

The Chinese pioneered the petroleum industry. Records predating the Old Testament refer to petroleum seeps. *Jo shui* (weak water) was the Chinese name for petroleum because, although it was liquid, nothing would float on it, a quality that must have been perplexing. Around A.D. 190 Thang Meng wrote, "there are certain rocks from which springs of 'water' arise. . . . This liquid is fatty and sticky like the juice of meat. It is viscous—like uncongealed grease. If one sets light to it, it burns with an exceedingly bright flame." Records from 300 B.C. mention using "oil water" as grease for axles and as a source of fire. According to Confucius, there are also records of drilling with bamboo poles for natural gas in Szechwan in 211 B.C.[53]

The Chinese did not randomly poke and drill in their search for minerals. They were among the first to develop a systematic technique of mineral exploration. They documented associations of different minerals, just as modern geologists depend on suites of mineral associations to identify drilling sites. The *Kuan Tzu* (or *Guan Zi*), a fourth-century B.C. text, lists such associations:

> Where there is cinnabar above, yellow gold will be found below. Where there is magnetite above, copper and gold will be found below. . . . Where there is haematite above, iron will be found below.

This ability to identify what may lie underground by what is found on the surface is significant,[54] and this knowledge is echoed again and again in later texts. *The Book of Mountains and Seas* (written somewhere between the seventh and fifth centuries B.C.) states that where one ore is found, another is likely to be beneath. An example cited involves pyrite and alum.[55]

The ancient Chinese kept detailed records of natural phenomena, including mineral classifications, which became more elaborate with time. Because of their use in medicines the Chinese started classifying minerals centuries before the birth of Christ. These classifications in the *pen tshao* (a series of natural histories of drugs that began in the fourth century B.C.) are elaborated on straight up to the *Pen Tshao Kang Mu* (Great Pharmacopoeia) of A.D. 1596. The *Shan Hai Jing* ("Classic of Mountains and Rivers"), written around the fifth century B.C., divides rocks into ores, nonmetals, special rocks, and clays. It indicates hardness, color, and luster, as well as shapes—clay lump, nugget, oval, grains, massive. It also tells whether the minerals can be smelted.[56]

Aside from geology's practical applications, mineralogical ideas in China also rose out of philosophy and became important in interpreting empirical observation and experience. The idea that *ch'i*—a Chinese word roughly meaning energy or emanation—was responsible for the formation of minerals exists as early as the second century B.C. This concept mirrors somewhat Aristotle's idea of the moist and dry "pneumas" of the earth.[57]

The Taoists believed that minerals undergo changes following a specific order, which led to the practice of alchemy. According to the second-century B.C. *Book of the Prince of Huai Nan, ch'i* turns into mercury and then mercury into gold after twenty-five hundred years. Gold eventually turns into water, clouds, thunder, and lightning.[58]

THE CHINESE puzzled over fossils but took an early interest in them. Fossils are traces left by earlier life-forms—sometimes remains of animals or plants themselves, but not necessarily. Footprints, for example, are fossils, as are the shapes mollusks leave in rock, even though the mollusk itself has dissolved and the void in the rock is filled with sediment. Petrified wood is a fossil, though the organic wood has been replaced by minerals cell by cell. According to the English scholar Joseph Needham, who extensively documented Chinese science,[59] the Chinese from the first century B.C. onward more consistently and accurately recognized fossils as evidence of once living creatures than did their Western counterparts. In A.D. 200 to 300, Chinese scholars living in a region of petrified pine forests believed that trees became petrified naturally after three thousand years.[60] By the fourth century A.D., however, Lo Han knew that fossils were petrified life.[61] In the fifth century A.D., fossilized brachiopod shells were recognized as shells turned to stone (called "stone-swallows" because they had winglike shells). The same fossils were not understood as fossils by Westerners until 1853.[62]

Much earlier Greek thinkers like Xenophanes (c. 560–478 B.C.) and Herodotus (c. 485–425 B.C.) believed fossils had been living creatures and theorized the need for gradual sea and ground-level changes to explain fossil origins. Aristotle (384–322 B.C.), Eratosthenes (c. 276–196 B.C.), and Strabo (63 B.C.–A.D. 25) all used fossils to develop theories of the great time frames involved in geologic processes. None of these ideas, however, were accepted by the Greeks or later Westerners for any length of time.[63] Aristotle said in his *De Respiratione,* "A great many fishes live in the earth motionless and are found when excavations are made.[64]

According to Needham, Chinese ideas about mountain building and the associated cycles of sea to land and vice versa also relate to Vedic Indian notions of the destruction and re-creation of the world. (Needham, great scholar that he was, was sometimes overenthusiastic in his interpretation of Chinese science.) A story is told in which "ashes"—probably bituminous materials—were found. The emperor Han Wu Ti consulted Buddhist monks about the origin of the ashes; the monks claimed, "These are the remains of the last cataclysm of heaven and Earth." Although the sources are suspect, and the story probably apocryphal, the idea existed, and probably dates from the first century A.D.[65]

THE ANCIENT CHINESE were clearly the first seismologists, and the first to measure the intensity of earthquakes.

Records of earthquakes exist from 1300 B.C.,[66] along with records of seismic sound. In Shanxi Province, according to the *History of the Wei Dynasty,* from 474 A.D., "There were sounds like thunder heard in Qicheng of Yanmen; they came from the west and sounded for a dozen times. When the sounds ended an earthquake occurred."[67] Needham, commenting on these frequent seismic events, says, "It was natural, therefore, that the Chinese should have kept extensive records of earthquakes, and these indeed constitute the longest and most complete series which we have for any part of the earth's surface." After an eighth-century B.C. earthquake, Chinese scholars theorized that it was caused by an imbalance of *ch'i* between heaven and earth. The imbalance arose, they said, when the yang was imprisoned or barred by the yin.[68] This is not exactly a scientific explanation. On the other hand, who else was trying to understand the problem?

The first seismograph was created in A.D. 132[69] by the scientist Chang Heng (78–139). His "earthquake watercock," according to Needham, had "a domed cover, and the outer surface was ornamented with antique seal-characters. . . . Inside there was a central column capable of lateral displacement along tracks in the eight directions, and so arranged [that it would operate] a closing and opening mechanism."[70] Outside this creation were eight dragon heads, each with a ball in its mouth, and eight toads directly underneath with open mouths. When an earthquake hit, the vibration would cause the dragon head in the direction of the quake to drop its ball into the toad's mouth. Some kind of inner pendulum set off the motion, and some inhibiting device ensured that only one head released its ball during the tremor.

Needham comments that modern seismologists have admired the Chang Heng mechanism because of its ability to hang on to the seven extraneous balls, allowing only the ball indicating the direction of the primary shock wave to drop. Every earthquake comprises, in addition to the major tremor, a number of lateral events that could cause many of the balls to drop, but the apparatus immobilized those balls immediately after the first shock wave. Once, a dragon dropped its ball even though no tremor was felt. Days later a messenger came with news of a distant earthquake in the direction indicated by the dragon head. Indeed, officials relied on the instrument to give them warnings of disasters in far districts, giving them time to prepare their response.[71]

The Chinese left us no working model of the Chang Heng mechanism, and no data about its innards. Modern researchers have had to guess as to its construction. Needham claims three reconstructions were attempted, but only one seems to have been built, at Tokyo University. It used an inverted pendulum, and the ball dropped only after the lateral waves hit instead of with the initial primary wave, except when the initial wave was unusually severe. Needham speculates that Chang Heng would have calibrated the device to be sensitive enough to drop the ball with the first longitudinal wave. Still, one would like more evidence than Needham finds satisfying.[72]

The geologist Edward J. Tarbuck suggests (perhaps a little smugly) that though the Chinese knew that the first earthquake shock had a directional component, the "weather cock" probably was unable to determine wave direction with any predictability, because of the complexity of the waves generated.[73] There is little doubt that we have improved on earthquake prediction in the past two thousand years. We can make fun of the ancient Chinese—and many have—for explaining quakes in terms of *ch'i.* Yet the earthquake watercock indicates that they understood some of the mechanics involved.

THE ANCIENT GREEK philosopher Thales of Miletus introduced the concept of magnetism to the West in the sixth century B.C. He humanized his hypothesis, explaining magnetism in terms of anthropomorphic attraction.[74] The Chinese may have known about it even earlier, as early as 1000 B.C., according to geologist S. Warren Carey. (The Central American Olmecs may have beat both, about which more later.) Early Chinese knowledge of magnetism included its attraction and repulsion properties and its pointing direction. Both the Chinese and the Greeks recognized the first property, but only the Chinese initially recognized the second, which, of course, became the key to navigation.[75]

According to Needham, the Chinese understanding of magnetism grew out of geomancy, which he quotes as being " 'the art of adapting residences of the living and tombs for the dead so as to cooperate and harmonize with the local currents of the cosmic breath.' " Geomancy allowed magicians to direct this flow for the benefit of the living.

What we now know as the compass began as a spoonlike device, the south-pointing ladle, or *sinan,* around 475–221 B.C. A lodestone (magnetite) was placed on a smooth board in such a way that allowed the spoon's handle to swing according to the earth's magnetic field. The device continued to appear in texts up to A.D. 907 indicating its use. Needham claimed that compass devices were not used for navigation until A.D. 618. However, according to a fourth-century B.C. essay in *Gui Gu Zi,* the instrument was taken on journeys so that travelers wouldn't lose their direction.[76] (Needham may have been referring to ocean navigation only.)

In addition to the lodestone spoon, we find descriptions of a "fish-shaped piece of wood to which a magnet was attached, so that it pointed north and south when it was set afloat." This was *chih-nan-yu,* the south-pointing fish. Later needles that were magnetized came into use around A.D. 300 to 400.[77] The first known true compass dates from A.D. 1080.[78]

DURING THE MIDDLE AGES, the Chinese continued to modify the compass and made the significant discovery of magnetic declination, the difference between true north and magnetic north. "I-Hsing measured the magnetic declination [the local angle between magnetic north and true north] in 750 A.D. during the T'ang dynasty," writes Carey. "Through the next eleven centuries the Chinese recorded the slow variation in the declination." In Europe, by contrast, the compass was in use by the twelfth century, but Europeans did not have the Chinese knowledge of declinations.[79]

In *Wu Ching Tsung Yao,* a compilation of military knowledge written in A.D. 1044, the making of the "south-pointing fish" from more sophisticated technologies is described. It was made from a flat, thin sheet of iron heated to red-hot, oriented to the north, and then rapidly cooled. This process allowed it to take on magnetic properties, after which it would orient itself in a north-south position.[80] Westerners traditionally stick an arrowhead on the north-pointing end of the needle. The Chinese were more interested in the south, and put a fish-head shape on that end. The magnetic needle is also described in writings by the astronomer Shen Kua in A.D. 1088:

> Magicians rub the point of a needle with the lodestone; then it is able to point to the south. But it always inclines slightly to the east, and does not point directly to the south. . . . It is best to suspend it by a single cocoon fibre of new silk attached to the centre of the needle by a piece of wax . . . then, hanging in a windless place, it will always point to the south.[81]

The passage clearly shows Shen Kua's knowledge of declination. The reasons behind the magnetism remained mysterious to the Chinese, who, as usual, credited it to *ch'i.* According to Needham, the magnetic needle was used extensively in navigation during the eleventh and twelfth centuries,[82] but he finds no mention of using it for navigation before A.D. 618 (Tang dynasty), and speculates it was used for geomancy alone.[83] As mentioned earlier, there is evidence Needham is wrong here, that the needle was used in the earlier centuries A.D. for navigation.

DID THE ANCIENT Chinese believe in a round earth? The answer is yes, but there is debate over when they came to this conclusion. Carey says the Chinese believed the world to be round by somewhere between A.D. 1 and 100, perhaps learning this from the earlier Aryan Indians.[84] Needham claims that popular belief held that the earth was a square set in the round heavens, though that idea was challenged. Between A.D. 1 and 200, Chinese scholars claimed that the earth was the yolk of an egg (a common idea in many cosmologies around the globe) and Yu Hsi (around A.D. 330) claimed the earth was round.[85]

Along with speculations about the earth's shape, the Chinese made intensive inroads into developing realistic maps and other records of natural features. Yu Kung made the first recorded realistic map of China around the fifth century B.C. The Greeks made similar efforts in their own land and were slightly ahead of Yu Kung, though his work was much more detailed.[86] Somewhere between the first century B.C.[87] and the third century A.D.,[88] Chinese cartographers categorized 137 rivers. True scientific cartography was started by Phei Hsiu (A.D. 224–271), who is credited with using graduated divisions for scale and a rectangular grid.[89]

CHINESE METEOROLOGY, like other Chinese science, combined a basic belief in the supernatural with rigorous and scientific methods of measuring and understanding how the supernatural makes the weather. Weather records in China exist from as early as 1400 B.C. Temperature records are found from the Han dynasty, and careful rainfall records were kept beginning in 1216 B.C., including "rain, sleet, snow, wind and direction."[90]

The ancient Chinese believed that immoderate weather indicated angry heavens, so record keeping was to some extent an attempt to learn the cause of heaven's displeasure. Very early on, the Chinese began to understand the earth's hydrological cycle. The book *Chi Ni Tzu,* by a fourth-century B.C. naturalist, states: "The wind is the *chii* of heaven, and the rain is the *chii* of Earth. Wind blows according to the seasons and rain falls in response to the wind. We can say that the *chii* of the heavens comes down and the *chii* of the Earth goes upwards." Later Han records are more down-to-earth and describe how evaporation from mountains causes water to rise and become clouds.[91]

Modern ideas of the earth's atmosphere can be inferred from Chinese concepts of *ch'i* and of vapors. Chiang Chi, an astronomer from around A.D. 400, writes: "The terrestrial vapours do not go up very high into the sky. This is the reason why the sun appears red in the morning and evening, while it looks white at midday. If the terrestrial vapours rose high into the sky, it would still look red then."[92]

The Middle East

During the Middle Ages the Arabs translated the learning of India, Persia, and ancient Greece, transferring it to Europe through the Moorish occupation of Spain. Ancient Romans had not kept alive the scientific inquiry of the Greeks, and medieval Europe, out of touch with the classics, had little scientific tradition of its own.[93]

The center of scientific knowledge was in Alexandria (on the edges of Byzantium, which later became Christian) and was built up by people speaking Arabic, Syrian, and Hebrew. In the sixth to eighth centuries, even Greek learning was forgotten in the Middle East, and Eastern learning predominated. Intellectual centers were found in the

Abbassid caliphate in Baghdad, Damascus, and Cairo. By the tenth century this flowering moved more toward Cordova and Moorish Spain.[94]

Much of Islamic scientific progress in the Middle Ages rested on the work of two men, tenth-century scholar-scientists al-Biruni and Avicenna. Al-Biruni (973–1048) was born in the eastern Persian state of Khwarizm, and grew up speaking the Khwarizmian dialect, Persian, and Arabic. He was educated by an astronomer-mathematician.[95] In northern Persia (Uzbekistan) al-Biruni learned Sanskrit, and he studied minerals from China and India to Byzantium. His work formed a critical link between Indian and Arabic knowledge.[96]

Born in Bukhara in central Asia, Avicenna, also known as Ibn-Sina (A.D. 980–1037), lived mostly in current Iran. At sixteen he became a physician. At his life's end, he is reputed to have commented that he learned "everything he knew" by the age of eighteen, when he studied psychology, chemistry, astronomy, and pharmacology.[97] He was a prolific translator of Aristotle and, for our purposes, is best known for his *De Congelatione et Conglutatione Lapidum* (On the Congelation and Conglutination of Stones), a commentary on Aristotle's earlier *Meteorologica.* Avicenna goes beyond Aristotle. He stated that meteorites come from space and fall to earth. Aristotle claimed that they originated on the earth and were projected into the heavens by wind.[98]

Both al-Biruni and Avicenna pioneered classification systems for minerals. Al-Biruni, in his *Collected Information on Precious Metals,* listed around a hundred known minerals. He determined the specific gravities of eighteen minerals by displacing water with the minerals and weighing the amount of water displaced, coming close to modern measurements.[99] Meanwhile, Avicenna classified rocks into four types: stone, metals, salts, and combustible sulfuric matter. This classification was widely used in the West into the 1750s.[100]

Natural oil was a familiar phenomenon in the Middle East during the Middle Ages. Burning natural gas wells were described by the Arab thinker al-Mas'udi (A.D. 915), who saw them in Baku (Iran), and by Mu'jamu'l-Buldan Yaqut (1179–1229), who was told that a merchant "saw there a bit of land from which fire does not cease issuing forth day and night. I think that a fire has fallen there from some person and that it does not cease because the [combustible] material sustains it."[101]

Ideas of where oil comes from were explored by the Brethren of Sincerity, a mysterious band of scholars living in Basra around A.D. 983, who connected Greek thought with the Koran.[102] They postulated that water and air grew through fire, and made "fiery sulphur" and "watery mercury." When these substances mixed with earth and were put to high temperatures, they formed minerals in the ground, including crude oils, asphalt, tar, and pitch. In fact, petroleum products do form under high temperature (caused by pressure). Because such products were full of oil and air, reasoned the Brethren of Sincerity, they could be liquefied and burned. Al Qazwini (c. 1275) expands on this idea to say that the condensation of the mercury and sulfur in deep mountain ravines caused the bituminous substances to form. This hints at the modern idea of the intense pressure needed to turn organic matter into petroleum.

BIBLICAL PASSAGES put the age of the earth at a few thousand years, not enough time for geological processes like uplift, weathering, and sedimentation to have taken place. The Europeans held with their orthodox Christian beliefs until the 1750s, dismissing huge sedimentary layers as relics of the Flood, and seeing fossils as creatures drowned in the Flood or tricks of the Devil or a kind of stone.[103]

In the medieval Middle Eastern world, modern concepts of these processes were more accepted, building on earlier theories from the Indians, Persians, and Greeks. Both al-Biruni and Avicenna were instrumental in developing and promoting these ideas. While visiting India, al-Biruni realized the Indus valley was once an alluvial plain that had been covered by the sea.[104] Here is a passage from *Alberuni's India,* translated by E. C. Sachua:

> One of these plains is India, limited in the south by the . . . Indian Ocean, and on all the other sides by the lofty mountains,

> the waters of which flow down to it. But if you have seen the soil of India with your own eyes . . . if you consider the rounded stones found in the earth however deeply you dig, stones that are huge near the mountains and where the rivers have a violent current; stones that are of smaller size at greater distance from the mountains, and where the streams flow more slowly; stones that appear pulverized in the shape of sand where the streams begin to stagnate near their mouths and near the sea . . . you could scarcely help thinking that India has once been a sea which by degrees has been filled up by the alluvium of the streams.[105]

Al-Biruni understands the sorting of sediment size that occurs during river erosion, from boulders down to the finest silt, as well as the rounded nature of rocks ground to smoothness by the action of stream water. He also demonstrates an understanding of the changing nature of both land and sea, as well as the huge periods of time involved in such changes.

Al-Biruni describes also how rivers have changed course since the time of Ptolemy's *Geographaia* in the first century A.D. He focused on the river of Balkh (Oxus), which ran between Jurjan and Khwarizm in the time of Ptolemy in a cultivated valley. During al-Biruni's time, nearly one thousand years later, the place was a desert, with rocks containing fossil "fish ears." Al-Biruni observed the evidence of water on a mountain, where the water must have built up as a lake (he postulated), then broken through the surrounding earth. In this fashion, he was able to follow the history of the lake until its present condition in his time as a salt marsh.[106]

Avicenna also knew about the incursion of the sea onto the land, the destruction of mountains, and the ebbing of the sea, and recognized that this had happened as cycles throughout history. Mirroring the eternal cycles of destruction and creation in ancient Vedic philosophies, he believed that the world had no end.[107] "As to the beginning of the sea," he states, "its clay is either sedimentary or primeval, the latter not being sedimentary. It is probable that the sedimentary clay was formed by the disintegration of the strata of mountains."[108] Modern geologists bear out this statement, since clay is a stable end-product of weathering, especially of shale outcrops.[109]

Avicenna identified processes involved in the formation of rock. Over the course of twenty-three years, he watched the clay on the Oxus River turn into a soft stone ("conglutination"). Observing how water dripping from caves turned into stone, he believed that water solidified in a process he called "congelation." His ideas seem bizarre on first glance. In this translation from Homeyard and Mandeville, Avicenna states:

> We know therefore that in that ground there must be a congealing petrifying virtue which converts the liquid to the solid. . . . If what is said concerning the petrifaction of animals and plants is true, the cause of this is a powerful mineralizing and petrifying virtue which arises in certain stony spots, or emanates suddenly from the earth during earthquakes and subsidences, and petrifies whatever comes into contact with it. As a matter of fact, the petrification of the bodies of plants and animals is not more extraordinary than the transformation of waters.[110]

Extraordinary indeed. In fact, the tenth-century Arab thinker al-Mas'udi developed a theory of evolution that bears some elements of this transformational process. Born near Baghdad, he claimed, in *The Book of Indication and Revision,* that minerals evolved to plants, plants to animals, and animals to man.[111]

Avicenna's belief that some force or quality in water was behind rock formation is not so far off from modern knowledge of mineral precipitation. Cave stalagmites and stalactites are common examples of how dripping calcium-rich water can form limestone precipitates (known as travertine) that can grow to considerable size.[112] However, knowledge of the cementing process of sediments under pressure (where deposits lithify both from compaction and from minerals in solution moving through the layers)[113] eluded him.[114] In the thirteenth century the Western medieval philosopher and scholar Albertus Magnus applied Avicenna's ideas to explain fossil branch imprints; he was an early Westerner to recognize that living things had turned to stone.[115] Meanwhile, a generation before, the Scottish philosopher John Duns Scotus had claimed that stones and metals were alive.[116] His name has since given us the word *dunce.*

As ideas concerning rock formation took form and were borne out by observation, ideas of the forces behind mountain building also took form. Avicenna exaggerated the role of wind and floods in carving large-scale features such as mountains, but he correctly identified earthquakes (although not the cause of earthquakes) as an essential mountain-building force. He theorized that earthquakes raise part of the ground and that winds and floods randomly erode one part of the earth, leaving as heights those they miss. This, Avicenna said, is the source of valleys bounded by mountains.[117]

Avicenna's account of wind and water erosion is accurate for certain smaller-scale processes, such as the weathering of rocks along the sea and potholes in rivers. It is also essential to understanding the much more gradual process of erosion as it actually occurs in mountains composed of granite and other crystalline rocks. In addition, it is accurate for butte formation of very soft sedimentary rock, where torrential downpours can indeed carve deep canyons and, over time, erode the land until only isolated remnants of layers are left standing. The badlands in North and South Dakota and Bryce Canyon in Utah are examples of fast erosion common in arid lands creating high, steep formations.[118]

Avicenna was closer to modern geology when he observed and described erosion and sedimentation:

> At present time, most mountains are in the stage of decay and disintegration, for they grew and were formed only during their gradual exposure by the waters. Now, however, they are in the grip of disintegration. . . . It is also possible that the sea may

> have happened to flow little by little over the land consisting of both plain and mountain and then have ebbed away from it. . . . It is possible that each time the land was exposed by the ebbing of the sea a layer was left, since we see that some mountains appear to have been piled up layer by layer, and it is therefore likely that the clay from which they were formed was itself at one time arranged in layers. One layer was formed first, then, at a different period, a further layer was formed and piled (upon the first, and so on). Over each layer there spread a substance of different material, which formed a partition between it and the next layer; but when petrifaction took place something occurred to the partition which caused it to break up and disintegrate from between the layers.[119]

Modern geology bears out his essential observation: many sedimentary rocks do result from the depositions of layers of sediment carried by water. The word *sedimentary,* according to geologist Tarbuck, is "derived from the Latin *sedimentum,* which means 'settling,' a reference to solid material settling out of a fluid."[120] He continues: "Sedimentary rocks form as layer upon layer of sediment accumulates in various depositional environments."[121] The sea also makes incursions back and forth onto land from both uplift and the action of glaciers, depositing silt with each cycle. Today this is considered a pretty basic geologic concept. However, Avicenna did not understand that while these actions cause the lowlands to fill with sediment, uplift is also necessary for the formation of most mountains.

WHILE EUROPE held fast to the concept of a flat earth, its Arab neighbors revived ancient Chinese, Indian, and Greek speculations of a round globe. The seventh-century Armenian Ananii Shirakatsi thought the universe was egg-shaped, with the sky being the shell, the air the white, and the earth the yolk.[122]

Al-Biruni believed the earth was round and calculated its size, coming up with figures close to present-day computations.[123] According to the science historian Seyyed Hossein Nasr, al-Biruni developed a method for measuring the antipodes and roundness of the earth, the height of cities, and latitudes and longitudes (the former more accurately than the latter).[124]

Geography was influenced by the Muslims because of their travels. Sailing trips, navigation, and cartography all expanded under them in the ninth to thirteenth centuries.[125] This, in turn, led to more sophisticated techniques for measuring the geographic features they saw, including the earth as a whole. In his tenth-century work *Meadows of Gold and Mines of Gems,* Al-Mas'udi refers to "half the circumference of the Earth . . . which, if reduced into miles amounts to thirteen thousand five hundred geographical miles."[126] Clearly, if the earth has a circumference, it is round. The actual total circumference of the earth is 24,902.4 miles, so al-Mas'udi's estimate of 27,000 miles (2 times 13,500) isn't bad. (In all fairness, the Greek Eratosthenes came up with a similar estimate some twelve hundred years earlier.)[127]

The sea presented a more puzzling problem. How could a liquid take on the form of a sphere? Yet sailors' daily experience bore out this idea. Al-Mas'udi knew that the ancient Hindus and Greeks believed the sea to be curved. He gives an account of sailors on the Caspian Sea approaching a volcano near Tehran:

> It is to be seen at a distance of one hundred farsangs, on account of its height. . . . The mountain is about twenty farsangs from the Caspian. If ships sail in this sea, and are very distant, they will not see it; but when they go towards the mountains of Taberistan, and are within a distance of one hundred farsangs, they perceive the north side of this mountain . . . and the nearer they come to the shore the more is seen of it. This is an evident proof of the spherical form of the water of the sea, which has the shape of a segment of a ball.[128]

Hence, Columbus sailed without fear.

Medieval India and China

The history of geology in medieval India is a short one. Two scholars dominate: Varahamihira and Vagbhatta. Varahamihira (A.D. 499–587) was born in Kapitthaka (present-day Kapitha), a great intellectual center in central India, and became a patron of the Ujjain ruler King Maharajadhiraja Dravyavardhana. Varahamihira's 106-chapter *Brhatsamhita* (Great Compendium) covers geography, meteorology, botany, agriculture, the calendar, and gemology.[129] Among other things, he theorized that the location of earthquakes was correlated to the time of day they struck and suggested that the moon was a factor.[130] He claimed the earth was round,[131] and said that precious stones formed from rocks through metamorphosis over time. Much later, Vagbhatta (A.D. 1300) developed a system of mineral classification that included precious stones, metals, and alloys.[132]

In the Middle Ages, Chinese philosophy continued to hold that *ch'i* was expressed in the earth: large quantities of *ch'i* could cause outcrops, rocks, and precipices, while smaller amounts became sand, silt, and other fine-grained material.[133] The Chinese, like other cultures, systematically identified and classified a wide range of minerals and pioneered methods of using plants and other biological features on the surface to find minerals. Once considered a dubious method, correlating rock type with vegetation is a technique that has been finely honed by some modern geologists.[134]

For example, according to sixth-century wisdom, if plants were yellow, there was copper below.[135] The A.D. 863 text *You Yang ZaZu* (Miscellany of the Youyang Mountains) elaborates:

> When in the mountains there is green onion, there is silver below. When in the mountains there is shallot, there is gold below. . . . If the mountain has precious jade, the branches of the trees on it are drooping.[136]

Today, certain species are known to grow on rocks of specific composition. It is not a major exploration tool, but there's more than a morsel of truth to this ancient technique, according to University of Massachusetts at Amherst geologist Sheila Seaman. Different plants thrive on different trace elements provided by different rock composition.[137]

The Chinese are known more for describing large-scale geologic processes. Concepts of geologic uplift are mentioned in medieval Chinese texts, some two hundred years before the great Arab thinkers al-Biruni and Avicenna examined this subject. The Chinese also recognized the meaning of fossils. According to Yen Chen-Chang, writing around A.D. 770, "Even in stones and rocks on lofty heights there are shells of oysters and clams. . . . Some think they were transformed from the groves and fields once under water."[138]

Throughout this era, fossilized plants and animals were generally recognized as ancient life (commonly considered to be "dragon's bones and teeth").[139] Li Tao-Yuan in the sixth century A.D., Yen Chen-Chang in the eighth century, Shen Kua in the eleventh century, and Chu-Hsi in the twelfth century all knew that fossils, such as footprints left behind by animals, were petrified remains of animal parts.[140] In A.D. 1133, one finds a detailed description of fossils in *Yun Lin Shih Phu.* What is notable is that this work describes where in the stratigraphic record of rock the fossils were found, a necessary step in developing the later theory of evolution. One can infer a time sequence from studying the location of sediments. When fossils are embedded in these sediments, one can infer an evolution of forms over time.[141]

A clearer understanding of the role of erosion in shaping mountains is expressed by Shen Kua around A.D. 1070 (some one hundred years after Avicenna):

> All its [Yen-Tang Shan mountain] lofty peaks are precipitous, abrupt, sharp and strange; its huge cliffs . . . are different from what one finds in other places. . . . Considering the reasons for these shapes, I think that (for centuries) the mountain torrents have rushed down, carrying away all sand and earth, thus leaving the hard rocks standing alone.
>
> In places like Ta Lung Chhiu, Hsiao Lung Chhiu, . . . one

> can see in the valleys whole caves scooped out by the forces of water.[142]

In an opposing process to erosion, Neo-Confucian writer Chu Hsi (A.D. 1130–1200) wrote about uplift as a mountain-building force. From his work *Chu Tzu Chhuan Shu,* we read:

> The waves roar and rock the world boundlessly, the frontiers of sea and land are always changing and moving, mountains suddenly arise and rivers are sunk and drowned. . . . I have seen on high mountains conchs and oyster shells, often embedded in the rocks. These rocks in ancient times were earth or mud, and the conchs and oysters lived in water. Subsequently everything that was at the bottom came to be at the top, and what was originally soft became solid and hard.

To which Seaman says, "Wow!" According to Needham, such ideas had been in the making for centuries and continued for centuries afterward. Meanwhile, to fifteenth-century Europeans, fossils found high in mountains indicated that the sea had once been at that level, supporting biblical ideas of the Flood. The modern concept of geologic time can be seen in an equivalent Taoist concept of *sang thien,* an ancient phrase that came to mean the huge period of time it took for the sea to recede from the land.[143]

The Chinese were engaged in a number of geological subdisciplines as well. They made wonderful maps in the Middle Ages and may have originated the phrase "It's not the heat, it's the humidity," being the first to devise a method of measuring humidity by weighing charcoal as it absorbs moisture from the air.[144] They had the tides figured out, too. Records of tides had been kept since A.D. 800[145] and shortly before that, Tou Shu-Meng had delved into the connection of the moon to tides in scientific detail.

China possesses one of the largest tidal bores in the world, on the Chhien-Thang River. Examining records of these tidal bores, one sees a clear correlation with full moons, and yet it was not until the first century A.D. that Wang Ch'ung wrote in *Lun Heng,* "Finally the rise of the wave follows the waxing and waning moon."[146]

By the eleventh century, Yu Ching declared that both the sun and the moon influenced tides, though he thought the moon more critical. In the same century Shen Kua studied the motion of the moon, and correlated the timing of high and low tides to lunar positions.[147] Take, by comparison, Galileo Galilei, whose "Treatise on the Tides" in the seventeenth century explained tides as the result of the earth's motion, the oceans sloshing about like water in a bathtub.[148]

Medieval Africa

Modern soil scientists speak of the "horizons" or layers of different kinds of soil as you dig down. A vertical section of these layers is called the "soil profile."[149] In 1936, Tanganyika soil scientist G. Milne documented a traditional African concept of a series of soil types, which he dubbed "catenas." Studying hills around Lake Victoria, he noted predictable sequences of soil types depending on where one was on the hill profile. The "primary" soil that has directly eroded from the bedrock is found near the top of the hills; it in turn becomes the basis for a deeper "red earth" soil at the hill bottom. There are seven different soil evolutions, which end with the *mbuga,* the clay bottomland horizon.

Local Africans, Milne noted, recognized these sequences, and had a highly developed classification, in their language (Sakuma), of the associated soils.[150] Milne writes: "African soils tend to be very old, and showed the influence of underlying rock types much less than in Europe. The concept of soil catena, by emphasizing topography and downplaying the role of geology, provided a much better guide to the way the soils had formed."[151] This African view of soil associations varies from modern conventional concepts, but it enabled Africans to make better choices of which crops to plant.

North American Indians

Unlike Europeans in the same era, Native Americans in the fifteenth century knew that the sea once covered the mountains, even those far inland. They were also aware of glaciers. For example, in Jackson Hole, Wyoming, site of the Grand Tetons, glacial ice encroached on the valley around 50,000 B.C. and slowly receded. The ice was gone from the valley around 4000 B.C., while the earliest evidence of people at Jackson Hole dates from 7500–6500 B.C. These glaciers left glacial moraines, which are piles of gravel and silt deposited as the glacier recedes. (Cape Cod began as a glacial moraine.) Also left were river terraces that once funneled the melting glacial water. According to anthropologist Gary Wright, "Many Indian tales deal with these geological processes as metaphor. There was Big Beaver whose eyes could melt the ice and Coyote Trickster who possessed a host of supernatural abilities. Recognizing these processes and understanding their effects on the landscape were crucial to the maintenance of the local group."[152]

The Shoshone, who came to Jackson Hole sometime in the fifteenth century A.D., had creation myths that referred to the sea, though the closest sea to Wyoming was the Pacific Ocean. It appears the Shoshone correctly induced by reading geological clues that the mountains had once been immersed in seawater. The Grand Tetons have limestone, sandstone, and dolomite deposits thousands of feet deep, evidence of past sea deposits. The Shoshone account is not written in the style of a peer-reviewed journal, but it is geology nonetheless. Here is a version of the Shoshone creation myth that Wright related originally:[153]

> Coyote walked with a maiden to a large lake [probably Jackson Lake or its predecessor] at the foot of the Grand Tetons. The maiden lived with her mother, whose name was Ocean Old Woman. Coyote had sex with both of them, and after both women gave birth, Old Ocean Woman put the children in a large jug of water to give to Coyote. Coyote carried the jug up to the Saline Valley, to Death Valley, Tin Mountain, and the Ash

Meadows. At each place he left babies, who founded the different Indian tribes.[154]

One possible interpretation: Ocean Old Woman is the ancient mother of the Shoshone race; thus the ocean was the mother of the land the Shoshone lived on. In a metaphorical sense, Ocean Old Woman rising up into the mountains (via the water jug) reflects the influence of the sea on rocks in the mountains.

The Indians were the first to tell James Wilkinson, the governor of the Louisiana Territory in 1805, about what is now Yellowstone National Park, a natural volcanic area. They showed the governor a map, drawn on a buffalo hide. It showed, said Wilkinson, "among other things a little incredible, a Volcano" distinctly described as being located at Yellowstone.[155] Wilkinson's tale suggests that the Indians understood that Yellowstone's geysers and hot springs had origins in volcanism. Explosive volcanic activity formed the giant Yellowstone caldera some 600,000 years ago. This was followed by a succession of lava flows occurring 500,000 years ago and then later between about 160,000 and 70,000 years ago.[156]

How did Native Americans know about this volcanic activity? Did they have firsthand knowledge? It has long been believed that the first people appeared in the Yellowstone area after 12,000 B.C.,[157] or fourteen thousand years ago. That figure is now being debated, but it is unlikely to be pushed back fifty-six thousand years, to the time of the last volcanic activity. Or perhaps the Indians made a lucky guess that there were volcanoes beneath Yellowstone, though they had never seen one in action. A more likely explanation is that Native Americans had learned about volcanoes and the volcanic landscape through other volcanic events: perhaps from more recent volcanism in the Cascade Mountains, or from trading with Indians farther south, in Mexico, where volcanoes are more common.

At Hells Canyon, the deepest gorge in North America, on the Snake River, which divides Oregon and Idaho, seven mountains, known as the Seven Devils, form a semicircle. A Nez Percé myth, paraphrased here, explains the formation as follows:

> Each year, seven giants traveled eastward destroying all in their path, including the Nez Percé. The chief of the Nez Percé got Coyote to help them. Coyote enlisted the help of his friend Fox who suggested they dig seven holes and fill them with boiling liquid. All the clawed animals helped dig and afterwards Coyote filled each hole with a reddish-yellow liquid kept boiling by hot rocks. When the giants moved eastward, they fell into the holes. "They fumed and roared and splashed. As they struggled, they scattered the reddish liquid around them as far as a man can travel in a day." The splashing of the giants turned into copper.[158]

These mountains were once volcanoes, which explains the boiling holes of reddish-yellow liquid and the "fuming" and "roaring." According to a geologist at the University of Idaho, "Eruptions blanketed the area with thick flows of lava. . . . Huge floods of lava from the west poured across the land."[159] The fact that the lava came from the west also corroborates the giants' march eastward.

White men often said that Indians were frightened of natural phenomena that had scientific explanations. In particular, Indians were reputed to be terrified of the geysers at Yellowstone. In a memoir, nineteenth-century fur trapper Warren Ferris mocked the Indians for refusing to stand as close to an erupting geyser as he did. Ferris said the Indians were "quite appalled" at his actions, and he attributed their fear to a belief in the supernatural. Ferris admitted that, having put his hand in the water, he "withdrew it instantly, for the heat . . . in this immense cauldron, was altogether too great for comfort."[160]

The Indians had substantial reverence for the Yellowstone area, which reflected their understanding of the universe. White people interpreted Indian prayers and gifts as a way to ward off evil spirits, but Human Wise, a modern descendent of the Bannock and the Wind River Shoshoni, says that native peoples traveled far expressly to be near geysers in order to pray and to bathe in the hot springs. Regarding the offerings of gifts, he said, "You take one element [i.e., the use of the hot springs], then you leave an element so that you are not disturbing the balance of nature."[161]

Central America

The Olmecs (1500–800 B.C.) lived along the Gulf Coast in southern Mexico. The Zapotecs (1150 B.C.–A.D. 1521) lived in Oaxaca in southern Mexico. The Maya lived in southern Mexico, Guatemala, and Belize (c. A.D. 1–900); the Toltecs (900–1150) lived in central Mexico; and the Aztecs lived in central and southern Mexico (c. 1150–1521). Dates for both Zapotec and Mayan civilizations vary with the source.

Like the North American Indians, civilizations in Central America carried on a vast trade around desirable rocks and minerals, which they used as jewelry but also as tools and materials. The Olmecs used a wide range of local rocks and minerals for implements: basalt, slate, pumice stone, obsidian, green quartz, serpentine, flint. Pumice stone and obsidian powder were used for abrasives.[162] They imported kaolin (clay used in china), quartz, serpentine, jadeite (all green stones) from contiguous civilizations; rock crystal and alabaster may have come from as far as Peru; and asphalt from the Gulf Coast.[163]

The Olmecs' contemporaries, the Zapotecs in Oaxaca, worked jade, turquoise, and iron ore and polished their mineral specimens with hematite powder. They used limestone and travertine brought in from distant quarries as building stones and carted in nearby soil fill and volcanic tuff for their terraces around buildings.[164]

Copper, found throughout Mesoamerica, was one of the first metals to be used.[165] Indications of western Mexican metalworking first appear around A.D., 600 in places where people had access to native copper, silver, gold, and lead, and ores of copper, sulfide, and silver.[166] A gold-copper-silver mixture was used from Mexico down to the Andes. Gold was found in streams as tiny nuggets or grains; people made it into sheet metal by pounding the nuggets with stones, smelting it, hammering it for hardness, and then annealing it to soften it.[167] Some Mexican metalsmiths cast silver and gold together, which demands a much fuller understanding of the melting temperatures of mixtures of the minerals. Toribio de Benavente, known as Motolinia (1495–1565), a Spanish explorer and Franciscan who wrote *History of the Indians of New Spain,* was impressed with native metalsmiths:

> They can cast a bird whose tongue, head and wings move, and they can mould a monkey . . . which moves its head, tongue, hands and feet, and in its hands they put little implements so that the figure seems to be dancing with them. What is even more remarkable, they can make a piece half in gold, half in silver, and cast a fish with all its scales, in gold and silver alternating.[168]

The Maya associated the god Hurricane with two Thunderbolt companions: Newborn and Sudden. "These two names refer not only to shafts of lightning but to fulgurites, glassy stones formed by lightning in sandy soil," writes the Mayan scholar Dennis Tedlock.[169] The Maya must have seen the formation of fulgurites, which are tubes formed by the fusion of silica sand due to lightning.[170]

The Aztecs were the last Central American civilization before the Spanish conquest, and they inherited centuries of knowledge from earlier societies. Their knowledge of stonework and metallurgy, in particular, owed a debt to the Toltecs who proceeded them.[171] According to the Florentine Codex, a book compiled by Bernardino de Sahagún, a Spanish priest who lived among the Indians and learned their language, the Aztecs meticulously indexed animals, birds, fish, reptiles, trees, herbs, stones, metals, water, soils, and food. The book was written in Nahuatl, the Aztec language, from dictation by Aztec nobles and with illustrations by Aztec scribes. This all-encompassing work of Aztec beliefs, knowledge, and practice took much of de Sahagún's life in "New Spain," from 1529 to 1590, to complete. De Sahagún established schools after the Spanish Conquest to educate the sons of the native aristocracy. The Florentine Codex was organized, overseen, and edited by de Sahagún, who was fluent in Nahuatl, but the actual writing was done by Aztec students. De Sahagún then translated the Nahuatl into Spanish. The codex now takes up fourteen large volumes.[172]

The codex lists thirty-nine different kinds of precious minerals, nonprecious minerals, and metals, by name, characteristics, and usage. In particular, three different forms of precious jade and two forms of turquoise are identified. In addition, the codex records stones by size, such as pebbles, river rocks, slate, sand, and building stone. For flint, the codex states:

> It is slick . . . it is asperous; it is rough, scabrous, concave, dented, hollowed out, bored. . . . [One] is white; one yellow, not really yellow, just blended. . . . One is shaded tawny. One is green, one transparent. . . . This [flint] has fire. When it is struck, sparks come out of it.[173]

This shows the Aztecs' recognition of the many forms of flint and its fire-making ability.

Like the Chinese, who used certain plant characteristics to identify the presence of copper or tin, the Aztecs knew where to find "green stone" (a common form of jade) because "[the herbs] always grow fresh. . . . They say this is the breath of the green stone. . . . It is an announcer of its qualities."[174]

The Aztecs recorded soil types in Mexico from A.D. 1500 to 1600. The Codex contain glyphs signifying ten different soil locations, from top soil down to sandy soil. The soil-classification system, recorded in hieroglyphic writing, was the most sophisticated in the non-Western world since the soil maps made in China thousands of years ago.[175]

MOUNTAINS, according to the Aztecs, were thin vessels with rock and earth only on the surface, the interior being filled with water. If the "mountains would dissolve; the whole world would flood."[176] We know today that this is wrong; mountains are not giant water tanks. However, the Aztecs may have been referring to the many crater lakes in the local volcanoes, to the liquid magma that oozed out of an active volcano, or to the underground water systems and the limestone cave and sinkhole (karst) formations common in the Yucatán. Regardless of their understanding of the interiors of volcanoes, volcanoes were important to the lives and geology of Central American peoples.

The Olmecs, one of the founding civilizations of Central America, put the center of creation in San Martín Pajapan, a volcano. They mined its basalt and must have seen the molten lava forming new rock. Volcanoes and "cleft mountains" figure strongly in their art. According to Mayan scholar David Freidel, "Volcanoes were, in the Olmec experience, the clearest example of the world being born out of the Otherworld below. No people who have seen the sky turn black in billowing clouds of eruption and then rain stony fire and desolation onto the fertile, surrounding countryside could doubt that mountains contain spiritual forces capable of dispensing prosperity or disaster in human lives. Perhaps because of this, volcano and cleft mountains are a prominent feature in Olmec art. They carry sprouting vegetation and represent openings between the earthly plane and the world below it." (Exceptions to Freidel's claim that no people who have seen a volcanic eruption doubt the spiritual forces within would obviously be the citizens of Washington State who witnessed the eruption of Mount St. Helens or other modern people who have seen eruptions. On the other hand, volcanic eruptions often stimulate a belief in the supernatural in otherwise modern minds.)

In 1897 came an amazing find: a statue having a headdress that featured a cleft-headed god and a corn plant was discovered on the edge of San Martín Pajapan's crater. These images appear in both Olmec and Mayan cosmology to reflect the planting of "the World Tree" and the creation. To re-create the divine order in the human one, the Olmecs, in around 1000 to 600 B.C., made a volcano pyramid in their town (now called La Venta), located in swampland along the coast. They carried, by human labor, tons of basalt quarried from inland volcanoes. Some pieces weighed a hundred tons. From southern Mexico, they brought greenstone and serpentine to make huge floors built up in layers. Another incarnation of the man-made volcano is later found in Mayan temples that resemble mountains.[177]

The global geology of the Olmecs and Maya was obviously flawed. They believed that the core of the earth was water. The Maya represented "First-True-Mountain," the mountain of the creation, with cleft openings out of which the first gods appear. Both the Olmec and the

Maya connected these mountains with water: pools like the sacred Mayan cenotes (wells) and the watery void out of which the world was formed.[178] On the other hand, these myths aren't woven entirely from the imagination. The Olmec and Maya were familiar with the Yucatán's limestone caves and associated water sources deep underground. Their representation of a watery interior earth was based on exploration.[179]

An almost exactly equivalent cleft-headed god from around 1150 B.C. appears in artifacts from the contemporaneous Zapotec civilization. This is Xoo, god of earthquakes and a symbol of the earth, whose fissures also sprout plants. Did the Olmecs, Zapotecs, and Maya recognize the connection between earthquakes and volcanism?[180]

In the Popul Vuh, the Mayan creation myth of Guatemala, the earthquake-volcano connection is again suggested. The cycle of destruction of world and land is described in terms of the two gods or dragons: Zipacna, who resembles a caiman, his name appearing to have origins in words meaning to slip or slide, and Earthquake. Both Earthquake and Zipacna seem to be combined in a Yucatán version called *itzam kab ayin* meaning "monster earth caiman."[181] The text reads:

> And this is Zipacna, this is the one to build up the great mountains: Fireplace, Hunahpu, Cave by the Water . . . [names of a belt of volcanoes running near Lake Atitlan and Guatemala City] as the names of the mountains that were there at the dawn are spoken. They were brought forth by Zipacna in a single night.
>
> And now this is Earthquake. The mountains are moved by him; the mountains, small and great are softened by him.[182]

Tedlock's contemporary Mayan sources interpret this last sentence as meaning either that the mountains are "overcooked" by Earthquake or that they are "soaked." The "soaked" reading associates Earthquake with landslides and the destruction of land by torrential rains.[183]

In 1966 anthropologists uncovered a piece of Olmec hematite that appeared to function as a compass. In 1973 astronomer John Carlson exhaustively tested the object and, by radiocarbon-dating methods, proved it to from before 1000 B.C., thus predating the Chinese lodestone spoons by a thousand years. Some researchers believe that the alignment of Mayan temple complexes, which changes slowly according to the age of the site, reflects a knowledge of compass directions and the change of the earth's magnetic field over time. Carlson, writing in the journal *Science,* found that Olmec ceremonial complexes were usually situated, in very rough accordance to standard declination, 7 to 12 degrees west of north.

Another researcher cited by Carlson (M. D. Coe, 1968) discovered that these buried artifacts lay exactly along the central axis of the ceremonial site. Carlson found tiny flat mirrors, highly polished drilled beads, and large, concave parabolizing mirrors up to ten centimeters in diameter. The parabolic mirrors could focus sunlight and were made by unknown methods.[184]

The grand prize, though, was the putative compass. It was a broken, unornamented, polished bar of almost pure hematite with a carefully carved rounded groove. It proved to be magnetic, with its orientation at about 35.5 degrees west of magnetic north. Discussing the directional discrepancy from actual magnetic north, Carlson notes that only a fragment of the object still remained, and that rectangular magnetic materials tend to grow increasingly polarized the longer and narrower they are. He theorized that directions could have been determined using the hematite bar by sighting down the groove. The groove's cylindrical shape has a slight but carefully consistent (2-degree) divergence from the bar's centerline, which Carlson suggests might have served to calibrate the object to magnetic north at the time it was made.[185]

In 1975, geographer Vincent Malstrom of Dartmouth College found further evidence of Mesoamerican knowledge of magnetism. On the western Pacific plain at Izapa in Chiapas, Mexico, he discovered two large basaltic sculptures of a snake and a turtle's head, one aligned southeast to the other, with an intervening stela. The turtle's head was carved of a single piece of basalt that proved to be magnetic. To make sure it wasn't a general condition of the area, Malstrom tested all the other exposed rock at the site for magnetic properties. None were detected. "This would suggest," says Malstron, "that the Izapans knew about magnetism in that they had reserved a basaltic boulder rich in iron for the carving of the turtle-head, and had executed it so carefully that the magnetic lines of force all came to focus in the snout of the turtle." Furthermore, the site yielded other representations of turtles, though none magnetic. Malstrom suggests this may indicate an Izapan association of the directional properties of magnetism with the known navigational instincts of sea turtles.[186]

In 1979 nine more sculptures were found on the same coastal plain, dispelling doubts that such magnetized works were unintentional. Seven statues of large rounded people (dubbed the "Fat Boys") had the magnetic poles on each side of the belly button; statues of heads had the poles on the temples. The Fat Boys appear to have been made as long ago as 2000 B.C., well before the Chinese discovered magnetism.

Finally, worth noting is the Mayans' rough method of measuring a landscape by using a rope about twenty yards, or twenty footsteps, long. They used this technique for measuring cornfields, but it is recorded in the Popul Vuh as the way the earth was laid out also, and the *Book of Chilam Balam* states that "the measurement of the world is described in terms of twenty footsteps . . . that span the twenty day names of the divinatory calender."[187] We mention this because such methods of land measurement, twenty steps at a time, are also described in far earlier texts from Sumer and in the Old Testament, as in Job 38:4–5:

> Where were you when I laid the earth's foundations?
> Tell me, since you are so well informed.
> Who decided its dimensions . . .
> Or who stretched the measuring line across it?

South America: the Inca Empire (A.D. 1100–1530)

At its height, the Inca empire stretched from northern Ecuador to central Chile. The Incans were accomplished metalworkers, experienced at coating silver with gold[188] and capable of working with platinum, which has an extremely high melting point (3216 degrees F) by mixing it with gold dust. Anthropologist Warwick Bray says this Incan technique anticipates modern metalworking. These ancient craftsmen found the platinum in the rivers of Ecuador, off the Pacific coast. Platinum was not identified as a metal in Europe until 1748.[189]

More remarkable was the Incas' identification of radioactive material. According to an essay by researcher Salvador Polomino on the Quechua people, descendants of Incas, the Incas knew about radioactive minerals, which they named *aya kachi,* meaning "stones or salts of the dead." Their teachings forbade the use of these minerals, so one assumes the Incas recognized their dangerous properties.[190]

Erosion and the power of water were critical forces in the high mountain areas of the Andes, and a mastery of the hydrologic cycle was essential to Incan survival. In the introductory essay to *The Huarochiri Manuscript,* a Spanish priest's 1608 recording of pre-Hispanic Inca belief, translator Frank Salomon speaks of myths about "the chasm-cutting power of rivers . . . or the storm and flash flood that create disastrous mudslides," and of entire villages being washed away.[191]

Mountains in the Andes range isolate valleys from each other, leaving little usable agricultural land. Quite suddenly, around 200 B.C., long before they were an empire, the Incas unified this land into what have been dubbed "vertical archipelagoes," which could run some 40 miles from bottom to top and represented a single interconnected irrigation system. This practice brought the Incas surplus food and allowed a growing population to withstand famine. They built elaborate systems of stone terraces to hold the soils they brought in, which they deposited in layers to allow for the proper drainage of water.[192]

Salomon says the *Huarochiri* defines the Incan concept of water as including everything from rainstorms, mudslides, and irrigation to the Milky Way. Water goes up to the heavens (in this account, inside a "celestial llama constellation [that] carries water up from below by drinking before ascending") and then back down to earth during storms and rain.[193] The Peruvian Andes have a six-month rainy season and a corresponding dry season. In their cosmology, the Incas viewed water on earth as moving from south-southwest to north-northeast and then rising into a heavenly river that flowed north to south. Water was thus seen as being in constant circulation, which supplied the reservoirs needed for the crops during the dry season.[194]

Water was associated with the giant water serpent Amaru, which manifested itself, among other ways, as a rainbow. Since rainbows seen from certain vantage points on high mountains are not arcs but circles, the Incas believed that the rainbow serpents come up out of the springs during the rainy season. When seen as arcs, the rainbow serpents were believed to have buried the rest of their bodies underground. The two ends of the rainbow on the ground were believed to be connected by water underground, completing the circle.[195] This reflects our modern understanding of the cycle of water: from interconnected underground sources, to surface water, to evaporation into clouds, and back to earth as rain again. Of course, today we've eliminated the part about the rainbows and snakes.

The politically incorrect Incas, writes Salomon, saw the watery portion of existence as male and the hard dry earth as female:

> The hydraulic embrace of moving water and enduring earth was imagined as sex. . . . When the lake-huaca [sacred being] Collquiri rushes downhill to his earth-lover Capyama, the bursting pressure of his virility squirts out of every channel and sprays destructive floods all over Capyama's people.[196]

The following myth documents the Incas' control of water and their irrigation system on a small lake called Yansa, probably the modern lake Yansacocha on the western slopes of the Andes. When Collquiri's excesses cause the lake to overflow, the villagers cry out:

"Hey, Collquiri! Hold back on the water!"

"Shut if off!"

Collquiri attempts to plug things up; finally he jumps into the lake and the water ceases, until soon there is not enough water—probably a metaphorical reference to the Andean six-month dry season and six-month rainy season. The people again complain. Collquiri, sitting in the bottom of the lake, instructs a servant to put rocks and earth into the lake to raise the water level until water again overflows the banks and can be used as irrigation. Once the bank is breached, Collquiri dams the water by building a wall without mortar at the lake's mouth.

This account, which Francisco de Avila extracted from *The Huarochiri Manuscript,* refers to an actual wall that still stood in 1608, when the manuscript was first read by Europeans. And, in fact, it was still in existence at the end of the twentieth century.[197] In the myth, Collquiri also instructs the people to mark the dam in five spots. When the waters reach these various points, at various times, the people are to release the water to the lower fields. In fact, the dam still contains five sluice gates at different heights, built with ancient stone-and-sod technology, and present-day residents still use the gates for irrigation.[198]

All this implies that the Incas understood the hydrologic cycle in the Andes and, like the ancient Egyptians and Sumerians, practiced a careful form of water-level measurement and control.

THROUGHOUT INCAN CULTURE, the water serpent Amaru is also a symbol of chaos, catastrophe, violence, and change. In one account, the "house of the false god" was destroyed because "a serpent was eating away at its joints."[199]

A document from the archbishopric of Lima (colonial times, post-1532) more specifically connects the snake with earthquakes:

> There are some giant serpents which move beneath the Earth and have a habit of making mountains fall, and when the said mountains topple and fall, they say that it is this guayarera [serpent] which demolished them.[200]

This belief was shared by the Chorti Maya along the borders of Guatemala, Honduras, and El Salvador, who attributed earthquakes to the movement of giant snakes under the earth.[201] William Sullivan, an Incan scholar, suggests we can equate the sinuous form of the serpent to the now well-studied S curve of an earthquake's shock waves. ("This equation seems like a reach," comments Seaman.) According to geologist Tarbuck, the secondary, lateral S waves of an earthquake travel through the earth, vibrating material at right angles to the wave. Surface waves, says Tarbuck, cause the ground both to rise and fall and to move side to side "similar to an S wave oriented in a horizontal plane."[202] This is further reflected in the fine stone masonry that the Incas mastered to build their walls that buttressed agricultural terraces. According to Sullivan:

> The masonry itself, a mosaic of interlocking, irregular polygons, represents tens of thousands of manhours of labor. . . . Over the centuries, these walls have proved invulnerable to earthquake, which is to say triumphant over the S-wave, or Serpent, moving beneath the earth. In an earthquake, the ashlars of Inca terrace walls lock into each other, allowing the whole wall simultaneously to flex and cohere.[203]

The walls per se are worthy of comment, being made of gigantic stones put together with no mortar and so tightly fitted that one cannot wedge a knife blade into the joints.[204]

Pacific Islanders

Pacific Islanders lived in an environment plagued by earthquakes, volcanoes, hurricanes, and floods. They must have had an intuitive sense of the impermanence of land, even of mountains, and the fluidity of life. Their migrations that settled the Pacific were carried out over two millennia. Samoa and Tonga were settled around 1100 B.C.; the Cook, Marqueses, and Society Islands around 500 B.C.–A.D. 300; Hawaii around A.D. 300–750; Easter Island, 300–900; and New Zealand, 1000. Modern researchers have re-created this migration using replicas of oceangoing double sailing canoes.[205]

David Malo was the first Hawaiian to write down the oral traditions of pre-European Hawaii. Born sometime around 1793 (Cook came in 1778), he witnessed the transformation of his native Hawaii by the Europeans. His conversion to Christianity certainly biased his judgment in places, but he scientifically listed Christian and pre-Christian explanations side by side in his writing. Early Hawaiians depended on oral histories, passed down through specific oral "genealogies" detailing human history, nature, and the creation.[206]

Malo reveals that his fellow Hawaiians clearly knew the difference between igneous and volcanic rocks: "In the mountains were found some very hard rocks which probably had never been melted by the volcanic fires of Pele [goddess of the volcano]." Noting that these rocks were used for axes, Malo adds that "all of these are very hard, superior to other stones in this respect, and not vesiculated [filled with gas-bubble holes] like the stone called *ala.*" However, "volcanic *pa-hoe-hoe* [lava] is a class of rocks that have been melted by the fires of Pele. Ele-ku and anna, pumice, are very light and porous rocks." Malo's translator, Nathaniel B. Emerson, disputes him in part, however, saying that Hawaiians used a fine-grained basalt, a volcanic rock, for axes. Hawaiians also distinguished between vesiculated rocks, coral, sandstone, and "a stone that is cast down from heaven by lightning." (Probably fulgurites.)[207]

Malo describes the origin of the Hawaiian Islands and land: "In the genealogy called *Kumu-lipo* it is said that the land grew up of itself, not that it was begotten, nor that it was made by hand."[208] The Hawaiians knew that their islands went down into the sea, and they believed there to be a sea floor, but they visualized the islands themselves as floating, unconnected to the floor.[209] This ties in with Malo's description of Hawaiian first reactions to the large ships of the Europeans: they called them *moku,* meaning "an island, a piece cut off." He writes, "A ship was like a section of the Earth quietly moving through the water."[210] The idea of landmasses, islands, floating over the ocean floor, presaged the theory of continental drift. "This is an impressive early statement of the gist of the continental drift idea,"[211] says Seaman. It shows, at least, that Pacific Islanders did not view the solid land as stable or stationary, though their view differs from present-day theory. We now know that the continents do not plow through the ocean floors but are part, along with the ocean floor, of rigid sixty-mile-thick plates that jostle against each other.[212] (A few modern geologists still do not accept the idea of continental drift, despite the evidence.)

Hawaiians also believed that the islands came from the ocean floor, even if unconnected. They saw the sea floor as slimy and silty, a place where coral grew. The story is told that Kapu-he'e-ua-nui (a demigod) fished until he pulled up a piece of coral. When he prayed over it, it turned into Hawaii. He kept fishing until all the Hawaiian Islands were pulled to the surface.[213] In fact, the Hawaiian Islands did rise out of the sea, and atoll coral reefs would have been very familiar to the Polynesians and the Hawaiians. Another genealogy, called the *Pali-ku* (Precipice) suggests the formation of land through earthquakes: "rising of the mountains by earthquake."[214]

THE ROCK-FORMING and rock-destroying properties of volcanoes must have been a fact of life for Hawaiians. Many myths tell of the Hawaiian volcano goddess's unpredictable violent love and temper. In the myth of Pele and Hiiaka, her beautiful younger sister, Hiiaka falls in love with Lohiau, a human. Pele therefore needs to punish him. She sends her fires on him, and he dies.

Hiiaka goes searching for him in the land of the dead. She rips through the ten strata of the land, down to Pele's domain: "She came at last to the tenth stratum with full purpose to break up this also and thus open the flood-gates of the great deep and submerge Pele and her whole domain in a flood of waters." This can be compared to the oceans pouring into a volcanic explosion, and it shows the Hawaiians' knowledge of the layers of earth, ending in the molten fires of a volcano.[215]

Meanwhile, the spirit of Lohiau is wandering in the caldera of Kilauea. Lohiau's friend hears Hiiaka searching for him. His chant suggests the current geologic view of an earth composed of a thin rocky crust underlain by fiery magma:

The world is convulsed; the Earth-plates sink
To the nether domain of Wakea;
Earth's rooted foundations are broken;
Flame-billows lift their heads to the sky. . . .
There's pelting of heads with falling stones
And loud the clang of the smitten plain,
Confused with the groan of the earthquake.
Yet this cools not the rock-eater's rage:
The Goddess grinds her teeth in the Pit.
Lo, tilted rock-plates melt like snow.[216]

Seaman calls the first line's reference to Earth-plates "very impressive."[217] The Hawaiians originally came from southeast Asia (probably Java) and moved from island to island until they went far into the Pacific. At that point, according to historian Michael Kioni Dudley, their world concept started to change. As they sailed the Pacific, they saw the islands disappearing from sight behind them. When they headed into the prevailing easterly winds, they viewed it as climbing. When sailing downwind to the southwest, they felt as if they were descending. "They possibly came to picture their world not as round," speculates Dudley, "but rather as inclined. At some time they also came to see it as bowed." Therefore, "the ancestral homes of the Hawaiians lie over the horizon and 'beneath' Hawaii." That is, to the south and west.[218]

The Mangaians, in the Cook Islands, pictured the world as perched on top of a coconut, inside of which were their ancestors, and ending with a great taproot at the bottom. The surface of the coconut was curved, as the sea's horizon is curved when seen from an island. Drawing this as a map, the Mangaians created diagrams that seem fantastic, unless one conceives them to be projections of a three-dimensional reality onto a two-dimensional surface.

Similarly, the Tuamotu islanders drew the universe in circular levels. They saw the places they had come from as lying below them—that is, over the horizon and down. Their drawings show people climbing up the "sides" of the world. Actually, these are navigational charts showing how the Tuamotu people originally came from Tahiti and other places, all "below the horizon."[219]

Australian Aborigines and New Guinea

The Australian Aboriginal version of geology is less than rigorous. To wit: the world was built by "Creative Ancestors" who shaped the surface of the earth by "their hunting, loving, and fighting."[220] What the Aborigines offer, on the other hand, is a phenomenal memory of the land. They are, by some accounts, the oldest continuous culture in the world. "So strong is the Aboriginal bond to the land," claims researcher Stanley Breeden, "that through oral history, some groups apparently still know the locations of sacred sites now under the sea that were exposed during an ice age."

People have lived in Arnhem Land in northern Australia, now encompassed by Kakadu National Park, for twenty-three thousand years, part of an older culture going back forty thousand years. Their ancestors putatively came from Southeast Asia when because of an ice age, the sea was lower. About 50,000 years ago, Australia and New Guinea were joined by a landmass.[221] Is it possible that the Aborigines knew that the land had once been above water? Did they remember the ice age?

The Murngin people in Arnhem Land recall their history by associating it with the landscape. Given that the Murngin had no written words or man-made monuments (as the Maya had) to aid their memory, they relied on the "script of sacred geology," according to two researchers, David Suzuki and Peter Knudtson: "To them, the land itself is a living, ever accessible repository of their memories of the Creation Time order. . . . Within their world view, place itself is the mnemonic of significant events and of personal and group history."[222] This can be viewed, perhaps, as a precursor to seeing the physical earth in historical terms, as it is etched in the stratigraphic rock record or in the formation of land features. The Aborigines seem to have a strong sense of the earth's history as reflected by the landscape.

Let us conclude with a view of the earth from the Huli, a tribe living in Papua New Guinea. They were indigenous to the region before whites arrived around the 1940s. Traditional Huli believe that the earth was created by a volcano from the sun, which threw volcanic stones around the universe. One of these became the earth. Their belief expands to say that all creatures are interrelated and all have their origin from the sun.[223] This creation myth is surely more sophisticated than early Western anthropomorphic stories or the biblical version that just pops the earth into being by God's word. The Huli myth implies the formation of the earth from a molten magma and implies the sun is at the center of the solar system. Current geologic theory indicates that the sun and the planets were formed from a giant spinning gas-and-dust cloud.[224] The Huli are not far off.

7

CHEMISTRY
Alchemy and Beyond

ANTOINE-LAURENT Lavoisier (1743–1794) was a financier, established a system of weights and measures that led to the metric system, lived through the early turmoil of the French Revolution, and was a pioneer in scientific agriculture. He married a fourteen-year-old girl and was beheaded during the Reign of Terror. He has been called the father of modern chemistry, and, in the course of his busy life, he brought Europe out of the dark ages of that science.

One of Lavoisier's early contributions resulted from his boiling water for long periods of time. In eighteenth-century Europe, many scientists believed in transmutation. They thought, for instance, that water could be transmuted into earth, among other things. Chief among the evidence for this was water boiling in a pot. Solid residue forms on the inside surface. Scientists proclaimed this to be water turning into a new element. Robert Boyle, the great seventeenth-century British chemist and physicist who flourished a hundred years before Lavoisier, believed in transmutation. Having watched plants grow by soaking up water, he concluded, as many had before him, that water can be transformed into leaves, flowers, and berries. In the words of chemist Harold Goldwhite, of California State University, Los Angeles, "Boyle was an active alchemist."

Lavoisier noticed that weight was the key, and that measurement was critical. He poured distilled water into a special "tea kettle" called a pelican, an enclosed pot with a spherical cap, which caught the water vapor and returned it to the base of the pot via two handlelike tubes. He boiled the water for 101 days and found substantial residue. He weighed the water, the residue, and the pelican. The water weighed exactly the same. The pelican weighed slightly less, an amount equal to the weight of the residue. Thus, the residue was not a transmutation, but part of the pot—dissolved glass, silica, and other matter.[1]

As scientists continued to believe that water was a basic element, Lavoisier performed another crucial experiment. He invented a device with two nozzles and squirted different gases from one into the other, to see what they made. One day, he mixed oxygen with hydrogen, expecting to get acid. He got water. He percolated the water through a gun barrel filled with hot iron rings, splitting the water back into hydrogen and oxygen and confirming that water was not an element.

Lavoisier measured everything, and on each occasion that he performed this experiment, he got the same numbers. Water always yielded oxygen and hydrogen in a weight proportion of 8 to 1. What Lavoisier saw was that nature paid strict attention to weight and proportion. Ounces or pounds of matter did not disappear or appear at random, and the same ratios of gases always yielded the same compounds. Nature was predictable . . . and therefore malleable.

Ancient Chinese alchemy, circa 300 to 200 B.C., was built around the concept of two opposing principles. These could be, for example, active and passive, male and female, or sun and moon. The alchemists saw nature as having a circular balance. Substances could be transformed from one principle to another, and then rendered back to their original state.

A prime example is cinnabar, known commonly today as mercuric sulfide, a heavy red mineral that is the principal ore of mercury. Using fire, these early alchemists decomposed cinnabar into mercury and sulfur dioxide. Then they found that mercury would combine with sulfur to form a black substance called metacinnabar, "which then can be sublimed into its original state, the bright red cinnabar, when once more heated," according to science historian Wang Kuike. Both mercury's liquid quality and the cyclic transformation from cinnabar to mercury and back again gave it magical qualities. Kuike calls mercury "*huandan,* a cyclically transformed regenerative elixir" associated with longevity. These ancient practitioners became familiar with the concept that substances could be transformed and then come full circle to their original state. They developed exact proportions of the amounts of mercury and sulfur, as well as recipes for the exact length and intensity of the heating required. Most important, according to Kuike, these operations could be performed "without the slightest loss of the total weight."[2]

It would appear that the ancient Chinese alchemists were empirically familiar with the conservation of mass fifteen hundred years before Lavoisier's experiment. He and his alchemist precursors discovered that the weight of the products in a chemical reaction equal the weight of the reactants.[3]

The earliest alchemic text is Wei Po-Yang's *Ts'an T'ung Ch'i* (Unification of the Three Principles), written around A.D. 140. The work describes an experiment, very likely the cinnabar-mercury-sulfur reaction described above. It is difficult to be certain, because the chemicals going into the fire are called by metaphorical names: White Tiger (probably mercury), Blue Dragon, and Gray Dragon (sulfur?).[4] More important is the vessel they used:

> On the sides [of the apparatus] there is the walled enclosure, shaped like a peng-hu pot. Closed on all sides, its interior is made up of intercommunicating labyrinths. The protection is so complete as to turn back all that is devilish or undesirable. . . . Like the moon lying on its back is the shape of the furnace and the pot. In it is heated the White Tiger. Mercury Sun is the flowing pearl, and with it the Blue Dragon. The east and west merge together, and the *hun* and *po* [two kinds of souls] control one another. . . . The Red Bird is the spirit of fire and dispenses a victory or defeat with justice. With the ascending of water comes the vanquishing of fire.[5]

The vessel is used for melting and subliming different metals. The instrument is similar to and more complex than Lavoisier's pelican, designed to "turn back" all products to ensure the conservation of mass.

THE HISTORY of chemistry, Western and non-Western, runs counter to the history of physics. The latter contains a cornucopia of theory, with experiment lagging far behind. In chemistry we see a fascination with empirical knowledge, experimentation with all variety of substances (liquids, solids, gases), using all sorts of methods (fire, boiling, distillation), but without a solid theoretical framework to guide the experimentation. The movie image of the frizzy-haired scientist in his lab sloshing together beakers full of brightly colored chemicals is not too far off. Chemistry has been a science of trial and error. The theory has not always been of the highest quality.

The West developed a coherent theory that predicted which elements will combine with each other and which will not, why some compounds are possible and others aren't, and what precisely will happen when one chemical is combined with another. In addition to Lavoisier there were two great pioneers in this area.

IN 1869, at the University of St. Petersburg, the Siberian-born Dmitry Mendeleyev could find no good chemistry textbook to assign to his classes. He began writing his own. Like Lavoisier and the ancient Chinese, he saw chemistry as "the science of mass." He enjoyed playing patience, a variety of solitaire, so he wrote the symbols of the elements with their atomic weights on note cards, one element per card, with its various properties listed (e.g., sodium: active metal; chlorine: reactive gas).

Mendeleyev arranged the cards in order of increasing atomic weight. He noted an obvious periodicity (hence, the "periodic table of the elements," as his arrangement came to be called). Elements with similar chemical properties were spaced eight cards apart. Lithium, sodium, and potassium, for example, are all active metals (they combine vigorously with other elements, such as oxygen and chlorine) and their positions are 3, 11, and 19. Hydrogen, fluorine, and chlorine are active gases, and hold positions 1, 9, and 17. Mendeleyev rearranged the cards in a grid of eight vertical columns. Reading across, the elements get heavier. Reading down, the elements in each column display similar properties.

Mendeleyev did not feel compelled to fill in all the slots in the grid, knowing that, as in solitaire, some of the cards remain hidden in the deck. If a slot in the table called for an element with particular properties and no such element existed, he left it blank. Mendeleyev was widely ridiculed for these gaps in the periodic table. Five years later, though, in 1875, gallium was discovered, and fit in the space beneath aluminum, with all the properties predicted by the table. In 1886 germanium was discovered, and fit in the space beneath silicon. No one has laughed since. Mendeleyev never won the Nobel Prize in chemistry, though he was alive and elgible during its early years. However, three chemists who found "gap" elements did win: William Ramsay, who discovered argon, krypton, neon and xenon; Henri Moissan for fluorine; Marie Curie for radium and polonium.[6]

Growing up in the 1950s and '60s, I, like other students of that era, spent many hours staring at Mendeleyev's periodic table, hung on classroom walls across the country. The periodic table is less in evidence today, which is unfortunate since it inculcates in even the slowest mind the importance of atomic number, an element's placement in the periodic table. The striking qualitative differences among the elements—carbon seems little like hydrogen, or lead like helium—are, on a basic level, differences in atomic number, which we now equate with charge on the nucleus.

The meaning of the periodic table and its regularities and repeating patterns remained hidden until the early twentieth century, when the atom was dissected, and physicists found electrons inside and a nucleus comprising protons and neutrons. Elements differed from one another because of the number of protons and neutrons in the nucleus and the number of electrons whizzing around them. Quantum theory ensued.

ONE OF THE PIONEERS of the quantum heyday (1900 to 1930) was Wolfgang Pauli. He didn't intend to solve the mystery of the periodic table; he was simply trying to understand the atom. Pauli was famous for his acerbic sense of humor. He spared no one. When the celebrated physicist Victor Weisskopf, at the time Pauli's assistant, presented him with a theoretical effort, Pauli said, "Ach, that isn't even wrong!" Pauli also sent a letter to Albert Einstein, recommending a student as an assistant. "Dear Einstein," Pauli wrote, "This student is good, but he does not clearly grasp the difference between mathematics and physics. On the other hand, you, dear Master, have long lost this distinction."

In 1924, Pauli announced the exclusion principle: no two electrons can occupy the same quantum state. It explained the order in Mendeleyev's table, and why we can use it to predict which elements combine with which and how. I won't go into the specifics of what constitutes a quantum state here. Suffice it to say that Pauli's exclusion principle limits the numbers of electrons in what we now call the "shells" of each atom: two in the first, eight in the second, eighteen in the third, and so on. Hydrogen, for example, has but one proton in its nucleus. To balance its single positive charge, we need one electron (negative charge), which occupies the lowest energy state, or orbit. Next in the table is helium. Its nucleus has two plus charges, so we need two electrons, which, according to Pauli's principle, fit together in the first shell.

When we get to lithium, and its three positive charges in the nucleus, we need three electrons. Two go in the first shell, but the third must be put in the second shell. This shell has a much larger radius than the first, and with only one of eight electron slots filled, we can see why lithium is an active metal, combining with other atoms with ease. When the outermost shells are filled, it is impossible to add an electron. The electromagnetic resistance is huge. When there are slots open, it's time to do business.[7]

The *Hindenburg* blimp is a prime example of this principle. Its tragic explosion over Lakehurst, New Jersey, in 1937 illustrates the Pauli principle. The United States had refused to export helium to Germany, so German dirigibles were inflated with hydrogen. Helium is safer because its two electrons fill its shell, making it an inert gas. Hydrogen has only one electron, making it an active gas, a fact that was evident when the *Hindenburg* went up in flames.

Hydrogen and helium are dissimilar despite differing by only 1 in atomic number. The vertical columns in the periodic table, on the other hand, contain elements whose outermost shells hold the same number of electrons, and thus these elements have similar chemical properties.

Thanks to Lavoisier, Mendeleyev, Pauli, and many others, seventh graders can do and understand experiments that would have seemed like magic to chemists working only a few centuries ago. Only in the past three-quarters of a century, thanks to Pauli, have we understood why chemicals mix and react as they do. It becomes clear to us why sodium and chlorine can combine to form salt, or hydrogen and oxygen to make water. I'm speaking theoretically. Not everyone grasps this valuable knowledge.

ALCHEMY IS NORMALLY associated with the pseudosciences and primitive, superstitious cultures. In a narrow sense, alchemy is the attempt to turn lead or other base metals into gold. Another goal of alchemists was to find an elixir of eternal youth. As we shall see here, alchemy can also be defined as an early form of chemistry.

The ambition to turn lead into gold is not so crazy. As we've seen, atomic number is the key to chemistry, and lead's number is similar to gold's, the elements being close to each other on the periodic table (atomic numbers 82 and 79, respectively), though, of course, the ancients didn't have a periodic table.

One of the first Nobel Prizes in chemistry was awarded in 1908 to Ernest Rutherford, who discovered that through radioactivity some elements transform into others. The elements are alterable. Listen to Rutherford's famous exchange with his collaborator Frederick Soddy.

> *Soddy:* "Rutherford, this is transmutation."
> *Rutherford:* "For Mike's sake, Soddy, don't call it transmutation. They'll have our heads off as alchemists."[8]

Rutherford went on to transmute elements in another way, bombarding them with particles to break off protons, "smashing the atoms" to turn them into lighter elements.[9]

In 1938, the Italian physicist Enrico Fermi won the Nobel Prize, in part for supposedly discovering new radioactive elements heavier than uranium. Fermi had bombarded uranium, atomic number 92 on the periodic table, with slow neutrons, and had produced two mysterious substances, which in his Nobel acceptance lecture he called "ausonium" and "hesperium," elements 93 and 94. Actually, Fermi had split the uranium atom into lighter elements, not added neutrons to make heavier elements. He had unknowingly produced fission. In 1939, Otto Hahn and Fritz Strassman, with some interpretative help from Lise Meitner and Otto Frisch, split the uranium atom, and realized they had achieved fission.[10] (Fermi's prize was well deserved; he was a great physicist, a "god" in the parlance, and had many other Nobel-level achievements.) In 1940, Edwin McMillan and Glenn Seaborg accomplished what Fermi's work had hinted at. They created the transuranic elements neptunium and plutonium via bombardment. They won the Nobel Prize in chemistry in 1951.[11]

We will meet ancient and medieval chemists who believed in alchemy and transmutation. Their ideas have been championed, if reluctantly, in our era by such men as Rutherford, Fermi, McMillan, and Seaborg.

Prehistoric peoples' first dabbling in chemistry involved any process that transformed initial ingredients: cooking, dressing hides, working metals, making medicines, painting and dyeing, and making pottery. Eduard Farber, a historian of chemistry, defines it as "the selection, separation, and substantial modification of materials.[12] In the ancient world, fire, the most visible agent of transformation, was at the heart of chemistry.[13] Water also holds an important role in early chemistry as the principal dissolving medium. An exact understanding of what early peoples knew about chemistry is hampered by the numerous names used for the same substance, by a single name denoting vastly different substances, and by the unknown amount of impurities involved in any process, even when the terms are clear.

Alchemy may have been denied its proper place in the history of chemistry because of an inability to interpret the religious, philosophic, and symbolic traditions that embodied its knowledge. Alchemists' fondness for metaphor, such as two lions being emblematic for sulfur, made it unintelligible to literal-minded people. In the chemist John Read's words, "It is easy to despise something which one makes no effort to understand."[14] Ages hence, one wonders what scholars will make of some of our "modern" terms in physics and astrophysics: Winos, WIMPS, quarks, dark matter, the eightfold way, superstrings, big bang, and the like.

Alchemists pioneered one technique that laid the foundations for much of modern chemistry: they experimented. The mystical-religious thinking surrounding alchemy also played a significant role, giving rise to beliefs that were later to become precepts of modern chemistry: conservation of matter, phase changes, and energy transfer. Alchemic mysticism connected the transformation of solid substances into liquids and vapors with the transformation of the human body into the soul. Sublimation, in which a solid converts directly into a vapor, in particular seemed analogous to the spirit leaving the body, and the magical recrystallizations out of a melt or vapor were connected with ideas of reincarnation and rebirth.

Alchemy was philosophy. Alternatively called hermeticism, alchemy had as its primary intention the regeneration of the human soul from its present sensory-dominated state into its original divine condition. It was about raising the life essence of things—metals in particular—to a nobler form.[15]

It's unclear where alchemy originated; some scholars say it began in Egypt and was filtered to China; others say it began in China. Alchemic ideas are seen in Hindu writings from 1000 B.C. in the Arthava Veda. In any case, alchemic ideas were seen far earlier in India, Egypt, and China than in Greece.[16] Let us start with Egypt.

Egypt

The Greeks thought of Egypt as the source of earliest alchemy, and they admired the ancient Egyptian skill in "enameling, glass-tinting, the extraction of plant oils, and dyeing—all dependent on chemical knowledge," according to John Read. "For such reasons, Egypt, or Khem, the country of dark soil . . . has often been pictured as the motherland of chemistry," he writes.[17] The word *chemeia,* Greek for a "preparation of silver and gold," may have roots in *Khem.* Other sources claim that *chemyia* has origins in word meaning to pour,[18] while scholar Bruce Bynum says that *Kem,* and *al-kemit* (as in "alchemy") referred not to Egypt's soil but to the earliest people of the upper Nile who established the Nubian, or Kemetic, civilization. They were black Africans. Hence, to the Greeks *Kem* came to mean "land of the blacks."[19] Other researchers believe that the Egyptian ideas came from Persian, Chaldean, and Hebrew sources.[20]

Third-century A.D. papyri from Thebes, copied from even earlier texts, may be the earliest records of alchemic schemes for turning base metals into silver and gold. The papyri contained little theory but much practical information on the alloys going into different metals.[21] Alchemy was, however, related to Egyptian philosophical and religious views, and, in a sense, to mummification, according to Eduard Farber:

> Egypt, the land of the black earth, was devoted to the cult of the dead. The god Osiris is revived from death after he has been ritually wrapped in bandages. To the Egyptian's mind, this indicated a valid analogy to the fact that minerals are bandaged and entombed in black lye to revive them into metals.[22]

Priests presided over this "embalming" of metals, which came from the dark earth to be transformed (they hoped) into gold. The Egyptian practice of embalming the dead was a logical extension of this belief and, according to Farber, an indication of how matter acted powerfully on the human spirit and life. Life, spirit, and physical elements were interconnected in ancient Egyptians' minds.[23]

This concept of material transformation must have fascinated early Egyptians as they watched metals change color and form after heating or undergoing other processes. The later Alexandrian alchemists (fourth through seventh centuries A.D.) emphasized a color progression in making gold: black was the first stage, from fusing "base metals" such as lead, tin, copper, and iron, or lead and copper with sulfur; bleaching was next, accomplished by firing the black compound with arsenic, silver, mercury, antimony, or tin. Next the substance was yellowed using gold or a lime-sulfur mixture. Finally the color violet prevailed. Violet-colored gold seems odd (and less than authentic) to us, but to the Egyptians this was a kind of heightened gold, the essence of gold, something that was seen as so powerful that it acted like "a yeast" to transform the metal into a spiritual substance.[24] The concept of yeast was seminal in ancient thinking, signifying something very tiny that causes huge changes. In a sense yeast is a precursor to chemical ideas of catalysts or enzymes and is related to the Chinese and later Arabic elixirs of life.

Keeping in mind the Egyptian connection of physical matter with spirit, the vapor of distillation during these processes was associated with spirit, while the remaining "base" material was the body, the corpse. This clearly relates to the chemical process of sublimation, in which solid matter directly turns to gas. The transformation of substances is an early way of thinking about the phase changes of matter from solid to liquid to gas.[25]

EMBALMING WAS the first step to take the human spirit from its dead body to reincarnation. The Greek Herodotus (fifth century B.C.) describes the process:

> First they draw out the brains through the nostrils with an iron hook, raking part of it out in this manner, the rest by the infusion of drugs. Then with a sharp stone they make an incision in

> the side, and take out all the bowels; and having cleansed the abdomen and rinsed it with palm wine, they next sprinkle it with pounded perfume. Then, having filled the belly with pure myrrh, cassia and other perfumes, they sew it up again; . . . they steep it in natron, leaving it under for seventy days. . . . At the expiration of seventy days they wash the corpse, and wrap the whole body in bandages of waxen cloth, smearing it with gum, which the Egyptians commonly use instead of glue.[26]

The natron in which the body is steeped occurs naturally in Egyptian lakes.[27] Chemists debate what natron was; some label it a sodium carbonate and bicarbonate precipitant,[28] others a sodium aluminum silicon oxygen salt.[29] It is now believed that the steeping in natron killed bacteria and dehydrated cells, while wrapping the corpse and sealing it in a tomb kept it from moisture and air. "All in all," write Cathy Cobb and Harold Goldwhite in their book *Creations of Fire: Chemistry's Lively History from Alchemy to the Atomic Age,* "the process was not much more mysterious than salting pork."[30] (The Hawaiians also preserved bodies by gutting and filling them with salt obtained from evaporated seawater. A body treated in this manner was termed *ia loa,* long fish.)[31]

As MUCH AS GOLD was revered in ancient Egypt, the occupation of goldsmith was not appealing to everyone. An ancient instruction book tells how Dua-Khety, a man living in the Middle Kingdom in Sile, wished to place his son Pepy in writing school rather than have him seek a profession as a goldsmith. "I have seen a smith at work before his furnace door, his fingers like [the claws of] crocodiles. He stinks more than fish roe."[32] The smell refers perhaps to the fumes resulting from the many chemical procedures done to gold, the curled fingers perhaps a symptom of heavy-metal poisoning. Still, this did not deter others from falling prey to the allure of gold.

The desired end product of alchemy, gold, is found pure in nature and dates from the Stone Age of Egypt. However, the earliest Egyptians could not separate gold from silver. Sometimes Egyptian gold was so rich in silver it seemed a different metal, variously dubbed white gold, asem, or electrum. The nineteenth-century French chemist Marcelin Berthelot analyzed artifacts from the Twelfth Dynasty (ca. 2000 B.C.) and found the metal to contain around 85 percent gold and 15 percent silver.[33] Later (circa 1300 B.C.) methods of separation involved heating the gold-silver alloy repeatedly with common salt, which eventually transformed the silver into silver chloride that would pour off into the slag.[34]

The two papyri of alchemic recipes found at Thebes were first translated into Greek and then into Latin, the version Berthelot analyzed. The papyri are a collection of chemical recipes for making metallic alloys; producing imitations of gold, silver, or electrum; dyeing; and other related arts.[35] The formulas are unabashed about their intended deception, so as to whether Egyptians saw no contradiction between "real" and "fake" gold, or whether the writer(s) were of a more practical nature than the priests, we can only speculate. Berthelot explains:

> The parts dealing with the metals are largely concerned with producing passable imitations of gold, silver or electrum from cheaper materials, or with giving an external or superficial color of gold or silver to cheaper metal. . . . There are often claims that the product will answer the usual tests for genuine products, or that they will deceive even the artisans.[36]

Here is a recipe for fake silver (amalgamated tin):

> Tin, 12 drachmas [3.411 grams]; quicksilver, 4 drachmas; earth of Chios [white clay], 2 drachmas. To the melted tin add the powdered earth, then add the mercury, stir with an iron, and put it into use"[37]

To make artificial pearls:

> Mordant [fix] or roughen crystal in the urine of a young boy and powdered alum, then dip it in "quicksilver" and woman's milk.

"Crystal," in this case, probably refers to softer, absorbent, transparent stones rather than to quartz, which would not soak up the solution. "Quicksilver" is probably fake mercury, made perhaps of mica or fish scales.[38] Today, of course, it is more problematic obtaining young boys' urine and woman's milk. Urine, an alkali, was mixed with alum (potassium aluminum sulfate) to fix colors. We're not sure of the purpose of woman's milk.

THE MAKING of skin creams and perfumed oils was highly developed in Egypt, as well as in Mesopotamia. Salves, ointments, oils, eye shadow, and nail paint protected the skin from the desert environment and had religious significance. In 2450 B.C., the sage Ptah-Hotep wrote, "If thou art a man of standing, thou shouldst found thy household and love thy wife at home as is fitting. Fill her belly; clothe her back. Ointment is the prescription of her body."[39]

Animal fat or vegetable oils from castor, colocynth, lettuce, linseed, olive, and safflower served as the base to which aromatic oils were added. Oil from anise, cedar, cinnamon, citron, mimosa, peppermint, rose, and rosemary were popular. The method of extracting the oils is uncertain but could have involved boiling mashed herbs in a pot covered with fat-impregnated cloth, the fat then soaking up the scent, a method still used by peoples along the Nile. Other methods include steeping flowers in fat until the odor was taken up by the fat, and dipping flowers in hot oils and straining off the liquid. Egyptians most likely expressed the oil by squeezing the ingredients in a cloth bag using wooden sticks. R. J. Forbes, a historian of technology, indicates that "milk, honey, salts and such aromatic gun-resins and oleo-resins" were included in beauty products, some of which probably fixed the volatile nature of the oils.[40]

Cosmetic recipes were set down in a sixteenth-century B.C. text, the Edwin Smith Surgical Papyrus, which apparently was a copy of a much earlier document. Its begins by stating, "Beginning of the Book of Transforming an Old Man into a Youth," a sentiment close to modern people's use of cosmetics.[41]

> Recipe for transforming the skin: Honey 1, red natron 1, northern salt 1. Triturate [pulverize] together and anoint therewith.

Honey and milk were common bases for cosmetics in Egypt and Mesopotamia. Forbes compares this recipe to modern skin lotions that include "alcohol, glycerine, lactic acid (85 percent), water and perfume." The Egyptians also practiced dandruff control, using potions composed of ground and roasted barley, bran powder, and soft grease, topped off with applications of fish oil and hippopotamus fat. Given shortages of hippo fat, modern dandruff shampoos rely on soft greases such as beeswax, lanolin, petrolatum, olive oil, or liquid petrolatum to dissolve the dry cuticle.[42] The Egyptians also used powdered antimony and green malachite for eye shadow.[43]

NATRON WAS USED both as a bleach for linen and as a kind of soap mixed with clay. The importance of soap is demonstrated by the fact that natron and other soaps made from soda and castor oil were overseen by the Egyptian authorities, who taxed the launderers' use of these materials.[44] Wool was cleaned with ashes (which would provide alkaline carbonates as reagents) and clay-water, which would work together as an abrasive substance.[45]

The Egyptians concocted a wide range of dyes, including purple, red, rose, yellow, green, and blue, made from safflower, orseille, woad,[46] mulberry juice, pomegranate blossoms,[47] cinnabar, and iron oxide.[48] The prize of the Egyptians was kermes, the dried pulverized bodies of female scale insects they used to make red and purple dyes. Of course, kermes's history predates its Egyptian use. Pre-eighteenth-century B.C. Persians dyed their rugs with kermes, which is the root of the words "crimson" and "carmine."[49] Blue dye was also popular, but we won't go into the details of its fabrication here. Suffice it to say that, once again, it involves urine.[50] Also pivotal to Egyptian dyeing was something called the Phrygian stone, which, according to Pliny (writing in the first century A.D.), was a pumicelike stone soaked in wine and then heated three times. When wool was boiled with the Phrygian stone, mixed with algae, and washed in seawater, it would turn purple.[51]

West Africa

We know less about West African chemistry than Egyptian. In both, philosophy and religion play a role. There is a baseline belief, says researcher John Mbiti, in "a power or energy in the universe which can be tapped by those who know how to do so, and then used for good or evil towards other people."[52]

The Yoruba people, occupying what is currently southwest Nigeria, compiled a body of oral knowledge known as the *Ifá,* which is both a set of scriptures and a deity—that is to say, a representation of God consciousness, through whom the Yoruba learned wisdom. *Ifá*'s origins are unknown and undoubtedly involve a mix of many African traditions. Some claim (this is controversial) that *Ifá* was brought to West Africa from Egypt sometime between 2000 and 500 B.C.[53]

The Yoruba have numerous deities, but all deities are considered to be the energetic vibrations of a single God. Each deity has a different set of herbs associated with it, whose energetic vibrations are seen as similar to the god's own energy.[54] We believe something very similar today. Every atom (or molecule) vibrates at specific frequencies, thus allowing us to identify chemicals through spectroscopy.

The *Ifá* belief system is reminiscent of Chinese yin-yang dualities and the four elements of the ancient world. The Yoruba see their universe as enclosed in a sphere or a calabash, the traditional bottle-gourd. In each quadrant are the four elements of the ancient system: earth, water, fire, and heaven (air). Heaven and fire are associated with light, masculine, positive qualities; water and earth with dark, feminine, and negative qualities. The lack of balance between these elements is seen as the cause of disharmony.[55] If one wishes to stretch the concept, we can connect it to modern ideas of the role of positive and negative ions in bonding chemical materials.

To another West African people, the Ga, who inhabited modern-day Ghana, numerous gods were known as *dzemanon*. Lesser beings connected with pharmacology and magic were called *won*. According to the researcher M. J. Field, a *won*, usually connected to a specific healing herb, "is something that can act but not be seen." These are nameless spirits that will act "for anyone, provided that the person has observed the proper ceremonies in becoming owner of the medicine, and provided he is careful about any taboos [contraindications?] attached to its use."[56]

West Africa has a rich textile tradition that draws on knowledge of dyes and dyestuffs. In Nigeria, traditional dye sources included guinea corn, teak, henna, African rosewood, kola nuts, palm roots, ash, and the leaves of many plants.[57]

One of the most celebrated African dyes is indigo blue, made from the leaves of several different plants including *Indigofera tinctoria*, which were fermented and then fixed with limestone or ash.[58] Yoruba dyers pounded and dried the leaves, pulverized them, and put them in a pot with a hole in the bottom. Ashes were added and then water percolated through the whole mixture, into a second pot below. The substance in the second pot was fermented until dye collected on the top of the fermentation vat.[59]

In the Sokoto, Kano, and Bida regions of Nigeria, the fermentation pits were often three feet wide and seven to twelve feet deep, able to take some three hundred gallons of dye bath. The pits were made watertight with a paste of ant heaps, ash, and goat hair, fermented locust bean pods, or clay smeared on the walls. The leaves were mixed with ash and left for a week. Then the fabric was added and retrieved, the water pressed out, and the cloth exposed in air to allow oxidation to occur.[60] The dyeing property of *Indigofera tinctoria* was known throughout much of the Middle and Far East as well. Used in Oriental rug dyeing in Baluchistan and Bengal, it represented one of the most complex processes, involving both fermentation and oxidation.[61]

Modern food processing, according to Nigerian scholar Richard

Okagbue, has its origins in ancient times. The main difference is the modern understanding of the biological and chemical principles behind the processing.[62] African food preparation represented a long and complex tradition of fermentation.

Cassava in its natural state is toxic (though not often fatal) because of the linamarin in the fresh tuber, yet it is a staple of Nigerian cuisine. Traditionally, cassava root is grated and placed in porous bags, then pressed by heavy blocks to dejuice it. The underlying biochemical process involves a bacterium that eats the starch to create organic acids, which then release the poisonous hydrogen cyanide from the linamarin. A second bacterium then creates aldehyde and esters that sweeten the flavor.[63] The prevailing theory is that people ate cassava only when safer food was unavailable. If cooks covered the unprepared root as it boiled, they stood a good chance of developing cyanide poisoning, which was chronic in peoples relying heavily on cassava.[64]

Ancient African physicians made no distinction between the characteristics of a healing plant and the effect of that plant. They described the medicinal qualities in qualitative terms: smell, taste, feel (dry, moist, hot, cold). If the plant produced these qualities, that was all that was important, not the mechanism behind the effect. In the eighteenth century, the Western world began to focus on measurable and objective criteria. Western pharmacology isolated the active chemical principle and believed that such a principle was entirely responsible for the cure.

Non-Western and ancient peoples took a more holistic approach, believing that the plants were inextricably connected to the world as a whole. It was a plant's "effective energy" or "energetics"—its ability to connect the patient with this larger whole—that was deemed the cause of the cure, not some inherent physical substance within them.[65]

The Ga people have a vast pharmacopoeia, compiled through the ages of herbs known to be useful for certain conditions, such as the *sese* root's ability to dull pain and cure insomnia. Chemical analysis of these herbs has shown clear pharmaceutical elements. Preparations include infusions, solution in fermented liquid (rum), vapor baths, and a kind of injection in which medicines are rubbed into cuts.[66]

The Ga people believe that the power of the herbs and mixtures used by the medicine men, however, derive not from their own qualities but from the invisible spirits emanating through them. There is no separation in Ga thinking between the biological action of food and drugs and the spiritual realm of healing. To the Ga, food keeps a person alive by pleasing the spirit of the person, his *kla*. If the *kla* does not wish to accept a certain food or medicine, the person will not heal, no matter how efficacious the substance is for others.[67] In Western terms, chemicals are thought to have the effects they do because of inherent characteristics. Westerners also believe it is the chemical essence or agent in medicinal plants that effect the healing. This "essence" is not so far from the Ga concept of the *won*.

With the Ga people, medicinal herbs are burned to a powder in a pot, then combined with various objects, such as clay or feathers, which are believed to control the *won*. When a patient is sick, the medicine man makes a tea of the same fresh herbs but sees the *won* as the source of the tea's power, not the tea per se. The tea merely makes clear the connection between the patient and the *won*.[68]

New medicines occur when the *won* speaks to the medicine man. M. J. Field writes, "Anyone will tell you that 'every herb is a medicine for something' but nobody knows *what* until the herb itself or a *won* talks and reveals. Everyone agrees also that only a clever medicine man can hear this talking. . . . [However] you gather that the 'talking' is only a metaphorical talking . . . that an object by its shape, colour, and function will suddenly suggest to the medicine man's mind connections between itself and the disease to be cured. For instance, seeds shaped like fingers, roots like hair . . . will suggest their use in doctoring whitlowed fingers, ringwormy hair."[69]

A modern African shaman's autobiographical account of his initiation into the ancient ways of the West African Dagara (who occupy modern Burkina Faso) claims that an expanded concept of consciousness, something that many Westerners would call hallucination or psychosis, allows shamans to identify the combinations of plants needed for healing:

> My perceptions had become hyperbolic. . . . I could see the different personalities of the trees. . . . Even their roots were visible to me. . . . I saw the medicine and the healing power in all of them.
>
> I remember the blind healer in the village who . . . was so skilled at conversing with trees that he baffled even his fellow medicine men. . . . The vegetal world would awake in the middle of darkness, every tree and every plant—all speaking to the man at once. . . . He would translate, telling each patient that such and such a tree said his fruit, dried and pounded and then mixed with salted water and drunk, would take care of the disease in question. Another plant would say that it couldn't do anything by itself, but that if the patient could talk to another plant (whose name the healer knew) and mix their substances together, their combined energies could kill such and such an illness.[70]

Other than the adding of salt to the dried fruit, which chemically can increase the amount of the active plant essence extracted, most scientists would not see a shred of logic in that description. However, the narrative illustrates well-known ties to human creativity and discovery. There is the classic tale of the German chemist Friedrich August Kekulé discovering the structure of benzene during a hypnogogic state. In 1862 he determined that benzene consisted of a cyclic ring composed of six carbon atoms each with a hydrogen atom spur.[71] His discovery was essential to the further development of organic chemistry, according to Pierre Thuillier in *Recherche.*[72] Kekulé's own description of the event:

> I turned the chair to the fireplace and sank into a half sleep. The atoms flittered before my eyes. Long rows, variously, more

> closely, united; all in movement wriggling and turning like snakes. . . . One of the snakes seized its own tail and the image whirled scornfully before my eyes. As though from a flash of lightning I awoke.[73]

As we shall see, South American shamans' use of psychoactive drugs parallels this experience even more insistently.

West Africans also understood stimulants. Cola comes from the West African plants *Cola nitida* and *Cola acuminata.* The seeds of the Kola nuts have been chewed for centuries by West Africans to counteract hunger and fatigue. Later, Westerners discovered the tree and transformed its extract into Coca-Cola, a significant source of caffeine in the Western diet. The cola seeds themselves have around 2 percent caffeine content. A later formula for Coca-Cola included not only *C. nitida* but also cocaine, derived from the Peruvian *Erythroxylon coca,* and wine.[74]

The poisonous Calabar bean (*Physostigma venenosum*) grows in West Africa and was used by the Efik people on the Calabar coast for executions, a kind of oral lethal injection. The Efik knew what dose would cause death, and knew as well how to take the dose so that vomiting would rid the body quickly of the poison.

Invading Europeans became interested in the toxin and in 1864 isolated physostigmine (eserine) as the active ingredient from the plant. Investigation proved that death was due to either paralysis or heart failure. Eventually they discovered physostigmine's use in treating glaucoma. In modern times the bean extract led to the discovery of acetylcholine, a neurotransmitter, and to the enzyme that activates the neurotransmitter that physostigmine blocks. Physostigmine has uses in treating myasthenia gravis, has potential use in reducing loss of memory in Alzheimer's patients, and was developed into a nerve gas by the German military in World War II.[75]

The Middle East

The best-publicized incident of Middle Eastern chemistry was performed by Moses, when he led his people out of Egypt and they wandered in the wilderness without water. According to Exodus 15:23–25:

> When they came to Marah, they could not drink the water . . . because it was bitter. . . . And the people complained against Moses, saying, "What shall we drink?" He cried out to the Lord, and the Lord showed him a piece of wood; he threw it into the water, and the water became sweet.

John W. Hill and Doris K. Kolb, in *Chemistry for Changing Times,* suggest a chemical reaction to account for the miracle. Desert water is often alkaline, and bases (alkali) taste bitter. A dead branch bleached by the desert sun would have undergone a chemical change so that the alcohol groups in cellulose would have oxidized to carboxylic acid groups. These acidic groups would neutralize the alkali in the water. Moses would not have known this chemistry, but his action implies that people of that era knew how to purify water and that certain agents were responsible for the transformation.[76] Granted, this is a flashy but somewhat trivial example. There is more substantial evidence of knowledge of chemistry in the Middle East, even theoretical chemistry.

The Sumerians, and later the Babylonians and Assyrians, were practical people, but they did speculate on alchemy and the nature of matter. A chemical connection between the spirit and physical matter can be seen in both seventeenth-century B.C. and seventh-century B.C. glass-making texts. Glass makers sacrificed human embryos in order to set up a confluence with the unformed souls in the other world to ensure that the glass and pottery process was completed successfully.[77] Similarly to the Egyptians, the Mesopotamians believed that taking metals and minerals from the ground disrupted the natural growth of the mineral in the earth's womb, and therefore required appeasement.[78] The Egyptians demonstrated this belief by burying dead children in pots and placing vases and food-filled pots in the graves to allow the child's spirit to eat the food on its way to reincarnation.[79]

Babylonian alchemists loved obscurity and secrecy. They used idioms and puns to describe their work. They were not popularizers. An exhortation written during the Kassite period (c. 1400–1155 B.C.) states, "Let him that knoweth show him that knoweth, (but) he that knoweth shall not show to him that knoweth not."[80]

Despite this secrecy, Sumerian cuneiform tablets show a clear use of prefixes to identify similar chemical and natural compounds. R. J. Forbes documents the naming of bituminous materials as follows: ESIR is the generic Sumerian term for bitumen or crude oil; ESIR.LAH the term for "white lake-asphalt"; ESIR.IGI for "shining . . . bitumen, asphaltite"; ESIR.UD.DA for "dry refined bitumen," and so on.[81] Here in the West, we are taught that Lavoisier and his collaborators were the first to invent a systematic method of naming chemicals, using prefixes such as *ox-* and *sulf-,* and suffixes such as *-ide* and *-ous. Martial ethiop* became *iron oxide; orpiment* became *arsenic sulfide.*[82]

THE ASSYRIANS, who displaced the Babylonians and in many ways continued their culture, were fascinated by acidic substances. They certainly knew of vinegar, and the term ZA.TU ("the acetated thing") was a prefix used as a descriptor of materials that would effervesce in contact with acid—for example, carbonate minerals. Calcite, argonite, white marble, and malachite are listed in Assyrian mineral lists, all with the ZA.TU prefix.[83] ZA.TU by itself refers specifically to white lead, a lead carbonate, which also would effervesce in vinegar.[84]

The Assyrian scholar R. Campbell Thompson posits that the Assyrians knew of sulfuric acid. Ancient peoples made early forms of sulfuric acid by heating iron sulfide and other sulfides with alum and then distilling the results. The residue from one particular sulfide was red vitriol, a cobalt sulfate. Thompson finds Assyrian texts that refer to "steaming green vitriol" (iron sulfate) in the same context as "decomposition of pyrites" (pyrite is a common form of iron sulfide), as well as terms for red vitriol. Putting all this together, he infers that the Assyrians recognized and could make sulfuric acid some two thousand years before the Arab alchemist Jabir ibn Hayyan is credited with the same discovery.[85] Forbes dismisses the notion that any early or "primitive" peoples knew of true distillation, which, he claims, was developed only beginning with the Alexandrian (i.e., Greek-influenced) alchemists.[86]

GLASS IS A FUSION of sand, soda, and lime. Glass beads and glazes are found in Egypt from 4000 B.C. but not until 1500 B.C. do the first Egyptian glass factories appear.[87] Forbes suggests that the sudden development in the 1500s was sparked by the communication between Egypt and ancient Sumer that dates from this time, and with the Sumerians' refinement of glass chemistry.

Glass is found in Sumer as early as 2500 B.C. The finds are both rare and of unusually pure glass content. Forbes reports that one glass artifact was mostly free of striations, unmelted quartz, or trapped impurities.[88] Mesopotamia did not have much in the way of varied mineralogy, but it did have clay, basically an aluminum silicate ideal for making pottery. Firing clay and water gives a hard porous substance; adding salts to the surface gives a glaze, a glasslike coating.[89] In Sumer, the earliest glass materials were pulverized quartz combined with blue and green minerals and then fired, causing a colorful permanent glaze.[90]

A 1600 B.C. tablet in Babylon includes a formula for pottery glazes. The tablet is in essence a glass text, written by Liballit-Marduk, the son of Ussur-an-Marduk, a priest of Marduk at Babylon.[91] The formula, basically a lead glaze with copper put over clay colored with green verdigris, varied little over the course of the next thousand years.[92] In fact, a method of dyeing clay green with copper and vinegar is still used by potters today.[93] A second and essentially similar text, written around the time of the demise of Nineveh (612 B.C.), is the only other glass text in Mesopotamian record.

The earlier Marduk text has been tested in modern times. The text gives formulas for two copper-lead glazes, each beginning with a basic Sumerian recipe for a colorless glass made from sand, alkali ash, and gum.[94] In 1948 H. Moore reported re-creating this complicated seventeenth-century B.C. recipe. He used a starter glass containing sand, dolomite, and soda ash. Moore had to reinterpret Sumerian chemicals to fit modern terminology. For example, he interpreted "copper" to mean malachite, "lead" as lead carbonate, and "lime" as limestone, all of which would have been accessible to the ancients. He fired the double glaze over the glass and produced a shining finish.[95]

The later, seventh-century B.C. Assyrian text details two kinds of furnaces that were used: a "furnace with a floor of eyes," which had holes in the floor for the fuel, and a "furnace of the arch," which had a door through which the metal was inserted.[96] Forbes says a third furnace was used in Assyrian glass factories, a fritting furnace for roughly fusing but not completely melting the sand grains. The "furnace with a floor of eyes" may well have reached temperatures up to 2012°F. Thus, Forbes says, this three-furnace system foreshadows later medieval European glass factories, which had three furnaces stacked one on top of another.[97]

The glazes prepared by the Sumerian method were then poured into molds or onto fired bricks, yielding the beautifully glazed brick-work renowned in Mesopotamian architecture. Seventh-century B.C. glass recipes exist for uncolored glass, red glass, blue frits, blue glass, purple glass, and even a red coral glass that included tiny particles of gold.[98]

Islam

The Arab peoples' contributions to chemistry have been minimized. At worst, they are credited only with preserving the knowledge of the ancient Greeks through the long intellectual drought of the Dark and Middle Ages. At best, they are credited with preserving ideas from India and China as well.

In fact, they did all of the above and more. Arabic texts took earlier ideas from India, China, and Greece and expanded on them, after which they filtered into Western Europe. The medieval Arabs can be credited with isolating many salts and acids. This, combined with their philosophical belief in balance, laid the framework for modern chemistry's acid-base ideas.

There is no significant distinction in Islam between chemistry and alchemy. The word *al-Kimya* covered such things as metallurgy, distillation of petroleum or perfumes, and the manufacture of dyes, inks, sugar, and glass. It also concerned matters of the cosmos and the spirit. Some scientists, Jabir ibn Hayyan and al-Razi, for instance, believed in the transmutation of metals, while Avicenna (Ibn Sina) and al-Kindi, two other equally distinguished scientists, did not. Yet all of them contributed to the advancement of chemistry. In fact, the two transmutationists, al-Razi and Jabir, became the most famous Islamic chemists in history, influencing Europeans centuries later. Scientific achievement does not always follow proper scientific ideology.[99]

The medieval Arab alchemists subscribed to a bevy of theories, many of them borrowed from the ancient Greeks. Al-Ruhawi built on Aristotle's simple elements of heat, cold, dryness, and moisture, stating that they were responsible for differences in metals: "Gold for example, contains more water than silver and so it is more malleable. Gold is also heavier than silver since its parts are more contracted. Gold is yellow and silver is white since the former contains more heat and the latter more cold."[100] There's lots of theory like this—balance is dwelled upon—but it's not fruitful to discuss it here at length. Islamic chemists were onto something with balance and stability, since we now know that atoms must be balanced in positive and negative charges in order to be stable. A fascination with hot and cold does indicate an understanding of the expansion and contraction of metals, and some scholars get excited over this.

Jabir ibn Hayyan (c. A.D. 721–815) of Kufa, Islam's renowned and historically shadowy alchemist, has many texts attributed to him.[101] His Arabic works include *Book of the Kingdom, Little Book of the Balances, Book of Mercury,* and *Book of Concentration*. Some scholars insist that Jabir was known to the West as Geber, whose Latin works appeared in the twelfth century and are supposedly translations of Jabir's.[102] Joseph Needham, known best for his histories of Chinese science and technology, disputes that Jabir and Geber are the same; Geber's work does not show evidence of translation from the Arabic, he writes, and Jabir was ignorant of much of what is in Geber.[103] Scholars also dispute that Jabir existed at all; many believe that his work was compiled by numerous Arab scientists.[104]

Aristotle said matter could be reduced to "four properties: hot, moist, cold, and dry; and four elements: fire, water, air and earth." In cyclical fashion, you could then, by this hypothesis, change each element into another that shared common properties: the hot-dry of fire would lead to the hot-moist of air, which would lead to the moist-cold of water, to the cold-dry of earth. Metals, Aristotle believed, arose from moist vapors coming off the earth.[105] Jabir took this idea but added a mercury and a sulfur stage that created the metals.[106]

To Jabir, mercury and sulfur were agents of transformation, rather than normal minerals. The perfection he saw in metals such as gold and silver resulted from the balance of mercury and sulfur with the proper heat. According to Roberts, he said that the imperfection in metals like "tin, lead, copper, and iron" resulted from the lack of balance in sulfur and mercury with proper heat. Gold came from "the most subtle, fixed and brightest mercury with a little clear, fixed, red sulphur. Silver is made from a combination of mercury and white sulphur, and the other metals from varying and less stable mixtures of mercury and sulphur, in less pure forms."[107] This shows that Arabs like Jabir did, in fact, preserve the "scholarship" of the ancient Greeks. In many cases, such as this one, it may not have been a blessing.

Other Middle Eastern scholars added to these ideas. The seminal tenth-century Islamic theorist Avicenna largely adopted both Aristotelian theory and the works of Jabir ibn Hayyan.[108] However, he held that all metals were separate substances that could not be changed into each other, an idea stretching back to ancient Indian thought.[109] He wrote:

> As to the claims of the alchemists, . . . it is not in their power to bring about any true change of species. They can, however produce excellent imitations, dyeing the red metal white so that it closely resembles silver, or dyeing it yellow that it closely resem-

> bles gold. . . . Yet in these the essential nature remains unchanged; they are merely so dominated by induced qualities that errors may be made concerning them.

This reflects the first principle of modern chemistry, that matter cannot be created or destroyed, and the modern chemical definition of what an "element" versus a "homogeneous mixture" is.

Another alchemist, Aidamir al-Jildaki (c. A.D. 1342) of Egypt, claimed that "substances react chemically by definite weights,"[110] according to the historian H. J. J. Winter. This essential concept underlies chemical reactions: certain precise amounts of reagents result in precise amounts of the end product.

This idea was a precursor to the hypothesis of Englishman John Dalton in the early 1800s. Building on the work of Lavoisier, Dalton stated that chemical compounds are made of atoms "combined in definite proportions."[111]

Al-Jildaki was, like many Arabic philosophers, largely concerned with balance. He used "numerology, the twenty-eight letters of the Arabic alphabet, and the numerical value of the name of the substance" to attempt to identify the precise ratio of heat to dry to cold to wet.[112] By modern standards, al-Jildaki based his ideas on an incomprehensible system, but he was on target in theorizing that chemical compounds react with one another in a quantifiable manner, and he set out to determine how. His theory was correct; his method was not.

IN WHAT IS NOW IRAQ, the thirteenth-century alchemist Muhammad ibn Ahmad abu al-Qasim wrote in *Knowledge Acquired Concerning the Cultivation of Gold,* "It appears, therefore, that these six metallic forms [gold, silver, copper, iron, lead, and tin] are all of one species, distinguished from one another by different 'accidents'; their extreme limit is reached when they become gold."[113] This idea had already been articulated by Jabir and al-Razi.

Such observations, though ultimately wrong (iron et al. do not become gold when "perfected"), reflect the chemical similarities of these elements. Copper, silver, and gold (in order of increasing weight) comprise the IB family of transition metals in the modern periodic table. That is, they have similar chemical properties and the same number of valence electrons in the outer energy level (which determines their chemical reactivity).

Furthermore, tin and lead make up the heaviest two elements of the IVA family, part of the carbon group, though still classified as metals. Iron and copper both occupy the fourth period.[114] Modern chemists have found many other similarities among these six elements, which we won't go into here. Perhaps the alchemists recognized the closeness of these elements without knowing why, making their desire to change silver into gold not so far-fetched. Mendeleyev, the father of the periodic table, also did not know why each element fell into the relationship it did in his table.[115]

Ancient and medieval alchemists recognized something fundamentally in common with the metals that they were able to isolate, even if they were mistaken in their belief that one could be changed into another. We credit John Dalton in the nineteenth century for recognizing that the each of the elements—hydrogen, carbon, oxygen, and so on—was made up of distinctive particles, which he called *atoms.* He borrowed the term of the fifth-century B.C. Greek Democritus, who used the word *atom* to mean that "which cannot be cut" (a-tom), an indivisible particle. We know today that what we and Dalton call atoms are filled with lots of smaller particles: electrons, protons, and neutrons, and the latter two comprise even smaller particles called quarks and gluons. Dalton erred in thinking that he had discovered Democritus's "a-tom," and his terminology has caused confusion ever since. Yet we do not vilify him in our textbooks but rather glorify him—and rightfully so—for declaring that each chemical element has a distinctive "atom." This paved the way for the periodic table and much of chemistry.

ON A PRACTICAL BASIS, the Arabs were demons in the lab. Jabir has been credited as the first chemist to distill sulfuric acid by first distilling alum, though, as we have seen, the Assyrians may have beaten Jabir to the punch. Jabir also made iron sulfate and potassium aluminum sulfate. He made aqua regia by dissolving sodium or ammonium chloride in aqua fortis. This acid is capable of dissolving gold. Jabir also distilled vinegar to form acetic acid; knew how to use manganese dioxide to make glass; and understood how to make pure arsenic and antimony from the raw sulfides.[116]

Al-Razi, or Rhazes, was less spiritually oriented than Jabir and was considered a better and purer chemist. According to Arthur Greenberg in the *Norton History of Chemistry,* "The preparation of pure hydrochloric, nitric, and sulfuric acids by Europeans in the thirteenth century depended crucially on the technology developed by Rhazes." Greenberg goes on to say that these acids were crucial in tipping off Europeans to the early ideas of oxidation.[117]

In his book *Secret of Secrets,* al-Razi sets down different chemical formulas and experiments: distillation, calcination (oxidation or reduction to ashes), crystallization. He also described a large number of chemical apparatuses: beakers, flasks, vials, casseroles, naphtha lamps, smelting furnaces, shears, tongues, alembics, pestles, mortars.[118] God seemed to help. Preparing sal-ammoniac (ammonium chloride) from hair to use as a dissolving reagent, al-Razi writes:

> Washed black hair put into an iron pan, covered with charcoal and burned until they are extinguished. Pour on the calcined residue of the hair 20 times its weight of distilled water of hair and let the mixture undergo coction [boiling] for one hour. Filter it, and coagulate with it the Spirits to whiten by means of coction. Or take the hair's water and oil separately and place over the residue an alembic and light a fire and fit the spout with a receiver wrapped in moistened felt. Then sal-ammoniac will be coagulated in the receiver, if it please Allah! The sal-ammoniac will crystallize in the upper receiver, and the oil will drop into the lower one.[119]

India

Was Indian chemistry ahead of Europe's in the Middle Ages? According to some historians, in 1100 to 1200, practical chemistry in India was more advanced than Europe's in the same period.[120] Will Durant, the modern pop historian, wrote that the Vedic Hindus were "ahead of Europe in industrial chemistry; they were masters of calcination, distillation, sublimation, steaming, fixation, the production of light without heat, the mixing of anesthetic and soporific powders, and the preparation of metallic salts, compounds and alloys."[121] Durant's documentation, however, is less than solid.

Ancient Indians did have alchemic ideas, but much of their writing is sober, more literal than allegorical. There are texts, though, that connect the elements of metal, sulfur, mercury, and fire. John Read writes, "In ancient India the Hindus held that the metals were born of the union of Hara (Shiva) and Parvati (the consort of Hara) through the help of Agni, the god of fire. Mercury was associated with the semen of Hara, sulfur with Agni, and earth (or crucible) with Parvati." Such alchemic ideas appear as early as 1000 B.C. in the Arthava Veda.[122]

THE JAINISTS were among the earliest Indian chemists, and they were not always on the mark. Describing a number of ways in which atoms combined to form matter, they included "body union." However, the Jainists predicted the importance of opposite electrical charges and even "spin," a quality of particles not discovered until the twentieth century.

The ancient Indians did some crude but interesting experiments with gas, and they showed their modernity by making the same major error that Western scientists later committed. A researcher named Udayana (his exact era and domain are unknown) filled balloons and bladders with air, smoke, and various gases and discovered that some gases are lighter than air. He also discovered that hot air rises. Udayana was a kind of Indian Robert Boyle, the British master of gases.[123]

The Indians' mistake: they believed in the "aether," an invisible medium that pervades the universe. Indian chemists described it as "infinite in extent, continuous, and external. It cannot be apprehended by the sense. . . . [It is] all-pervasive, occupying the same space that is occupied by the various forms of matter."[124] In the nineteenth century, many physicists, including the Scottish physicist James Clerk Maxwell, of Maxwell's equations fame, hypothesized that an invisible medium, called the ether, pervaded all of space, through which light waves and other electromagnetic radiation could propagate. Isaac Newton had also subscribed to this theory. One of Albert Einstein's important insights was to divine that no ether is needed for the propagation of light, just space itself.[125] In 1907 American physicist Albert Michelson won the Nobel Prize in physics for his optical studies that cast doubt on the existence of the ether.[126] This didn't end the matter, points out Goldwhite, as the debate continued long after 1907, with some scientists maintaining their belief in this invisible field.

China

The ancient Chinese alchemists were in step with modern science on the relationship between theory and experimentation. The alchemist and Taoist writer Ko Hung (c. A.D. 281–361) spelled out the importance of basing theories on empirical evidence in his text *Pao P'u Tzu:* "Indeed the diversity is boundless, and some things which appear different are in fact the same. Sweeping laws should not be formulated too soon. . . . If a generalization is driven too far it always ends in error."[127] However, Chinese alchemists applied this principle to realms alien to us.

Chinese alchemy had three objectives: 1) the search for the elixir of life using semichemical methods; 2) production of artificial gold and silver, but for therapy, not for wealth; and 3) pharmacology and botanical research.[128] Ko Hung's aims differed slightly: 1) the preparation of a liquid gold, producing longevity; 2) the production of artificial cinnabar, the "life-giving" red pigment, for use in gold making; and 3) transmutation of base metals into gold.[129]

Tied to Taoism, alchemic writing shows up around 500–400 B.C. The Chinese believed that jade, pearl, and cinnabar had life-giving properties. The Chinese text *Zhan Guo Ce* (Records of the Warring States), from around 475–400 B.C. states:

> The king of the state of Chu was presented with an "elixir of deathlessness" by thaumaturgical technicians. Qin Shi Huang, the First Emperor of the Qin Qynasty (221–207 B.C.) . . . recruited a number of thaumaturgists in making a "miraculous drug ensuring immortality," and dispatched thousands of virgin boys and maidens headed by an adept Xu Fu over the seas in quest of it."[130]

As in other cultures, the Chinese believed in the idea of contraries, or what they called "two principles"—active and passive, male and female, and so on—which represented earlier (and modern) ideas of attraction and repulsion, mimicking Sumerian and Egyptian concepts. These opposites emerge later as yin and yang. Beginning in the eighth century A.D., Chang Yin Chiu divided elements according to their yin and yang in a compendium of alchemy called the *Tshan Thung Chhi Wu Hsiang Lei* (The Similarities and Categories of the Five [Substances] in the Kinship of the Three). Rules were set down on how yin and yang materials react to each other ("Mercury might behave as a Yin substance to sulfur, but as a Yang substance to silver"). Needham, perhaps overenthusiastically, claims that such rules foreshadow the modern "electro-chemical series of the elements."[131]

Essential to the Chinese alchemist was the desire to create drinkable gold. Their motivation was immortality, not wealth. Gold was associated with fire, with yang, with life and *ch'i,* the Chinese concept of energy and spirit.[132] Chang Yin-Chiu wrote,

> Gold is the seminal essence of the sun, corresponding to the sovereign, and the principal *chii* of Thai Yang. Mercury is the *pho* soul of the moon, and the principal *chii* of Thai Yin. When

> they are combined and absorbed into a man's body he cannot die. . . . The ancients said, "If one ingests gold one will be like gold. . . . The nature of gold is endurance and resilience. When heated it does not crack or soften, when buried it does not rust, when placed in the fire it will not burn. Hence it is a medicine which can make man live [forever].[133]

In the twelfth century A.D., Wu Tsheng commented on an earlier text, saying "that if the elixir succeeds it will appear as an impalpable powder like bright window dust. If such an elixir . . . is ingested, it will irrigate the three Red Regions of the body of man." A phrase translated as "bright window dust," referring to the dust motes caught by sunlight, is used as a metaphor for potable gold. Needham waxes poetic on the subject of elixirs able to pass through a body like "incense smoke."[134] There is no proof, however, that the ancient or medieval Chinese ever produced such golden milkshakes or, if so, that they imbued everlasting life on the drinker.

IN CONDUCTING chemical experiments, the ancient Chinese used essentially two methods: fire and water. Fire was the more important.

To heat chemicals, the Chinese had sophisticated equipment by ancient world standards. The elaborate peng-hu pot was used to foment a number of chemical reactions. The alchemists were focused on discovering an elixir for eternal youth, and this makes the whole process suspect to our modern eyes. Yet, along the way, these ancient chemists discovered a great deal about heat as they roasted mercury, cinnabar, silver, and other substances.

Around A.D. 140, Wei Po-Yang wrote about "the five dazzling colors" that emerge when cooking metals. "One after another they appear to form an array as irregular as a dog's teeth." It is unclear whether the changing colors refer to the fire or the melting metals, but we may be seeing here the crude beginnings of spectroscopy, or the analysis of the spectrum of colors in blackbody radiation.[135]

A Buddhist monk tells of visiting China's Tse-chou region during the Tang dynasty (A.D. 618–907). The monk and his hosts collected saltpeter, and "on burning it emitted copious purple flames." The Sogdian (from Bokhara) monk said, "This is a marvelous substance which can produce changes in the five metals, and when the various minerals are brought into contact with it they are completely transmuted into liquid form."[136]

Chinese alchemists used water and other liquids to dissolve metals. *The Chapter on Grotto Gods of Dao Zang,* a text dating from approximately the third century A.D., records fifty-four recipes for dissolving thirty-four minerals. "The chief solvent was concentrated vinegar with saltpeter and other chemicals dissolved in it. Saltpeter is called *xiaoshi* (stone dissolvent) in Chinese alchemy classics, since it was believed to be able to dissolve seventy-two kinds of stones." Chemically, this solution would result in weak nitric acid, which could oxidize many metals and metallic ores, a process still in use today.[137]

The ultimate use of solvents, though, was to dissolve one mineral, gold, so that one could drink it. Wang Kuike, a leading modern expert on ancient Chinese alchemy, claims that gold, though chemically soluble and stable, will nonetheless react to various acids and mercury.

A Tang dynasty recipe, perhaps by Ko Hung, calls for "xuan ming long gao [mysterious bright dragon's fat]," meaning mercury or vinegar, and raw raspberry. The formula called for gold to be sealed in a vessel in the aqueous solution for one hundred days. Kuike insists that if "dragon's fat" meant mercury, gold will dissolve in it. If "dragon's fat" meant raspberry, raspberry juice will create cyanide ions, which in the presence of vinegar transform into hydrocyanic acid. With saltpeter, alkali ions such as those of potassium and sodium will form, producing a solution in which gold will dissolve, albeit slowly (hence the one hundred days).[138]

Many of these elixirs were poisonous, and the Chinese alchemists developed yet another passion: intense experimentation to detoxify sulfur, mercury, and other metals. Let's not get into that here. Meanwhile, in the New World, even stranger things were afoot.

South America

I shall limit my discussion of ancient chemistry in the New World to South America and Mesoamerica. Whereas the Old World concentrated on alchemy, the new world focused on pharmacology, psychopharmacology, and what might be called toxic pharmacology. And although all Amerindians worked in these areas, the most remarkable discoveries occurred in South America and Mesoamerica. Previously, I said we would not be getting into non-Western medicine to any great depth. Here we will touch upon medicine, but only its chemical underpinnings.

Chemical knowledge with regard to medicine pervades all ancient and indigenous cultures. The Chinese documented herbal cures beginning in around 3000 B.C.; the Assyrians collected some one thousand medicinal plants over their fifteen-hundred-year civilization; the Indians, Egyptians, and later the Greeks left the world a huge legacy of both herbal cures and techniques. Much of this knowledge has laid the foundation for Western medicine.[139] For example, the use of willow bark, which contains analgesic and antipyretic salicylic acid, was documented in the ancient Egyptian Ebers papyrus to combat fever. Salicylic acid combined with acetic anhydride is today known as aspirin (acetylsalicylic acid).[140] The native peoples of the Amazon—considered primitive and superstitious by most Westerners—inhabit one of the most ecologically rich regions of the earth and are notable for the sheer quantity of their pharmacological knowledge.

The Amazon Basin includes some 2,722,000 square miles of the most diverse plant life in the world, more than eighty thousand species of higher plants, almost 15 percent of the total plant life on earth. It spans much of Brazil and parts of Bolivia, Colombia, Ecuador, Peru, and Venezuela. Human habitation is also diverse, supporting native peoples speaking more than five hundred languages. These peoples developed an immense and time-proven knowledge of the healing powers of local plants. Some sixteen hundred species used by local people have been identified.[141] In a pioneering study, Richard Evans Schultes and Robert

F. Raffauf compiled an exhaustive compendium of 1,516 poisonous and medicinal plants of the Amazon, with slightly fewer than half of them having now drawn the interest of Western scientists, researchers, and drug companies for pharmacological and commercial use.[142]

Caffeine, which is of both commercial and medicinal use, has been isolated in the bark of yocco, a tropical forest vine from which Indians make a stimulating drink, and in *Ilex guayusa,* used by Ecuadorian tribes as a stimulant and as a vomiting agent.[143] (Coffee, by the way, is not native to South America but appears to have originated somewhere around Ethopia). The greatest stimulant of them all, however, has its origins in Peru. Cocaine, *Erythroxylon coca,* probably came from the Amazon but was well established in the mountain kingdoms of the Incas at the time of Pizarro. John Mann, the author of the 1991 book *Murder, Magic, and Medicine,* writes:

> The Incas believed that the Gods presented coca to the people to satisfy their hunger, to provide them with new vigour, and to help them forget their miseries. . . . It was intimately involved in their religious ceremonies and in the various initiation rites; and that shamans used it to induce a trance-like state in order to commune with the spirits. It was a far too important commodity to be used by the common Indians, and their exposure to coca was very limited before the invasion by Pizarro and his *conquistadores.*[144]

Only after the Spanish destroyed the Incan civilization did cocaine abuse become common. Later Westerners discovered cocaine's numbing effects and used it as an anesthetic. In modern times cocaine has been replaced by synthetic anesthesis, of which Novocain is probably the most familiar.[145]

Which is not to say that recreational drugs were unpopular. The hallucinogenic drink prepared by peoples throughout the Amazon and into Central America is variously known as *ayahuasca* ("vine of the soul"), *caapi, natema, pinde,* and *yajé,* the Latin name being *Banisteriopsis caapi.* The effects of this double-twined forest vine include soft blue hallucinations, which can be heightened by the addition of numerous toxic additives. In particular, the leaves of *Psychotria viridis* and *Diplopterys cabrerana* are used to extend the time and intensify the color and images of the visions. Both contain the toxic and highly hallucinatory alkaloid dimethyltryptamine (DMT),[146] which has no effect if drunk without a monoamine oxidase inhibitor. However, *B. caapi* contains substances (B-carbolines) that effectively inhibit the monoamine oxidase enzyme.[147]

Richard Shultes comments, "One wonders how peoples in primitive societies, with no knowledge of chemistry or physiology, ever hit upon a solution to the activation of an alkaloid by a monoamine oxidase inhibitor. Pure experimentation? Perhaps not."[148] How did the native people learn this effect from the staggering number of plant choices at their disposal? Why would people have sampled the drink to begin with? According to Schultes and Raffauf, "The beverage is extremely, sometimes nauseatingly, bitter, and vomiting usually accompanies the first draught. It almost always causes diarrhea."[149] Additionally *P. viridis* and *D. cabrerana* are poisonous by themselves, so again, what would have promoted native Amazonians to try to use them?[150]

THE ARROW POISON of the Amazon, curare, is prepared from the genus *Strychnos Linnaeus,* usually from the bark and roots of any of twenty-eight species of this vine, many of which have edible fruits. The Spaniards got firsthand experience with the potency of this poison as they entered the Amazon. Francisco de Orellana wrote, "The Indians killed another companion of ours . . . and in truth, the arrow did not penetrate half a finger, but, as it had poison on it, he gave up his soul to our Lord."[151]

The poison is also made from members of the moonseed family (Menispermaceae), which includes the genuses *Abuta, Anomospermum, Chondodendron,* and *Curarea,* among others. Methods and mixtures used in each recipe vary from tribe to tribe. Some plants are used individually; others in combination with plants of the same genus; others in combination with different plants altogether.[152] *Strychnos* and the moonseed family are rich in alkaloids. (As its name suggests, *Strychnos* is also the source of strychnine).[153]

Curare brings on paralysis of the central nervous system[154] and is used medicinally as a muscle relaxant. Tubocurarine, an active ingredient in curare, was explored for use in medicine in the 1800s and has been used in surgery so that the amount of anesthesia required is reduced. Another surgically important constituent, C-toxiferine, is extracted from *Strychnos toxifera* and is far stronger than tubocuranine in relaxing muscles during surgery. Both drugs were developed with curare as their base, and both have now been supplanted by artificial drugs.[155]

The complex process of preparing some curares argues that native discoveries of medical cures were not happenstance. Jeremy Narby, an anthropologist who studied the Peruvian Amazonian Indians, relates the process of production:

> There are forty types of curares in the Amazon, made from seventy plant species. The kind used in modern medicine comes from the Western Amazon. To produce it, it is necessary to combine several plants and boil them for seventy-two hours, while avoiding the fragrant but mortal vapors emitted by the broth. The final product is a paste that is inactive unless injected under the skin. If swallowed, it has no effect.[156]

In his notes, Narby tells how a German zoologist failed to heed the prohibition of inhaling the vapors of the boiling mixture and died.[157]

Despite its deadly application in curare, the moonseed family is also highly valued by the Amazonian Indians for its many curative properties. *Abuta grandifolia* is used by the Sions, the Karijonas, the Makunas, the Andoke, and the Taiwanos for curare; yet it is also prepared as a tea for bleeding during childbirth and given to children suffering from colic or nervousness. The Andoke name, *o-je-ji-ka-ka,* means "poison of the toad called *oje.*" The roots and stems of *Chondodendron tomentosum,* whose native name translates as "poison vine," are used as a diuretic, a fever reducer, and an aid for menstrual difficulties. It is applied externally to help heal bruises and is taken internally for mental illness, or "madness," as Schultes and Raffauf put it.[158]

In another beneficial application of potential poison, the Waika of Brazil use a hallucinogenic snuff from the resin found in the bark of the several plants of the *Virola* genus. The chemically active compound is a form of tryptamine. The resin is also used as a mild poisonous curare for hunting, and is effective in treating external fungal infections. Western scientists have since studied *Virola* and confirmed that it inhibits the growth of fungus.[159]

In the West's search for cures for cancer and AIDS, drug companies and research organizations are attempting to exploit indigenous healers' knowledge of the diverse tropical rain-forest flora. A National Cancer Institute (NCI) article describes how plants considered "powerful" by native shamans—that is, those having multiple medicinal uses in native pharmacopoeias—yielded greater medicinal activity (25 percent) than plants selected at random (6 percent) for use against the HIV virus. However, the percentage fell after the NCI followed routine pharmaceutical procedure in "dereplicating" the plant extracts—the separation out of raw plant substances such as tannins and polysaccharides to isolate what the NCI assumed were the active ingredients. After dereplication, the effectiveness of the plants fell to almost the same percentage as that of the plants selected at random. Despite admitting that dereplication "removes compounds that are known to have immunostimulatory activity" and that antiviral agents can be found in tannins, Michael Balick, of the Institute of Economic Botany at the New York Botanical Gardens, stated, "Thus . . . general ethnobotanical collection does not appear to be advantageous in developing leads for HIV treatment."[160]

M. M. Iwu, of the Department of Pharmacognosy at the University of Nigeria, says that compounds like polysaccharides and tannins "leach over time into the circulatory system and are gradually released from the proteins or macromolecules. . . . Dereplication might be a suitable method for our current state of knowledge, but once we understand how these compounds work, these assays may no longer be appropriate. . . . We should not have a strict *a priori* rule about which types of compounds are responsible for bioactivity."[161]

Walter Lewis and Memory Elvin-Lewis at the Washington University Department of Biology come to the same conclusion, citing several examples of "real" medically active plants used by the South American Jivaro Indians. "A few examples of individually targeted plants used medicinally by the Jivaro for which the specific utility has been verified scientifically are worth examining," they write. Lewis and Elvin-Lewis cite the holly *Ilex quayusa,* which is used as a stimulant by the Jivaro and has high concentrations of caffeine, along with traces of theobromine and theophylline, known appetite suppressors (which are useful if you're not sure where your next meal is coming from). They also mention *Cyperus articulatus* and *C. prolixus,* used by Indians to brew a tea that encourages contractions and lessens bleeding during childbirth.[162]

Indigenous pharmaceutical knowledge appears to be rooted in empirical experience. After a year living with the Achaninca in the Peruvian Amazon, anthropologist Narby described how he came to trust their empirical approach to developing new medicines:

> People in Quirishari taught by example, rather than by explanation. When an idea seemed really bad, they would say dismissively, "*Es pura teoría*" ["That's pure theory"]. The two key words that cropped up over and over in conversations were *práctica* and *táctica,* "practice" and "tactics"—no doubt because they are requirements for living in the rainforest. . . . After about a year in Quirishari, I had come to see that my hosts' practical sense was much more reliable in their environment than my academically informed understanding of reality. Their empirical knowledge was undeniable.[163]

Narby got firsthand experience. Suffering from chronic back pain since adolescence, he had received cortisone and heat treatments, to no avail. In 1985, a Quirhishari shaman prescribed half a cup of *sanango* tea, after warning him of the side effects (a feeling of coldness, lack of coordination, and hallucinations). Three days later the side effects vanished, along with the back pain. In 1998, when he wrote his book, the pain had not returned.[164]

When asked the source of their knowledge, South American shamans reply with more or less the same answer as the West African shamans: the plants tell us.[165] The shaman partakes of hallucinogenic plants, which are believed to guide him in identifying which plants will heal what illnesses. The plants are called *doctores* because the shaman is taught by them.[166] Schultes speaks of Salvador Chindoy, a renowned medicine man in the Valley of Sibundoy in western Colombia, "who insists that his knowledge of the medicinal value of plants has been taught to him by the plants themselves through the hallucinations he has experienced in his long lifetime as a medicine man."[167] In ancient cultures as in the modern, drug use and physicians go hand in hand.

Mesoamerica: the Maya and Aztecs

The Maya believed that natural phenomena and objects had an inherent sacredness to them, the *ch'ulel,* or soul, whereas man-made objects and buildings had to have this sacredness added by ceremony. *Ch'ulel* was found in blood (the biological life fluid) and was associated with red pigments. Thus Mayan buildings often included hidden collections of precious objects under the floors, including, among other things, cinnabar, specular hematite (both of which were red), and mercury. The Maya must have known that cinnabar yielded mercury when heated. Possibly they saw this happening in volcanic areas and duplicated it in their own fires.[168]

Related to *ch'ulel* was the *itz,* the excretion or secretion of living beings, or, as it's sometimes called, "cosmic sap." Tree resins and other forms of sap were harvested as *itz.* To the modern Maya, *itz* refers to the secretions of the human body. But, according to Mayan scholar Linda Schele Freidel, it also "can refer to morning dew; flower nectar; the secretions of trees, like sap, rubber, and gum; and melting wax on candles."[169]

Given the spiritual significance of blood in Mayan and Aztec life, the color red was important. The Mayans derived red dye from a variety of sources: the annatto tree (*Bixa orellana*), which they called *ork'uxu,*[170] and the cochineal croton tree *Croton sanguifluus* (or *kaqché,* for "red tree"). Dye from the "red tree" was used as a substitute for human blood in sacrifices. Cochineal, a bright red dye, was also obtained from scale insects.[171] The Aztecs used red cochineal in a purified form but also mixed an uncolored variety of it with ashes or chalk or flour for different effects.[172]

In order to get the red color, the dye had to be precipitated with alum.[173] One presumes the Aztecs knew enough chemistry to obtain fairly pure potassium aluminum sulfate (alum), an acid made by heating potassium sulfate with aluminum sulfate until the chemicals dissolve and then cooling the mixture. Conceivably they also found reasonably pure varieties of alum occurring in nature.

The Aztec scholar Bernardino de Sahagún reports it as a kind of "earth," or dirt. The Aztec name for alum translates as "sour fruit earth," referring to its sour taste. "It causes one to salivate; it deadens one's teeth . . . it makes one acid," reports de Sahagún. The Indians used it to refine colors and as a cleaning agent, and presumably as a fixitive.[174] In any case, the Aztec and Maya had copious brilliant dyes—reds, blues, yellows, purples—that the conquering Spaniards found equal or superior to any seen in Europe. As one wrote, "The Indians make many colors from flowers, and when the painters wish to change from one color to another they lick the brush clean, for the paints are made from the juice of flowers."[175]

Fray Toribio de Benavente, a Franciscan friar and Spanish explorer called Motolinía ("Poor") by the Aztecs because of his dilapidated clothing, documented turpentine among the Aztecs. In *A History of the Indians of New Spain,* finished in 1541, Motolinía writes of the Indians tapping a certain tree (perhaps pine) and allowing the sap to catch on maguey leaves, in a process not so different from tapping maple syrup today. There it hardened into copal and was later mixed with oil, producing "a very good turpentine."[176]

Balsam was another resin extracted from trees by the Aztecs. Today balsam is used as a base for cough syrups, other medications, and perfumes. Motolinía reports that the Aztecs were making balsam before the Spanish invasion and used it to treat diseases. He also reports on a tree in the sweet gum genus called liquidambar by the Spanish (from the Latin for "liquid amber"). The Aztecs extracted sap from the sweet gum and then combined the sap with its bark and sometimes bitumen to use in solid form as perfume or medication.[177]

Chewing gum was also pioneered by the Aztecs, who transformed chicle, the milky juice of the sapodilla, into the gum's primary component. Aztec women also used the bitumen that occasionally washed up on the beach as a chewing gum, because of its "sweet scent." It made their mouths smell good, although they complained later of headaches.[178] And the gummy latex exuded by the *Castilla elastica* tree was transformed by the Aztecs into hard rubber balls, a topic we shall take up in chapter 8.

Finally, salt was an important commodity among the Aztecs, who often extracted it from urine. The Spanish explorer Bernal Díaz del Castillo reported seeing canoes filled with urine for the purpose of making salt and for tanning leather.[179] Today leather is first cured with salt and then soaked in water. Since urine contains salt and ammonia and also serves as a cleanser, it may have been an all-purpose tanning agent: salt for curing, ammonia for removing the hair or fur, and the liquid in general as a final solution and cleansing agent. The ancient Chinese used a similar process.[180]

For one skeptical of non-Western science, there is plenty of ammunition in this chapter: alchemy, elixirs for everlasting youth, plants that talk to shamans . . . the list goes on. One could stack the evidence to show that these chemists accomplished nothing but voodoo. Even when they made their many significant discoveries in chemistry, one can always discount their achievements since they did not fully understand the mechanisms. This had been true of Western chemistry before the introduction of the Pauli exclusion principle.

Consider this passage from Friedrich Nietzsche:

> Do you believe then that the sciences would ever have arisen and become great if there had not beforehand been magicians, alchemists, astrologers, and wizards who thirsted and hungered after abscondite and forbidden powers?[181]

8

TECHNOLOGY
Machines as a Measure of Man

NOWHERE is there more phony information than in the area of technology. Eager to establish legitimacy, some multicultural scholars have manufactured dubious claims: Chinese warriors in the eleventh century firing away with machine guns, Incas frolicking above the Nasca plains in hot-air balloons.[1] My favorite theory is liquid stone. This supposedly explains the location of the Gate of the Sun, a ten-ton carved gate that stands alone on an isolated plateau at Tiahuanaco, Bolivia, thirteen thousand feet above sea level. How did this heavy structure get there? Joseph Davidovits, of the Geopolymer Institute in St.-Quentin, France, claims that the gate was not constructed and carried to its destination stone by stone. After conducting an electrochemical analysis of stone fragments, he concluded that the original Huanka builders used oxalic acid extracted from rhubarb leaves to dissolve the stone at the quarry, then transported the fluid plastic form to the site, where they poured it into molds.

Davidovits proposes the same hypothesis for the building of the giant Olmec stone heads in central Mexico, and even the Egyptian pyramids. He presented his findings at the International Symposium on Archaeometry at the Brookhaven National Laboratory. Brookhaven chemist Edward V. Sayre said, "It's intriguing but definitely controversial."[2]

Let's not dwell on such speculation. There's plenty of good material. In *The New Instruments,* Francis Bacon wrote that three inventions—gunpowder, the magnetic compass, and paper and printing—transformed the modern world and marked it off from antiquity and the Middle Ages. They all came from China.

Iron suspension bridges came from Kashmir; papermaking was commonplace in China, Tibet, India, and Baghdad centuries before it existed in Europe. Movable type was invented by Pi Sheng in 1041, long before Gutenberg. The Quechuan Indians of Peru were the first to vulcanize rubber; Andean farmers, the first to freeze-dry potatoes. European explorers depended heavily on Indian and Filipino shipbuilders, collecting maps and sea charts from Javanese and Arab merchants. (Vásco da Gama learned navigational techniques from his pilot, Mhmqad Ibn Majid.) The lack of quality manufactured goods in the West was a stimulus for European navigation, driving English, Dutch, and Portuguese merchants to sail off in search of superior Asian products, such as textiles from India. In the New World, meanwhile, Peruvian artisans produced 109 distinct hues, using natural dyes of such strength and brilliance that museums still exhibit brightly colored Peruvian textiles over two thousand years old.

Alfred W. Crosby, a historian at the University of Texas and the author of *Ecological Imperialism: The Biological Expansion of Europe, 900–1900,* claims that two centers of invention transformed history. The first was the Middle East—the Sumerians and their successors. The second was central Mexico—the Olmecs and others. Europe was simply a transfer station. "Think of the dozen most important things ever invented. The wheel, the stirrup, movable type, metallurgy, things like that. Not one of them was invented in Europe," he writes.[3]

We shall follow the development of technology in an unorthodox manner. It is usual to begin with the Chinese, since they were the most spectacular techno-masters. They may not have had machine guns, but they did in fact use flame throwers and a variety of other horrifying weapons. Chinese technology, however, has been so well popularized—by sinologist Joseph Needham and others—that we shall leave it for last, and begin with the two birthplaces of technology, Mesopotamia and Mesoamerica.

Mesopotamia

Mesopotamia is a geographical area, not a civilization. It was controlled over the several-thousand-year period we are examining by a diffuse and diverse number of peoples: Sumerians, Hittites, Assyrians, Arabs, and others. Yet technology transferred and evolved among these civilizations within Mesopotamia almost as if it were one coherent society.

We can easily trace the path of technology back to the Sumerians, to what may well be the mother civilization of humans outside of the Americas. The Sumerians were probably tribes who came from the east, the mountains of Elam, perhaps as early as 8000 B.C. They settled near the swamps at the head of the Persian Gulf between the Tigris and Euphrates Rivers. The Greeks later called this land between the rivers, at the eastern end of the Fertile Crescent, Mesopotamia. It stretched from the Persian Gulf to the Mediterranean Sea. The Fertile Crescent became the crossroads of the Eurasian world—the starter kit, for most, if not all, subsequent cultures in the Eastern Hemisphere.

Sumer arose some time before 5000 B.C., and there the written chronicle of humanity begins. "If we compare the Sumerians with the hunter-gatherers who preceded them," writes Crosby, "we see that the contrast between these dawn people of civilization and any Stone Age people is greater than the contrast between the Sumerians and ourselves." In looking at the Sumerians, Akkadians, Egyptians, Israelites, and Babylonians, "we are looking into a very old, very dusty mirror."[4]

The technological era in the Middle East began, according to Crosby, when humans started to grind and polish, rather than chip, their stone tools and ended as they learned to smelt metal and work it into superior implements. In between, our ancestors domesticated "all the animals of our barnyard and meadow, learned to write, build cities, and created civilization."[5] Columbus and his European contemporaries owe as much to the civilizations of the ancient Middle East, adds Crosby, as they do to all that was invented in Europe.

The Sumerians started a textile industry, working wool into cloth and flax into linen. They built canals and dikes to control the river water and carry its surplus to their fields. They invented the wheel in around

3500 B.C. The wheel made pottery wheels and carts possible, made moving things easier, and allowed for chariots and other engines of war. The first objects with glazed surfaces appeared around 4000 B.C.; the first freestanding glass objects around 2500 B.C. in both Mesopotamia and Egypt. The Sumerians started to develop writing about the same time as the wheel—around 3500 B.C. Researchers have long thought that writing evolved to keep track of property and the exchange of goods. Tens of thousands of cuneiform texts record Sumerian poetry, account ledgers, property records, lullabies, lists of astronomical events, animals, and medicinal plants.

Sumerians imported metals for making bronze, as well as the techniques for working with metal, from the mountains of Iran and Turkey. By land and sea, Sumerians imported metal, wood, lapis lazuli, and other stones, and exported textiles, jewelry, and weapons. Before 3000 B.C., Sumer's temple accountants had devised sets of standard weights for conducting business; they set large penalties for anyone trying to cheat with false weights.

By 3000 B.C. cities prospered throughout the region. The earliest may have been Uruk, in what is now Iraq, with a population of fifty thousand. The Sumerians built their cities from sun-dried clay bricks. They designed arches—curved structures over an opening—capable of supporting weight, then domes or vaults of rounded roofing covered with arches of bricks. In the center of cities they erected temples, called ziggurats, constructed on artificial mountains of brick and built in layers, each one smaller, usually in seven stories with a temple or shrine on the top. One ziggurat at Ur took thousands of bricks, and had weep holes at regular intervals to allow water to drain. Much of two stages of this temple survives today. Ur, in around 2500 B.C. had the earliest known arches.[6]

The usual processes of war and conquest ensued. Around 2350 B.C., following several centuries of intermittent warfare between the Sumerians and the Semitic-speaking Akkadians, Sargon the Great of Akkad challenged the Sumerian hegemony. Sumerian Ur fell in about 2025 B.C. Sargon proceeded to build Babylon on the Euphrates in southern Mesopotamia, a region described (later) as the location of the Garden of

Eden. The city-states regained their independence briefly, but by 1900 B.C. the Akkadians and the old Sumerian elements had merged to form the First (or Old) Babylonian Empire. King Hammurabi (c. 1775 B.C.) fashioned his code of laws, reflecting rules and usages going back to Sumerian times. The Babylonians established urban complexes and trade routes in the river valleys of the Egyptians, the Tigris-Euphrates, and the Indus. This was the first real "empire" in Mesopotamia, stretching from the Mediterranean to the Persian Gulf.

AROUND 1600 B.C., the Hittites, Indo-European immigrants from the Black and Caspians seas to the north, invaded the Tigris-Euphrates. They were probably the first to smelt iron, and they tried, unsuccessfully, to keep secret its process. They mined copper and silver and traded metal around the Fertile Crescent. The Hittites' use of the wheel in military machines probably initiated the gear-and-axle construction of the water-powered corn mill a few centuries later.

In the wave of empires rising and crashing, the Hittites, in around 1100 B.C., succumbed to the Assyrians, who settled along the Tigris northwest of Babylonia in the city state of Assur. The Assyrians acquired ironworking from the Hittites and became the first to outfit entire armies with weapons of iron. Aggressive and fierce, the Assyrians invented movable towers, battering rams, and other siege engines and made good use of the latest chariot designs. Their army was large and well equipped. At the battle of Qarqar in 853 B.C., King Shalmaneser III led an army of 52,900 foot soldiers, 3,940 chariots, 1,900 cavalrymen, and 100 camel riders against a coalition of Syrian, Lebanese, and Palestinian kings.[7]

The Assyrians conquered Babylon in around 700 B.C. and razed the city, changing the course of the Euphrates to flow over it. They built roads for the movement of their troops and developed a postal service for army communications. The Assyrians built Nineveh on the Tigris from taxes collected in conquered lands. They wanted the most splendid capital in the world and built a great library containing clay tablets from all the peoples of the Fertile Crescent. Nineveh was fortified with double walls fifty feet thick and one hundred feet high; it had fifteen gates. Still, in 612 B.C., Chaldeans (nomads from the Arabian desert), Medes, and Persians joined forces and destroyed Nineveh, eventually conquering the whole Assyrian empire.

IN 616 B.C. the Chaldeans seized Babylon and rebuilt it as their capital. The new Babylonian king, Nebuchadnezzar (605–562 B.C.), went on to conquer most of the Fertile Crescent. He is best known for building the Hanging Gardens of Babylon, designed for one of his wives, a daughter of a Median king, to assuage her homesickness for her homeland. The garden, suspended in air and with a quarter-mile-square base, looked from a distance like an ornately terraced hillside.

Horticulturists planted each terrace with large trees and shrubs. The topmost terrace was a massive roof so one could walk underneath the gardens. Below was a system of galleries supported by walls twenty-two feet thick, separated by passageways ten feet wide. The bases of the terraces were massive slabs of stone covered with plants, cantilevered so that each level projected over the one below. Holes in the structure allowed light to penetrate the galleries. The roof had a base of lead to prevent water from draining to the lower level and was covered with two layers of brick, a layer of reed matting set in asphalt, and a thick soil layer in which the trees were planted.

The garden was kept moist with screwlike devices that lifted water from the Euphrates. These machines may have preceded the screw of Archimedes—the greatest inventor of the classical world—by over seven hundred years. The remains of the Hanging Gardens have never been found, but Babylon was so enormous that it has never been completely excavated.[8] Our account of the Hanging Gardens comes not from the Babylonians but from Diodorus Siculus, a Greek historian.

With a population of around a half million during its apogee, Babylon seemed the center of the human universe; everything in it was on a grand scale. Its ziggurat, the temple of Marduk, seen and described by Herodotus in around 450 B.C., was decorated in gold; its eight levels were connected by a spiral staircase, with seats where climbers could rest. It was two hundred feet high, and may have been the inspiration for the biblical Tower of Babel. After Nebuchadnezzar's reign and civil war, Babylon was captured once again, by the Persians in 539 B.C.

Thus the narrative of humankind's seed civilizations is already an old story: replete with technological and social advances, imperialism, conquest; then the wars, the social and technological disintegration, and the sack. Each succeeding empire, while destroying much of the previous society, retained and advanced some of its technologies. The Greeks and Romans took some. The North Africans carried away some to Egypt and its great city, Alexandria. But much of the science and technology packages put together by the peoples of the ancient Middle East remained, tended to and enhanced, first by the Persians, and later the Muslims.

IN THE ARID ECOLOGY of the Middle East, water was a constant preoccupation. Given an annual rainfall in Iran that averages six to eight inches, for example, it is unsurprising that the supreme large-scale technology of the ancient Middle East was hydrology. Although the pyramids of Egypt were among the biggest construction projects ever undertaken (c. 2000 B.C.), as much labor and ingenuity went into constructing Mesopotamian embankments and canals, built at about the same time for flood control and irrigation.[9] Water management remained the most important technology from the ancient civilizations through the brilliant centuries of medieval Islam. Providing people with water, Muhammad is said to have observed, is the act of greatest value.[10] (See Frank Herbert's planet Dune and its Fremen culture.)

One of the first irrigation systems in recorded history was in Jericho, where water tanks have been dated to around 6000 B.C.[11] One of the first canals on record, the Al-Gharrif waterway from the Tigris, was cut by the governor of the Sumerian city of Lagash before 2500 B.C. (The work of the Sumerians wasn't all positive. Alfred Crosby points out that they ruined much of the farmland of the Middle East by irrigating their fields with water from their rivers. The water evaporated, leaving salt behind, the "same process," says Crosby, "going on in our southwest, where, as in

Sumeria, there are fields white with salt." Still, says Crosby, the difference between the Sumerians and the Stone Age people who preceded them is greater than the contrast between the Sumerians and ourselves.)[12] The Egyptians had a department of irrigation as early as 2800 B.C. The dam Sadd-al-Kafra, twenty miles south of Cairo, was built in 2500 B.C. The remains survive. In 690 B.C. the Assyrian king Sennacherib constructed a masonry dam on the Atrush River and a thirty-six-mile long canal to Ninevah.[13] Around 100 B.C. the Nabateans of south Jordan and the Negev Desert of Israel built seventeen thousand dams.

The Persians inherited this mastery of dam, canal, and underground waterworks. In one instance, the Persians captured an entire Roman army and put them to work building a dam.[14] In a huge engineering feat between A.D. 530 and 580, the Persians constructed two dams that diverted water from the Tigris River into the Nahrwan Canal. After their conquest of the Middle East in the seventh century A.D., the Muslims adapted the inherited techniques, and enormously extended the application of mechanical and hydraulic technology.

Near the city of Basra, founded in the seventh century A.D., Muslims built up a vast network of dams and dam-fed canals. A dam built over the Kor River in Iran between Shiraz and Persepolis in A.D. 960 irrigated three hundred villages with more than ten water-raising wheels and ten water mills. It still exists.[15] South of Qum in Iran, thirteenth-century Muslims built the first known example of a true arch dam. Unlike most dams, it did not depend on gravity for its resistance. Instead it was constructed as an arch laid on its side, its convexity pointing upstream, the sides anchored into the rocky banks of a gorge, where the forces of water pressure against it transferred to abutments.

During the Islamic expansion, Muslims built many dams on the Iberian peninsula, including a dam at Cordova that was fourteen hundred feet long and a series of eight dams on the Turia River in Valencia, with associated canals. These canals had a total capacity that was slightly less than that of the river, suggesting that the engineers were able to gauge a river and then design dams and canals to match it.[16]

Water mills were a variation on the theme to exploit water for power. Their origin is unclear, but some were present in the pre-Islamic Middle East. By the Islamic medieval period, three types of water mills were in use: undershot, vertical and horizontal overshot. They were used for flour production, papermaking, cloth making, and the crushing of sugarcane and metallic ores.

There was a remarkable diversity of machines for milling in Iran and Iraq. In Baghdad, with a population approaching one million inhabitants, conventional milling wheels could not keep up with demand, so Baghdadis carried out corn milling using a series of floating water mills on the Tigris that operated continuously on twenty-four-hour shifts, using undershot wheels that drove the millstones through wooden gears. By A.D. 1000 smaller horizontal, turbinelike wheels with the millstone mounted on the same shaft, directly above, were used throughout Eurasia, from western Europe to China. Near Basra, ten mills operated by the ebb and flow of tides about a century before the first mention of tidal mills in Europe.[17]

Windmills were also invented in the Middle East, where water for power was scarce. Records from eastern Iran date windmills in that area from around A.D. 950. Some of those are still operating today. According to legend, the inventor of the windmill lived in Iran during the time it was conquered by the Muslims in the mid–seventh century. The second caliph, Omat, levied heavy taxes on windmills, and, according to the story, the inventor was so irate that he murdered Omat. Nonetheless, windmills spread throughout the Islamic world, then to India and perhaps China. The technology reached England in the mid–twelfth century.[18]

AFTER THE COLLAPSE of the Roman Empire, the burning of the library at Alexandria (the Birdcage of the Muses) in A.D. 640 and the European Dark Ages, the Islamic Middle East preserved engineering technologies as well as pure science. Such fields included building construction, mirrors, weights (gravity physics), surveying, hydraulics, military technology, navigation, and the designing of ingenious machines. Often technology and engineering were given as much merit as pure science. The word *handasa* in Arabic, for example, means architecture and engineering as well as geometry. The distinction was not rigidly made between scientists and technicians. Many men were both.[19]

The obvious Islamic paragons of this type were the three Banu Musa brothers of ninth-century Baghdad. They were astronomers and mathematicians as well as engineers. Around 850 they wrote a compendium, *Kitab al-Hiyal* (The Book of Ingenious Devices, or On Mechanical Devices).[20]

The Banu Musa brothers designed increasingly complex waterwheels and other sophisticated water-drawing systems. Although they were influenced by the clever Alexandrine inventors of Hellenistic Egypt, whose work was translated into Arabic during their lifetimes, the Banu Musa made many advances. They designed a device for providing hot and cold water, dredging machines for harvesting jewels from sea and river bottoms,[21] and an oil lamp that raised its own wick and fed itself more oil. The brothers built elaborate fountains. They are credited with the earliest use of a crank as part of a machine (the crank wasn't employed in Europe until the fifteenth century) and the first use of suction pipes.[22]

Fun was an important element. The Arabic word *hiyal* can denote almost any mechanical object, from a small toy to a siege engine. The leisure class took its toys seriously, and Islamic courtly circles funded engineers. Consequently, many of the most advanced Arab designs were both toylike and useful. The Banu Musa brothers designed eighty-three "trick vessels." There are pitchers from which pouring cannot be resumed after it has been interrupted; vessels that replenish themselves if a small amount of water is removed; vessels into which a mixture of liquids can be poured together, yet discharged separately. The components usually included variations on conical valves, siphons, airholes, balances, pulleys, gears, miniature waterwheels, floats, and cranks.[23]

Automata, or self-operating mechanisms, were very popular. A prominent designer of automata was Badi al-Zaman al-Jazari, a twelfth- and thirteenth-century engineer who may have been in the employ of the southeastern Turkish Artuqid dynasty. Al-Jazari devised most of his large-scale automata to collect and transport water, and was known for his gear systems, one of which showed up two centuries later in Europe in Giovanni de' Dondi's mechanical clock. One of al-Jazari's automata toys was a mechanical boat with drinking men, designed to amuse guests at a drinking party. When activated, the *hiyal* came alive in a counterpoint of sailors rowing and musicians playing.[24]

The caliphs of Baghdad exploited this richness of invention and engineering to build private playgrounds. Perhaps memories of Babylon inspired Islam's monumental gardens, models of paradise on earth. A report from the early tenth century describes a pond-filled garden. In the middle of one pond stood a tree with a silver and gold mechanical whistling bird. Another of the garden's ponds was filled with mercury, upon which floated gold boats. Around the ponds were automata of singing birds, roaring lions, and other moving animals.[25] A thousand years ago the Arabs were experimenting with animatronics.

Many of the basic building blocks of European technology originated in the ancient Middle Eastern river-valley civilizations. Medieval Islam's central location in Eurasia allowed it to acquire inventions from India and China as well as make crucial advances on technology inherited from ancient Greece and Hellenistic Egypt. In time, the technological knowledge of the Middle East was transferred to Europe via Spain, and to Asia and Africa. Muslim engineers contributed enormously to the technology of medieval Europe, and Europeans may have feared the dominance of Middle Eastern technology and learning. Dante reveals European animosity toward Islamic culture in the *Commedia.* In Canto VIII of *The Inferno,* the Florentine poet places the mosques in the city of Dis, and in Canto XXVIII he puts Muhammad in the eighth circle.

Mesoamerica

If the Eurasians of the Old World were the first great techno-masters, the first peoples of the New World were the greatest explorers, traversing the Bering Strait and quickly migrating throughout the Western

Hemisphere. When Europeans arrived in the New World, indigenous peoples of the Western Hemisphere had not yet developed iron tools, except for a very few of meteorite iron. Nor did they have the wheel or any animals that could be ridden or used to pull a plow. Yet they were cultivators of highly subtle crop varieties. The agricultural impact of the New World on the Old World was enormous. New World farmers were, perhaps, the world's greatest plant breeders. They had experimented with and exploited the many varieties of the wild ancestors of their crop plants. When they reached Europe, Asia, and Africa, these crop varieties ignited an agricultural revolution. Maize (corn) and potatoes, for example, were regarded as "miracle crops" after their introduction in Europe.[26] Some experts estimate that Native Americans gave the world three-fifths of the crops now in cultivation.[27] In Mesoamerica the cacao bean, the source of chocolate, was so valuable that it was used as currency and only the elite were able to drink the hot chocolate mixed with honey. Mayan kings had royal chocolate makers.[28]

At the time of the European invasions in the early sixteenth century, the only New World peoples with urban technologies in place were those in Mesoamerica and in the Andes. Mesoamerica is a cultural area extending from north-central Mexico to Pacific Costa Rica, and includes southern Mexico, Guatemala, Belize, and Honduras. It began its cultural identity with the spread of farming villages around 2000 B.C.[29] The scholarly consensus is that most if not all the great cultures of Mesoamerica had as their mother civilization the Olmecs of southwestern Mexico. The Olmecs, who lived between 1500 and 600 B.C., erected massive basalt (a dark volcanic rock) sculptures and public monuments that date to 1400 B.C.

Mesoamerican peoples shared certain cultural traits, probably of Olmec origin, that were absent or rare elsewhere in the New World. These include hieroglyphic writing, books of fig-bark paper or deerskin that were folded like accordions, a complex calendar, knowledge of the movements of the planets (especially Venus) against the background of the stars, a game played with a rubber ball in a court (called *chaah* by the Maya), highly specialized markets, human sacrifice by head or heart removal, a general emphasis on self-sacrifice by blood drawn from the ears, tongue, or penis, and a pantheistic religion that included nature gods as well as deities emblematic of royal descent.

They shared an urban design ethic in which cities were built around a central temple-pyramid. The typical tapered shape of these structures helped them withstand earthquakes.[30] Although there were many differences among cultures, the base Mesoamerican diet included the "ancient foursome"—maize, beans, chili peppers, and squash. According to the Mayan foundation epic, the Popul Vuh, the ancestors of the Maya were created from maize dough. Mesoamerican agricultural strategies were good enough to support a preconquest population of eight to ten million people in the Mayan lowlands. Aerial surveys detect evidence of a virtually continuous occupation of the Yucatán from about A.D. 750.[31] (Pre-Columbian population figures in the New World are inexact and controversial. According to Charles C. Mann, a correspondent for the *Atlantic Monthly* who surveyed anthropological demographers, figures can fluctuate by a factor of ten.)[32]

SPORT WAS the inspiration of a significant technology, what is called an *oopart,* an acronym for *out-of-place artifact.* The Maya were obsessed with their ball game, and most towns of any size had ball courts located near the temple pyramids. The earliest ball courts of the Olmecs were simple earthen retaining walls. By the time the ball court was finished at the city of Copán (before A.D. 800), the walls were of stucco-faced masonry with sloping playing surfaces. Archaeologists consider Copán to be the most perfect of the ball courts, with tenon-jointed sculptures in the shape of macaw heads as markers.[33] The Aztecs carried on the tradition from 1200 onward.

The courts somewhat resemble today's handball courts, and while the exact rules of the game are undetermined, we do know that, given the occasional consequences of death or mutilation, losing was to be avoided. The ball-court wall reliefs at Chichén Itzá show the decapitation of a player. Doubtless, the game there was played "for keeps," says archaeologist Michael Coe, and the losers would end up with their heads on the *tzompantli,* or skull racks.

Various forms of the ball game may have been played as early as the second millennium B.C., from the northern part of South America to the American Southwest. As mentioned, we are not sure how the Classic Maya played the game, but we have some insights into the Aztec version of the sport. Cortés was so intrigued with the game that he took a troupe of Aztec players to Europe in 1528, where they performed before royal courts.

The Aztec played with an eight-inch-diameter solid rubber ball. It was almost as large as a basketball, but solid and heavy. One to four players on a team controlled the ball Hacky Sack style (no hands), using the lower leg, thigh, torso, and arms to keep the ball from hitting the ground. They attempted to aim the ball through or against stone rings or other markers along the walls of the court or to shoot it into a goal. Scoring was difficult, and the ball itself, because of its mass, could disable a player. Gamblers were omnipresent.[34]

More important, for our purposes, was the nature of the ball used. While Western scholars today accept the Spanish conquerors' tales of the Amerindians' obsession with sport, their description of the amazing ball was hard to believe. The Spaniards told of a sphere that bounced many feet in the air when it was dropped on the ground or hit off a court wall. There was nothing like it in Europe. In fact, there was nothing like it anywhere outside of Mesoamerica until 1839.

Rubber is tapped from a variety of rubber trees, dripping out as sap, or latex. In its natural form, when dried, rubber is soft, sticky, and not very elastic. In 1839 Charles Goodyear developed the vulcanization process, mixing the latex with sulfur and heating it, thus making the rubber hard and bouncy. Reports that the Amerindians were able to do the same thing to rubber in the sixteenth century, and perhaps as early as 1600 B.C., left scholars incredulous.

The raw latex for most Mesoamerican rubber came from the native *Castilla elastica* tree. Its viscous white liquid becomes brittle when dried. Sixteenth-century Spaniards reported that the Mesoamericans mixed the latex with juice from a species of morning glory vine that wraps itself around the latex trees. Recently, MIT archaeologist Dorothy Hosler and undergraduate Michael Tarkanian rediscovered the ancient technique. When Hosler and Tarkanian went to Chiapas to gather raw latex material for testing back in the lab, they saw to their surprise that farmers were collecting latex from *C. elastica* by the same method described in the old documents. About ten minutes after the farmers mixed the latex and vine juice, a mass of rubber rose to the surface. The farmers showed the scientists how to form it into a ball that easily bounced six feet.[35]

The two brought the rubber ball, raw latex, and vine juice back to MIT, where, with materials scientist Sandra Burkett, they analyzed the substances with nuclear magnetic-resonance spectroscopy. The scientists found unidentified organic compounds in the latex that, after the process was completed, were no longer present in the rubber. This suggested that the mystery substances might be plasticizers that keep the latex runny by preventing its polymers from cross-linking. (Today's rubber is made by cross-linking polymers.)

If the vine juice dissolved the plasticizers, they theorized, the polymer molecules of the latex would be more likely to entangle and form a rubbery mass. In the morning glory juice they found traces of sulfonyl chlorides and sulfonic acids, components that react with polymers, stiffening polymer segments and making them more likely to interact. Only a few such entanglements, they said, would be enough to give the mixture its rubberness. Hosler plans to test rubber made with differing amounts of vine juice to see if the Mesoamericans could have designed rubber with specific elasticities.[36]

The Olmecs, Maya, Aztecs, and others made solid and hollow rubber human figurines and wide rubber bands to fasten stone-headed axes to wooden handles. They painted with rubber and used it for lip balm. Above all, they used solid rubber balls in the sacred and sporting ball games that were central to Mesoamerican societies. According to one sixteenth-century source, the "sports authority" of the Aztec capital Tenochtitlán ordered sixteen thousand rubber balls each year as tribute from one province.[37]

Along with rubber, obsidian was a unique Mesoamerican tooling technology. The superhard volcanic glass was to Mesoamerican civilizations what steel is to the modern world. (Except, obviously, glass wasn't used as a building material.) People brought obsidian down from the mountains and fashioned it into knives, lances, dart points, and prismatic blades for woodworking and shaving. The blades were used for cutting out hearts.

It took archaeologists years to figure out how the Mesoamericans made the famous prismatic blades. Basically, through precise chipping, the Indians refined a piece of obsidian into a symmetrical blade core. The upper flat surface of the core was then ground with a basalt tool to roughen it up, and the actual blade was removed from the core by applying a steady force to a small area of the core. Known as pressure flaking, this required a force greater than a person's arm strength, and researchers are not yet sure how the Amerindians did this. Microscopic studies show that obsidian blades have the sharpest edges of any tool, ancient or modern. The edge of a well-made prismatic blade can be sharper than a surgeon's scalpel, and, in fact, some modern surgeons are beginning to experiment with obsidian scalpels.[38]

THE BEST KNOWN of the classical civilizations of Mesoamerica, the Maya originated in the Yucatán around 2000 B.C., emerged from simple pleasant life, and grew to prominence around A.D. 250 in what is now southern Mexico, Guatemala, Belize, and western Honduras. During the Classic Period the Maya lowlands encompassed hundreds of cities and towns, which are now generally buried under an almost unbroken canopy of tropical forest.[39]

Architecture was pivotal to the Mayan civilization. In the lowlands, limestone was everywhere and easily quarried. Since it hardened only when exposed to the air, it was easily worked with Stone Age technology.[40] Lowland Maya always erected their temples atop older ones, and over time, the earliest constructions were deeply buried within the towering accretions of the Classic Period (A.D. 250–900). Much earlier in Petén Yucatán ("Maya Land": Petén is now in northern Guatemala, the Yucatán in Mexico and Belize)—where there was an abundance of limestone and flint tools with which to work it—the Maya had discovered that if lime-stone fragments were burning, the resulting powder mixed with water made a plaster of great durability. They quickly saw the structural value of a concretelike fill made with limestone rubble and marl (a crumbly mixture of clay, sand, and limestone). With these construction elements, they were able to build their cities.[41] The Maya constructed magnificent ceremonial buildings, including the highly ornamented temple-pyramids, palaces, and observatories that are the most striking feature of a Classic Maya city. Built from the hand-cut limestone block on platforms with tiers and stairways, they towered over the surrounding buildings.

The Classic Maya central city included a series of these temple-pyramids of stepped platforms topped by masonry superstructures arranged around broad plazas. In the largest cities, such as Tikal, several such building complexes would be interconnected by causeways. Although the temples usually contained more than one room, the rooms were so narrow they could have been used only on ceremonial occasions, and were not for commoners' eyes. The palaces, which served as administrative centers, were single-story structures built along the same lines as the temple-pyramids, but on much lower platforms and with as many as a dozen plastered rooms.[42] Citizens lived in simple dwellings, also built up on low rectangular mounds of earth and stone to avoid summer floods.

A unique Mayan temple feature is the corbel arch. The corbel vault has no keystone, as Old World arches do, and appears more like a narrow triangle than an archway. It seems to have originated in the roofs of tombs in older periods. Successive courses of stones were set in overlapping rows up to the vault summit, which was capped by flat stones. The great thrust from above is absorbed by the massive walls and the strength of the rubble cement fill.[43] Perhaps the Maya designed the corbel arch because they never had the keystone technology. Or perhaps the lack of keystone was deliberate: the Mayan temple had nine stone layers, representing the nine layers of the underworld. A keystone would have created a tenth layer, violating Mayan cosmology.

The Maya were known for massive structures. El Marador, in northern Petén, Guatemala, featured a network of causeways, raised earthen roads, and connecting groups of monuments, including the Canta pyramid, which reached a height of 230 feet, and with its smaller superstructures constituted in overall bulk, what was possibly the largest structure in Mesoamerica. It was completed before A.D. 200 and perhaps begun as early as 200 B.C.[44] During the Mayan golden age Tikal was huge. It had a six-square-mile area, with an estimated population of ten to ninety thousand inhabitants, giving it a population density several times greater than that of the average European city in the same period.

This drive for mass affected the ball courts as well. In the waning days of the Mayan and during the rise of the Toltec civilization, the two cultures fused to a certain extent. The Toltec-Maya Ball Court at Chichén Itzá, in the northern Yucatán region of Mexico, with walls 27 feet high, 545 feet long, and 225 feet wide, stands today as the largest ball court in Mesoamerica. One sophisticated aspect of the ball-court architecture is the acoustics. Standing in the "end zone," one can clearly hear a whisper from the opposite end.

Spectacular acoustics can also be heard in other Mayan structures—for example, in the famed Castillo, or Temple of Kukulkan, also at Chichén Itzá. With a hand clap at the bottom of one of the massive staircases, a visitor to the temple can produce a piercing shriek reverberating from the top. To some people, the echo sounds like the cry of the quetzal, the gorgeous, exotically feathered bird whose plumery was so prized in Mesoamerica that the animal was on its way to extinction by the ninth century.

The Aztec knew a lot about human anatomy, had names for all of the organs, and understood the circulatory system long before William Harvey made his studies in the seventeenth century. There was a reason for this.

In 1978, power-company workmen in the heart of Mexico City accidentally uncovered a gigantic stone sculpture, instigating a huge excavation of the Temple Mayor, which was probably built during the early years of Tenochtitlán, the Aztec capital city.[45] There were many levels to the structure, but the main temple was often referred to as Coatepec, or "Serpent Hill." It was the site of thousands of human sacrifices. The bloodstained twin stairways of Temple Mayor loomed over this sacred precinct. Below the central temple was a low circular platform dedicated to Ehecatl, the wind god. Below the platform was the *tzompantli,* the skull rack. One of Cortés's soldiers wrote that the *tzompantli* held 136,000 skulls, but this, according to contemporary estimates, is probably an exaggeration. Below the skull rack was the ball court.[46]

Captives were brought through the streets of Tenochtitlán and led up the stairs to the sacrificial altar of the great temple. To the accompaniement of drums, the victim was held down on the altar, and in an instant the obsidian knife of the priest opened his or her chest. The priest then reached in, grabbed the heart, and held it high for all to witness. The Aztec sacrifice machine was nearly insatiable. During one famine year the priests sacrificed more than ten thousand victims, most of whom had been captured in wars. (During the Renaissance, Italian physicians convinced authorities to let them perform autopsies on the bodies of executed criminals to aid in the study of anatomy.)

The Aztecs were more warlike than the Maya or even the aggressive Toltecs. The Aztecs warred because they thought the lands of Mexico were theirs by divine right. They warred to obtain more goods and land to support their ever burgeoning population. And they warred to supply themselves with fresh victims for the great ritual sacrifices.

The Aztecs were the nouveaux riches of Mesoamerica, their dominion due less to inherent genius than to the technology and architectural traditions of the Olmecs, Toltecs, and other vanished peoples who had constructed the city of Teotihuacán (100 B.C. to A.D. 900) near what is now Mexico City. At its apex, between A.D. 450 and 650, Teotihuacán embraced 150,000 citizens and spread over twenty-one square miles, making it one of the largest cities in the world at the time. The Teotihuacáns were conquerors, amassing one of the earliest empires in Mesoamerica, one that came to an end in the seventh and eighth centuries when Teotihuacán burned and was abandoned, for reasons unknown.[47] The Aztecs imitated Teotihuacán and Toltec design early and often.

The Aztecs arrived in the central Mexican valley between A.D. 1200 and 1267, after a hundred-year migration from possibly as far north as the southwestern United States. During the centuries that the Mayan city-states rose and fell and the Toltecs and other cultures flourished, the Aztecs (or Mexica peoples) were hapless nomads, virtual outcasts, wandering in the wilderness. Theirs is a Mesoamerican rags-to-riches story, made possible largely because of the extraordinary engineering and urban planning of their capital, Tenochtitlán.[48] At the time of the Aztecs, the Valley of Mexico, the site of Mexico City today, was a large basin ringed by volcanic mountains. It had deep rich soils with a system of shallow swamps and saline lakes in the center, full of fish, turtles, insect larvae, blue-green algae—all part of a rich diet for the taking.

In the thirteenth and fourteenth centuries the Aztecs built Tenochtitlán in the middle of Lake Texcoco on five islands connected to the mainland by causeways. In many areas canals took the place of streets, and people got about by canoe. When the Spaniards saw Tenochtitlán they called it "the Venice of the New World." Originally, Tenochtitlán was a typical Mesoamerican city arranged around a central sacred zone of plazas. After the victorious border wars of the early fifteenth century, the Aztecs came into their own, and had the power and the resources to rebuild Tenochtitlán to outdo Teotihuacán.

By the 1519 arrival of the Spaniards, the Aztecs had transformed central Mexico into an entirely social landscape, with Tenochtitlán and the agricultural arms of the lake system engineered into the core of an empire with more than a million inhabitants. The Aztecs appropriated past urban design principles to re-create Tenochtitlán in the image of Teotihuacán and other abandoned cities. They applied a grid layout to establish a common alignment for all buildings. Using another Toltec city, Tula, as a template, they redesigned the downtown and civic buildings. They built a walled compound around the sacred buildings, creating a holy city in the heart of Tenochtitlán that was about five hundred meters on a side and covered about thirty-five acres. Limiting public access was part of the Aztec plan to raise their religion to a mystical state.[49]

The glory of Tenochtitlán was tied to the Aztecs' new agricultural strategy. They created large-scale raised fields, called *chinampas,* which covered many miles of what is now Mexico City in long narrow ridges arranged in a grid. (A few fragments remain; for a fee one can canoe through the old canals.) Farmers traveled to the fields via canoe and germinated plants in seedbeds built of floating reed rafts that they towed to individual *chinampas* for replanting. The region was the breadbasket of the Valley of Mexico. However, it's possible that by the time Cortés and his army arrived, in 1519, the valley had already reached its carrying capacity.[50]

THE INCAS were an early people in South America. Our evidence dates to the first century A.D., when they began moving into the Valley of Cuzco, Peru. In the twelfth century they began to dominate other societies, and by the fifteenth century they had conquered more than twenty-one different societies, exercising control over more territory than any other people had done in the pre-Columbian Western Hemisphere. This final empire, Tahuantinsuyo, "Land of the Four Quarters," lasted only about a century, but stretched more than 2200 miles, as far north as today's Ecuador and the southern edges of Colombia, and as far south as the southern Andes of Chile and Argentina. The Incas conquered vast territories without wheeled vehicles or animals that could be ridden. Their empire embraced an enormous variety of environments, as well as diverse cultures. From 1438 to 1532, the Incas incorporated more than a million subjects into a strongly hierarchical sociopolitical system. Based on an organization sometimes compared to communism, this tightly controlled system collapsed after its emperor, Atahuallpa, was garroted by Spanish invaders in 1533. How the Incas incorporated such a diversity of sociopolitical entities into an empire is a poorly understood wonder.

Of all the urban peoples of the New World, the Incas were the most brilliant engineers. The Inca Road represents the essence of their organizational and engineering talents harnessed to their will for empire. Consisting of approximately twelve thousand miles of roadways, the Inca highway constituted a transportation network comparable only to the Romans' for its time. Since they did not use the wheel, the roads served only foot travelers and llama pack animals. Two royal Incan roads were built, on which Incan rulers traveled the length of the empire. One royal road followed the Pacific coast from northern Peru to northern Chile; the other ran through the Andean highlands from modern Colombia to northern Chile. The Incan "highway authority" built tunnels and bridges through the mountains, erected levees across swamps, and carved steps into steep rock slopes.

The royal Incan highways were fed by arterial routes and tributaries used for transporting maize, coca, metals, and gems. Even today, archaeologist Thomas Lynch writes about one of the most desolate high desert regions in South America: "One can follow the Inca Road through the Gran Despoblado, by means of its turquoise litter, about as easily as Hansel and Gretel traced their route by bread crumbs."[51]

The Incas used the road system for a number of social purposes: communication, restructuring political boundaries between ethnic groups, imposing the Quechua language of Cuzco on conquered subjects, and raising revenues.[52] No one could enter Cuzco without a tribute relative to their status. No matter how highly placed a nobleman, no one could appear empty-handed before the ruling Incas.[53]

The principal lord of a province used a counting instrument called a quipu, a knotted string devised for mathematics, to tote up the tributes. It was an effective toll. When the lord caught subjects cheating, he meted out punishment. Even slight offenses were punishable by an unpleasant and slow death.[54] (The quipu was more than a counting device. Alfred Crosby says that it was also a memory aid, a precursor to writing.)[55]

Africa

Africa is the only continent to extend from the northern to the southern temperate zone. It also is the most likely birthplace of *Homo sapiens* and where, in the gorges at Olduvai in Tanzania and on the shore of Lake Turkana at the Great Rift Valley, our precursors began to fashion their first tools.[56]

Africa is a continent with two stories, one detailed, one mysterious. Though the history of North Africa is well documented, the long history of sub-Saharan Africa is shrouded in mystery, partly because few of its cultures developed a written language and partly because the historical sciences are just now beginning to learn how to cope with its history. The old European bias of the "Dark Continent" collides painfully with newer, perhaps more egalitarian methods of examining African civilizations' unique contributions to the world. The Africans' influence on the "classical" cultural expansions to the north is an area of huge controversy, embroiled in the politics of race and historical accuracy, as exemplified by the "Black Athena" theory of Martin Bernal.[57] It is becoming clear that some African cultures were major players on the ancient stage and were technologically well in advance of Europe during certain eras, if not more advanced than the Middle East and Asia as well.

From around the third century B.C. to the third century A.D., Egypt shared Africa with two other impressive civilizations: the Cush and Axum. From the early part of the second millennium B.C. to the end of the fourth century A.D., the land of Cush, including the territories in the valley of the Upper Nile in the land now called Nubia, was inhabited by various groups. By the first millennium they had unified almost the entire Nile valley, from the border of contemporary Egypt south to Khartoum, the capital of today's Republic of Sudan. These peoples established a hydraulic culture, based on irrigation agriculture, making Sudan today one of the world's oldest continuous centers of civilized life.[58]

The Cushites were responsible for introducing to the Upper Nile valley region a tradition of iron metallurgy that still exists there. Cush's last and legendary capital was Meroë, located near the junction of the Atbara and Nile Rivers in central Sudan, an area the Greeks called Aithiopia ("Land of the Burnt-faced people").[59]

Cush, with its powerful military, ruled Egypt for a half century, from the eighth to the seventh century B.C. Cush created a renaissance in Egyptian culture and made Egypt an active player again. Cush armies helped delay the conquest by the Assyrians of the smaller cultures of the eastern Mediterranean basin, such as the Jews and Phoenicians.[60] Good trading relationships and increased demand for the African goods that Cush could supply brought wealth to its rulers and a great expansion of population to the Upper Nile. Cush reached its zenith about 200 B.C.

The first metals Africans used were gold and copper, which could be more easily worked than iron. Neolithic people of lower Egypt knew of gold—*nub,* in the old Egyptian tongue—from Nubia before 4000 B.C.[61] Although gold deposits could be found throughout the Upper Nile valley, the most productive mines were located in the deserts of the Nile in lower Nubia. Gold-processing equipment, abandoned mine shafts, and ruins of ancient mining settlements are not uncommon in southern Egypt and northern Sudan. The nightmarish conditions under which criminals and political prisoners worked the gold mines are described by the second-century B.C. historian Agatharchides of Cnidus.[62]

Meroë may have been the mother civilization of the Iron Age in Africa, and iron is the key to understanding Africa from the mid–first millennium B.C. onward. The Cushites probably had their first confrontation with iron when they tasted the tempered iron weapons of the Assyrian invaders, sometime around 600 B.C. The Egyptians knew about iron for centuries but ignored it for the most part. It remained rare in Egypt, until the Meroites initiated smelting works on a huge scale, by the middle of the first century B.C.; they would develop the largest iron-smelting industry on the continent. Modern archaeologist Basil Davidson has called Meroë "the Birmingham of ancient Africa."[63]

Knowledge of ironworking technology could have traveled south to west-central Africa before the first century B.C. The earliest secure dates begin around 500 B.C. in the Nigeria, Cameroon, and Gabon area and in the Great Lakes region in the Great Rift Valley. The technology moved in a north-to-south trajectory into southern Africa. It was slowed by the transdesert distances and by the fact that Cushite rulers tried to keep the craft of ironworking a closely guarded secret of the royal or priestly castes. The slag heaps of Meroë lie only a few hundred yards from the ruins of the Temple of the Sun.[64]

Cush fell in the fourth century A.D. to a combination of attacks by nomadic peoples from the east and invasions by the forces of the growing empire of Axum, which became a political giant in Africa.[65] Throughout the kingdoms of Cush and Axum, iron smelting and international trade, especially in gold and iron, were major growth factors. Peoples of the African forest belt had iron technology sometime before the eleventh century, and the earliest origins of ironworking south of the Sahara can probably be stretched back to before 300 B.C.

By 1067, the year after William of Normandy crossed the English Channel, a report came out of Ghana describing its capital region, Kumba Salah. Long vanished, it was said to be filled with grand houses, with pavilions gilded in gold, and with kings whose dogs had collars of gold and silver. The Ghanaian king not only knew the trading value of gold but also understood the concept of reserving it, allowing most of the gold dust to remain with the village miners to keep it from flooding the market. One cannot help but make comparisons, notes Basil Davidson, with the grim cities of Europe at the time.[66]

Miners in West Africa in the Middle Ages employed methods of shaft-sinking, extraction, and refinement that were sui generis. In southeastern African, documentation reveals a flourishing iron and gold industry. In 947, a Baghdadi, Abdul Hassan ibn Hussen, wrote a travel book entitled *The Meadows of Gold and Mines of Gems,* about an East African coastal people he called the Zanj, who had a capital on the Zambesi River. The Zanj were skilled workers in metal who prized iron over gold. They transported the iron to India, where they sold it for a good price. The Indians turned it into their *wootz* (a high-quality steel), which was sold throughout the medieval world and forged into Damascus blades, among other things. The Crusaders encountered the sharp end of Saracen weapons, which were made of steel mined in Africa, forged in southwestern India, and fashioned in Persia and the Middle East. Thus southeast Africa was an integral part of the widest trading circuit in the medieval world.[67]

THE AFRICAN CONTINENT, especially sub-Saharan Africa, has yet to give up its best secrets. Only an hour's drive from Lagos, in the rain forests of southeastern Nigeria, British archaeologist Peter Darling is studying the remains of one of the largest monuments in sub-Saharan Africa: a hundred-mile-long wall that was built sometime between A.D. 800 and 1400. (A carbon analysis of part of the ramparts dates to the tenth century.)[68]

This nearly eight-hundred-square-mile complex of earthern ramparts may be the world's second-largest human-made structure (after China's Great Wall). Stretched end to end, the wall, which comprises more than five hundred interconnected communal enclosures, would measure ten thousand miles long. Its reddish banks rise seventy feet in the air, and its builders would have had to move more material than the builders of Egypt's largest pyramid. "There is nothing whatever like this in Europe," writes Crosby.[69] With the help of a team of Nigerian and British colleagues, Darling estimates that most of the complex was constructed over a 450- to 650-year period, when much of the area was conquered by the power later known as the Benin Kingdom. The ruins suggest a highly organized culture. This monument, called Sungbo's Eredo, may have functioned as boundaries separating communities of related people who were forbidden to intermarry, Darling suggests. He also suspects that the walls and moats served less as a physical barrier than a spiritual one, dividing the real and spirit worlds. "It's like a yellow line for not passing," he says.[70]

India

In the third millennium B.C. in the Indus valley, a little-understood civilization sprang up that may have been as large as or larger than the empires of the Fertile Crescent. Its people seem to have placed more value on well-organized cities than on awe-inspiring temples, palaces, and conquest. Stone Age farming towns seemed to leap almost overnight into complex, well-engineered, even elegant urban environments. Small, immaculately constructed objects like seals and beads appear to have been the most highly prized cultural objects. Terra-cotta ceramics were fired at high temperatures to make stoneware, using technologies reinvented only centuries later in China. There is little evidence of warfare. The same weights and measures were used for more than a thousand years, an incredible feat in the Bronze Age.

Situated in the western part of southern Asia, in what is today Pakistan and western India, this culture is often referred to as the Harappan civilization, after Harappa, a city first rediscovered five thousand years later. Harappan traders set up flourishing colonies in the Persian Gulf and Mesopotamia. They may have been the top exporters of their time.

Excavations show that Harappa was far larger than once thought, perhaps supporting a population of fifty thousand at certain periods. One of the other principal cities, Mohenjo-Daro, is in Sind, Pakistan, next to the Indus. Here, in the 1920s, archaeologists uncovered the Great Bath, uniform buildings and weights, hidden drains, and other hallmarks of the civilization.

Mohenjo-Daro was dominated by a large citadel built on an artificial hill. On the same summit was the Great Bath, twenty-nine feet long, twenty-three feet wide, and ten feet deep. The pool was lined with bricks laid in two layers with an inch-thick layer of bitumen in between to hold the water. An outlet in one corner of the bath led to a large drain. Whether the Great Bath had a ritual purpose is uncertain, but washing was clearly a central part of life in the Indus valley cities.[71]

The third millennium B.C. was the "Age of Cleanliness." Toilets and sewers were invented in several parts of the world, and Mohenjo-Daro circa 2800 B.C. had some of the most advanced, with lavatories built into the outer walls of houses. These were "Western-style" toilets made from bricks with wooden seats on top. They had vertical chutes, through which waste fell into street drains or cesspits. Sir Mortimer Wheeler, the director general of archaeology in India from 1944 to 1948, wrote, "The high quality of the sanitary arrangements could well be envied in many parts of the world today."

Nearly all of the hundreds of houses excavated had their own bathing rooms. Generally located on the ground floor, the bath was made of brick, sometimes with a surrounding curb to sit on. The water drained away through a hole in the floor, down chutes or pottery pipes in the walls, into the municipal drainage system. Even the fastidious Egyptians rarely had special bathrooms.[72]

The Indus architects designed sewage disposal systems on a large scale, building networks of brick effluent drains following the lines of the streets. The drains were seven to ten feet wide, cut at two feet below ground level with U-shaped bottoms lined with loose brick easily taken up for cleaning. At the intersection of two drains, the sewage planners installed cesspools with steps leading down into them, for periodic cleaning. By 2700 B.C., these cities had standardized earthenware plumbing pipes with broad flanges for easy joining with asphalt to stop leaks.[73]

The Harappan civilization links the ancient Indus valley and the modern subcontinent. While the details of the civilization's demise are unknown, many of its elements have survived, from prototypes of Hindu gods to current Punjabi burial practices. Even the rupee monetary system, used until recently from Dhaka to Peshawar, seems to stem from base-eight Indus mathematics.[74]

THE INDIAN SUBCONTINENT was never famed for mechanical invention—machines with gears, pulleys, cranks, or cams were not big items there. The Indians followed a different trail, compared with the technical developments of China or the Middle East. The Indian subcontinent made contributions to the world's technology in such things as food crops, minerals, and metals.[75]

India began the organized hunt for diamonds (the Sanskrit word translates to "thunderbolt") and for centuries exported them to Europe for jewelry and talismans. For generations Indian craftsmen had whirled diamond bits to drill holes in beads of extremely hard stone.[76] Some of the highest-quality steel in the world was also made in India, as early as A.D. 600. Known as *wootz,* it was limited primarily to use in swords and knife blades. *Wootz* is most famous, and most valuable, for its role as the raw material of Damascus swords, made in Syria. Nineteenth-century European metallurgists were puzzled by the unique quality of *wootz* steel. One factor was the quality of the African ores used by Indian steelmakers, but more important was their technique. *Wootz* was produced only in small quantities and at great cost of fuel and labor.[77] The smelting technique was crude by the standards of the Chinese, yet the Chinese tried without success to imitate the quality of steel it produced.

THE INDIAN TECHNOLOGY that had the most profound effect on the West was textiles. Almost every house in medieval India had a spinning wheel, yet we do not know with certainty that India invented the device. The spinning wheel is of particular interest to historians of technology because it features an early application of an endless belt drive and may demonstrate the first use of the flywheel in machines. There is still controversy about whether the spinning wheel first appeared in China, Persia, or India. It is first mentioned in India around A.D. 500, and in Europe in the thirteenth century, introduced into Spain by the Arabs.[78]

In the first millennium A.D. India's workers were kept busy manufacturing luxury textiles for the royal palaces. One *karkhana,* or royal factory, in New Delhi employed four thousand silk workers. Some of them may have been conscripted. Such factories were highly efficient in meeting the courts' demands for luxury goods but could not respond to a varying market demand. By medieval times, every city, town, and village manufactured textiles, and some were famous for a particular kind. Dacca muslin, for example, was the finest. It was said that a piece twenty yards long and one yard wide could pass through a finger ring. Buhranpur calicoes were exported to Persia, Turkey, and the rest of the world.[79]

Their colors did not fade when washed, and their brilliance of color and pattern was coveted everywhere.

Cotton textiles originated in India, and their production, like so much else in India, was labor-intensive. It involved much beating and cleaning. Indian cottons were Z-spun (revolving left to right), rather than, as was more common, S-spun (revolving in right to left). Yarns can be washed with less damage with Z-spinning. Women covered their fingers with chalk to keep their sweat off the fine but strong yarn. The weaving process was laborious. "Specialists" who did the bleaching, for example, belonged to a separate caste. They boiled the cloth in lime, lemon, and special soap and took it to a river or pond to beat it vigorously on stone slabs; they then spread it out in the open to dry and steamed it. This process produced a calico, for example, of a very close fabric, though the bleaching process also inevitably caused damage to some batches.[80]

The superiority of Indian fabrics continued through the Middle Ages and into the 1700s, by which time India had long been the world's leading exporter of textiles, sending its silks and cottons to Europe, Africa, and Asia.[81]

Indian textile technology had a profound influence in Britain during the industrial revolution, stimulating inventors there to devise methods to attain similar results—the brightness and permanence of the colors, the delicacy of the cotton yarn—with machines. The British had little success in attaining the quality of handmade Indian textiles. British spinners showed little interest in how their Indian counterparts achieved the high quality of their textiles and would have been disappointed had they known. The secret was painstaking and laborious hand spinning.[82] The Indians focused on the quality of the final product with little emphasis on the production process. The machines used for the production of yarns and cloth were always of considerable interest to Western minds, while in India the equipment was seen mainly as a means to an end.

The Western bias toward things mechanical was crucial for the industrial revolution, but paradoxically, even in the late nineteenth century the superior production equipment of Europe produced poorer quality textiles. Yet the economics of the subcontinent stultified its textile industry. Indian fabric processes were operated on a large scale, using massive labor and minimal equipment. Wages were low, so banks and merchants had little impetus to invest their capital in better equipment, and instead spent their money on ships to transport the exports.[83]

China

The three inventions Bacon considered world transformers—paper and printing, the magnetic compass, and gunpowder—were also cited by Karl Marx as the inventions that prefigured capitalist economics. Bacon regarded the origins of these inventions as "obscure and inglorious." They all came from China.

At the beginning of the second millennium A.D., China was an advanced scientific and technological society, and would continue to dominate for another three or four centuries. To a visitor from another continent it might seem that China had invented everything anyone could ever need and beyond. Besides Bacon's big three, other Chinese technological feats included cast iron, porcelain, sternpost rudders for ships, canal lock gates, stirrups and harnesses for horses, fishing reels, hot-air balloons, the seismograph, whiskey, gimbals, the umbrella, crank handles, kites, mechanical clocks, paper money, convertible bank notes, and many agricultural innovations, such as row cultivation, the iron plow, and the seed drill. The Chinese also spun off, with glorious abandon, oddities such as the south-pointing carriage, fantastical fireworks, magic mirrors, and a rocket-propelled toy called an "earth rat."

THE INVENTION we most associate with ancient China is gunpowder. In the ninth century A.D., during the Tang dynasty, Chinese priests described a new compound they'd created by combining charcoal, saltpeter, and sulfur in the proper proportions. Long before the first written observations of these investigations, the Taoist alchemists were down in the basement mixing up variations of these ingredients, often blowing themselves to smithereens. Later Taoist literature strongly recommends that investigators not mix these chemicals, especially with arsenic, since some who had done so set their beards on fire, seared their fingers, and burned down the house.[84]

One hypothesis holds that gunpowder was invented by alchemists searching for a drug of deathlessness, or for the metallurgical key to the making (and faking) of gold. One can imagine, wrote Joseph Needham, these alchemical adepts "mixing everything off the shelves in all kinds of permutations and combinations to see what would happen, whether perchance an elixir of life would be formed."[85]

Saltpeter was recognized and isolated at least by A.D. 500. It seemed almost inevitable, wrote Needham, that "the first compounding of an explosive mixture would arise in the course of a systematic exploration of the chemical and pharmaceutical properties of the substance."[86]

In *Science Since Babylon,* Derek de Solla Price says that while science must follow what seems to be a dictate of nature rather than a property of our mental perspective, technology is an arbitrary property of a civilization.[87] A technology evolves within a culture and its particular demands and preoccupations, intertwined with that society's particular environment. That being so, it is not surprising that the Chinese were the first to invent gunpowder.

The Chinese were fascinated and preoccupied with preparations of perfumes, gases, airborne poisons, noxious bombs, explosions, and flaming eruptions. From the Ch'in and Han dynasties onward (221 B.C.–A.D. 220) they burned incense; fumigated for health reasons, to rid their houses and books of insects and pests; and produced smoke ritually to drive out demon spirits. Smoke, detonations, and loud explosions were intrinsically associated with the spirit world. Militarily, they used toxic smoke screens generated by pumps and furnaces in siege warfare from the fourth century B.C., or perhaps earlier.[88]

The Chinese did (and do) love fireworks, and created them in a huge variety of Catherine wheels, Roman candles, and many other styles. Fireworks flourished at the dynastic courts, with colored lights and balls of flame. Rockets and rocket-composition gunpowder must have been used in these displays as soon as they were discovered.

Around 1040 Tseng Kung-Lang published a gunpowder formula to be used in a variety of weapons, including an incendiary arrow, an incendiary bullet, a burning bomb with a hook to catch on wood, a bomb to be hurled by a trebuchet (a Chinese version of the catapult), and a hand grenade. By the mid–tenth century, the fire lance, or fire spear, had appeared.

The oldest image of a fire lance and a grenade is on a silk banner from Tun-huang from about A.D. 950 now hanging in the Musée Guimet in Paris. The banner depicts the meditating Buddha. Surrounding him are Mara the Tempter and her minions, who hurl things at the Buddha in an attempt to distract him from attaining enlightenment. One of her demons, sporting a headdress of three striking snakes, aims a cylinder from which flames spout forth horizontally. Another is in the act of throwing a weak-casing bomb from which flames are starting to fly.[89]

The fire lance consisted of a tube mounted on the shaft of a lance and filled with a mix of gunpowder, toxic chemicals, lead pellets, and pottery fragments. When ignited it spouted flame and sparks for about five minutes, frying the enemy in streams of fire.[90] Made first from bamboo tubing, the fire lance used homegrown materials. Like the natural abundance of saltpeter in the ground, the plentiful growth of bamboo was a factor in the development of firearms. As a natural tubing, Needham maintains, the stem of the bamboo is the ancestor of all barrel guns and cannons.[91] Later the tube was made of cast iron and bronze.

The fire lance played a large role in the wars between the Sung and the Juchen Tatars from around 1100 onward. By the middle of the thirteenth century, the Sung and Mongols were locked in combat, and by 1230 we find written descriptions of destructive explosions in the campaigns, and accounts of continuing advances in the development of barrel guns and cannons. At first, soldiers held fire lances. The southern Sung made them in much larger diameter, perhaps a foot across, and mounted on legs with wheels. It is with these that the first bronze or iron barrels appeared, using high-nitrate gunpowder and a projectile—a cannonball or bullet—that completely filled the barrel. The true gun or cannon probably appeared in the 1280s, three and a half centuries after the invention of flamethrowers.

By 1288, Chinese soldiers under Mongol command were using weapons that had made the transition from fire lance to gun. A bronze barrel found at a battle site in Manchuria was meant to fit on the end of a wooden shaft. It was designed for an explosion at the base of the barrel, not for slow burning from the barrel mouth. The bronze has thicker walls and a touchhole in the area where the explosion would occur. The thickening of gun barrel walls around the point of the explosion became a distinctive characteristic of Chinese guns. Another prototype, designed for mounting in a fortification, looked like a vase or bottle.[92]

The array of gunpowder weaponry developed by the Chinese starting in the ninth century is of Strangelovian proportions: the "thunder-fire whip," a fire lance in the shape of a three-foot-long sword that discharged lead balls the size of coin; the "vast-as-heaven enemy exterminating Yin-Yang shovel," with a broad crescent-shaped blade that emitted poison as well as lead pellets and flames. There was a huge battery of fire lances called "the ingenious mobile ever-victorious poison-fire-rack." Later there came the "cartwheel gun," which had thirty-six barrels radiating from its center like the spokes of a wheel but was small enough that a mule could carry two.[93]

For mortars you had "the flying, smashing, and bursting bomb-cannon." By the eleventh century there was the "thunderclap bomb," hurled from a trebuchet, that terrified enemies' horses while starting fires. Thunderclaps were also made in the form of grenades that could be hurled by hand. A new improved bomb in the twelfth century was the "thundercrash bomb," with an iron casing to cause maximum shrapnel damage. The Chinese were just getting started. They let a thousand bomb varieties bloom: some packed with anti-personnel material, poison bombs, gaseous bombs, bombs filled with human excrement. There was also the "bone-burning and bruising fire-oil magic bomb," the "magic fire meteoric bomb that goes against the wind," the "dropping-from-heaven bomb," and the "bees-swarm bomb releasing ten thousand fires."

By 1277 the Chinese had developed land mines; one was called "the ground-thunder explosive camp." Some of the trigger mechanisms of these land mines were kept secret until the seventeenth century. The *Fire-Drake Artillery Manual,* published in 1412, describes the "submarine dragon-king," a complex wrought-iron sea mine carried on a submerged wooden board. This device for blowing up ships featured a burning joss stick floating above the water that determined the fuse ignition time.[94]

In 1245 Pope Innocent IV sent an ambassador to the great khan's capitol in Mongolia, most likely to check out the fabled firepower the Mongols had picked up from their enemies to the south. Soon thereafter other Europeans visited, including one Willem van Ruysbroeck, a Franciscan who returned to Europe in 1257 and told his associates about gunpowder weapons. The following year, Europeans began experimenting with gunpowder. Other Westerners discovered gunpowder the hard way, in their warring with Islamic nations. In 1249, Crusaders ran into an Islamic counterattack of incendiary devices and grenades in Palestine. The effect was horrific.[95]

The Europeans learned quickly. A picture of the bombard, a small bulbous cannon that fired arrows, appears in a 1327 manuscript, *On the Majesty, Wisdom, and Prudence of Kings,* in the Bodleian Library at Oxford. Chinese drawings of bombards reveal sets of them mounted on a carriage, similar to the first European ones. Copies? "If so, it would mean the purely propellant phase of gunpowder and shot, [the] culminating stage of all gunpowder uses, was attained in China with bottle-shaped bombards before any knowledge of gunpowder itself reached Europe," says Needham. It appears that the entire line of development took place in China first, and passed to Islamic nations and then to Europe.[96] The export of gunpowder and guns to the West led to the utter transformation of Europe.[97]

THIS WAS NOT the first time inventions from China had revolutionized Europe. The widespread use of the Chinese stirrup in the early Middle

Ages had given birth to the knight, a warrior now able to stabilize himself on his horse. The advent of gunpowder blew away that knight, perched like a big immobile target on his horse. Gunpowder that could punch holes in the heaviest fortifications signed the death warrant for the castle and Europe's aristocratic military feudalism.[98]

While Europe was broken into hundreds or thousands of small economic and social units, the Chinese *usually* lived under a powerful centralized administrative authority with close internal commerce and a unified language, writing, and religion. (The operative word is "usually." In between stretches of order, barbarians kept barging in from the north, and there were, according to Alfred Crosby, "periods of godawful instability.")[99] Maintaining stability required military strength, hydraulic control, transportation systems, a calendar, land measurement, technology, map drawing, palace building, and other construction technologies to display the images of imperial power.[100]

Metallurgy and metal manufacturing was a unifying technology. The Chinese "industrial-dynastic-military complex" was a voracious consumer of iron and steel products. Records from the eleventh century show a single order for nineteen thousand tons of iron just to make coins. The million-men-plus army maintained by the Sungs was a giant maw for iron and steel: two government arsenals manufactured thirty-two thousand suits of armor a year.[101]

A superb bronze and cast-iron metallurgy was part of what the physiologist Jared Diamond calls an autocatalytic process, one that catalyzes itself in a positive-feedback cycle, proceeding ever faster once it is started.[102] Long before iron and bronze casting provided the receptacles for gunpowder weapons, the early mastery of cast iron led to the sharp axes that opened up vast areas to forestry; it provided craftsmen with honed chisels, awls, saws, and other tools of a firmness previously unknown. Cast iron allowed new kinds of construction for buildings and bridges and the hard rotary bits for a deep-drilling industry not seen in the West until the seventeenth century. From around the sixth century B.C., the Chinese were adept in cast-iron forging in special vertical blast furnaces. With the vertical furnace, iron and steel technology in China diverged from that of other regions of the world and followed a unique path.[103]

The Chinese were blessed with clays with high refractory qualities, which they used for the walls of their blast furnaces, thus intensifying the heat. They discovered that phosphorus reduced the temperature at which iron melts. By the fourth century B.C., the Chinese were able to cast iron into ornamental and functional shapes.[104] In the West, blast furnaces are known to have existed in Scandinavia by the late eighth century A.D., but cast iron was not widely available in Europe before 1380.

By the third century B.C., the Chinese had discovered annealing (heating then cooling) techniques for making a malleable, nonshattering cast iron. Plowshares could survive hitting large rocks; swords could clang with impunity. So plowshares, longer swords, and even buildings were eventually made of iron. During the Han dynasty (206 B.C.–A.D. 220), iron was of such interest to the officials that in A.D. 119 the rulers nationalized all cast-iron manufacture. During the Han there were forty-six Imperial Iron-Casting Bureaus throughout the country where bureaucrats supervised the mass production of cast-iron goods.[105]

Chinese iron making inspired a continuous stream of inventions. First were the agricultural tools: cast-iron hoes in the sixth century B.C. and a new model in the first century B.C. called the "swan-neck" hoe capable of weeding around plants without damaging them; the moldboard plow was invented in the third century B.C. Called the *kuan,* it was made of malleable cast iron, with a central ridge ending in a sharp point to cut the soil, and with wings that sloped gently up toward the center to throw the soil off the plow to reduce friction.

Again, the introduction of Chinese iron agricultural tools to the West revolutionized European culture. Intensive hoeing and the iron plow were perhaps the greatest technological advantages China held over the rest of the world. "Nothing underlines the backwardness of the west more than the fact that for thousands of years, millions of human beings plowed the earth in a manner that was so inefficient, so wasteful of effort, and so utterly exhausting, that this deficiency of sensible plowing may rank as mankind's single greatest waste of time and energy," writes sinologist Robert Temple. Throughout the first millennium B.C., the Chinese refined the iron plow. When the newfangled plow (along with the Chinese seed drill) finally arrived in the Netherlands and England in the seventeenth century, it instigated an agricultural revolution.[106]

THE CHINESE were making steel by the second century B.C., although they were probably not the first civilization to do so. They furthered metallurgical technology with at least two inventions that were to be reinvented centuries later in the West. One is what we call the Bessemer steel process today, invented in England by Sir Henry Bessemer in 1856. Bessemer's work had been anticipated a few years earlier by William Kelly, who brought four Chinese steel experts to a small town near Eddyville, Kentucky, in 1845. The experts taught Kelly the secrets of steel production that had been used in China for more than two thousand years.[107]

In short, the Bessemer process is the removal of carbon from iron. Cast iron is brittle because it contains a large amount of carbon, about 4.5 percent. To get steel, one removes most of the carbon. (For wrought iron, nearly all the carbon is removed.) As carbon is removed, the metal gets more supple. Steel with high carbon is strong but is more brittle than lower-carbon steel. The Chinese used different carbon contents to great effect. For example, the back, blunt edge of a saber might be made of wrought iron, for elasticity, while the cutting edge would made of harder steel. The Chinese removed carbon from cast iron by blowing oxygen on it, a technique similar to the one "discovered" by Henry Bessemer in the nineteenth century. The Chinese technique is described in the classic work *Huai Nan Tzu,* published in about 120 B.C.[108]

In the fifth century A.D., the Chinese invented another steel-manufacturing process, in which cast iron and wrought iron were melted together to yield steel. In the modern world this is called the Siemens process, invented in 1863 in England. The Chinese were doing it fourteen hundred years earlier. It is more properly called the Ch'iwu

Huai Wen process, in honor of the metallurgist who made sabers of "over-night iron" by baking wrought and cast iron together for several days and nights.[109]

With a variety of irons and steels of differing hardness and flexibility, the Chinese did more than build spiffy swords. They used wrought iron, for example, to construct the world's first suspension bridges, possibly as early as the first century A.D., using chains of wrought-iron links instead of woven bamboo. By comparison, the first suspension bridge in the West of any size was built in 1809 across the Merrimack River in Massachusetts.[110]

CHINESE METALLURGICAL advances made possible a whole range of innovation. In A.D. 976, for example, an engineer named Chang Ssu-Hsun invented the chain drive for use in a large mechanical clock. The Chinese were fascinated with chains and clocks. Since the first century A.D., they had used iron-linked chain pumps and the common sprocket chain to transmit power in clocks and elsewhere.

Chang Ssu-Hsun's successor, the even more famous clockmaker Su Sung, also adopted the chain drive for his huge astronomical clock, in 1090, calling it the "celestial ladder." The first European chain drives were made in the eighteenth century, and in 1897, chain drives became the basis of the bicycle. It is ironic, Temple comments, given that bicycles are a leading form of transportation in China, that only a few Chinese have any idea that the chain drive was a native invention nine hundred years in advance of its application in Europe for the bike.[111]

THE FIRST completely printed book is thought to be the *Buddhist Diamond Sutra,* completed in A.D. 868 and now preserved in perfect condition in the British Museum. A scroll 17.5 feet long and 10.5 inches wide, it contains the text of a Sanskrit work translated into Chinese. There were also large print runs for ordinary books. Calendars and horoscopes were as popular then as now. In fact, so many astrological calendars were being privately printed that in 858 the governor of the Szechwan province tried to ban them. They were sold under the counter in marketplaces before the Board of Astronomers could approve and issue them. The prohibition spurred sales of these calendars, which contained weather forecasts, prophesies for lucky and unlucky days, edifying sayings, and other Farmers' Almanac types of things.[112]

Writing is the unification technology par excellence of civilization. Chinese writing is preserved from the second millennium B.C. but probably began earlier. The Hsia dynasty, c. 2205–1766 B.C. and shrouded in legend, may have had rudiments of literacy. Inscriptions from the Chou dynasty from 1100 to 221 B.C. record the conquest and absorption of non-Chinese-speaking populations by the Chinese states. (Anthropologist Claude Lévi-Strauss wrote that ancient writing's main function was to "facilitate the enslavement of the other human beings.")[113]

Although writing evolved around the same time in Egypt and Mesopotamia, the Chinese writing of 1300 B.C. had unique signs and principles that lead most scholars to think it evolved independently. The preserved writing of those times consists of religious divination and ritual inscription about dynastic affairs incised into "oracle bones."[114] Before paper's invention, words were written on various materials—on grass stalks by the Egyptians, earthen plates by the Mesopotamians, tree leaves by the Indians, sheep skins by the Europeans, and even on tortoise shells and shoulder blades of oxen by the early Chinese. Then the Chinese invented paper.

The oldest surviving piece of paper in the world comes from a tomb near Sian, in Shensi Province. It was made sometime between 140 and 87 B.C. from pounded and disintegrated hemp fibers.[115] From this and other fragmentary evidence it is clear the Chinese knew the general mechanics of papermaking one thousand years or more before the Europeans. (Paper is not that complicated. It's a layer of disintegrated fibers in a watery solution pressed onto a flat mold. The water is drained away, the layer is dried, and you have paper.)

Although most early Chinese paper was made of hemp, in the second century A.D. a court official named Cai Lun produced a new kind of paper from a mix of bark, rags, wheat stalks, and other things. Perhaps the first recycled paper, it was also the first modern paper. It was fairly cheap, thin, light, strong, and suitable for brush strokes. The Chinese also used paper for clothing, shoes, and toilet tissue, which amazed the Europeans when they first saw it. They invented wallpaper, kites, umbrellas, paper money, the paper-folding art of origami, and more. Paper reached India in the seventh century, and the Islamic nations a hundred years later. For five hundred years the Arabs jealously guarded the secret of papermaking from the Europeans, but sold paper to them at a hefty profit. Paper manufacturing did not come to Europe until the thirteenth century, when the Italians took it up.[116]

THE BEGINNINGS of printing are lost in history. About two thousand years ago in the Western Han dynasty (206 B.C.–A.D. 28), stone-tablet rubbing was the favored way to spread Confucian texts or Buddhist sutras. The practice of block printing began in the Sui dynasty (A.D. 581–618): one engraved writing or pictures on a wooden board, smeared the board with ink, then printed the image on pieces of silk (or, later, paper) page by page. During the Tang dynasty (618–907), the technology spread to Korea, Japan, Vietnam, and the Philippines.

Block printing was cumbersome, with boards that were sometimes useless after one printing. A single mistake in carving could ruin a whole block. Between 1041 and 1048, Pi Sheng (sometimes called Bi Sheng) invented movable type. He carved single characters on pieces of fine clay as thin as the edge of a copper coin, which he slow-baked until extremely hard. He then set the type in an iron frame and stuck it to an iron plate with a mixture of resin, wax, and paper ash melted over fire. A plate thus prepared could print hundreds or thousands of sheets of paper. Each piece of type could be removed to be used again.

The first record of Bi Sheng's invention is found in the 1086 book *Dream Pool Essays* by the scientist-encyclopedist Shen Kua. It was not unusual for a chronicler to own fifty thousand books, he wrote. To publish books with Chinese characters, a printer might need up to 360,000 pieces of type. In the centuries that followed, the Chinese used wood, enamelware, or metal type more commonly than clay.

The American physicist and essayist Philip Morrison noted in 1974 that when Gutenberg first set the Mainz Bible in print, "Chinese libraries already held editions of printed books older than Gutenberg's product is now." For every *Book of Songs* or *Analects* the West has, wrote Morrison, there are ten thousand printed texts from every period of China.[117] The Mongol armies pressing into Russia, Poland, and Hungary in the thirteenth century reached the borders of Germany not long before printing surfaced there. Johannes Gutenberg printed his now famous Bible using movable type in 1456.

Perhaps the non-Western world peaked too soon, technologically speaking. By inventing a method of vulcanizing rubber a thousand years before Goodrich or originating the Bessemerizaton of iron a thousand years before Bessemer, these ancient inventors may have given the West a chance to "reinvent" and rename their innovations. Today we view technologically oriented societies as being superior. We see exploration and the ability to conquer as exponents of superiority.

There is an old skit from the TV comedy show *Saturday Night Live* in which extraterrestrials land their spaceship on earth and demand that humans bow down to them. It becomes quickly apparent that the extraterrestrials are stupid and ignorant. They eventually admit that they didn't invent their spaceship; they found it. Imagine the reaction of the Aztecs to the conquering Spaniards, treating their wounds by pouring hot oil on them and praying, while the "backward" Amerindians used early antibiotics. Cortés and his men had guns; they had found them in China. As *New York Times* writer Gail Collins put it, "The Chinese . . . had toothpaste, while people in Europe barely had teeth."[118]

The seafaring ways of the Europeans have often been attributed to superior technology, but, in fact, the Chinese invented a staggering number of shipbuilding advances—fore-and-aft rigging, the lateen sail, the sternpost rudder, and watertight bulkheads, to name a few. With those advances and the compass, the Chinese could have theoretically gone anywhere the Europeans did—and long before. Indeed, while Columbus was making the rounds of the courts of Europe seeking funding for his adventures, Chinese maritime technology was advanced enough for Chen Ho, chief admiral and eunuch of the Ming emperor, to send to India and then to East Africa fleets of vessels armed with cannons and manned with thousands of sailors and passengers.

It is this admiral, suggests Alfred Crosby, who should be acknowledged as the greatest explorer in the age of exploration. "If political changes and cultural endogeny had not stifled the ambitions of Chinese sailors," writes Crosby, "then it is likely that history's greatest imperialists would have been far easterners, not Europeans."[119] The Chinese could have made arduous journeys around the world on any seas they wanted, had they had a reason to do so. Western European economies offered nothing China could not acquire much closer to home at much less cost.[120]

So as it happened Chen Ho did not sail east, and Christopher Columbus sailed west, "greedy to find the gold of Cathay and the courts of the Grand Khan as described by his countryman Marco Polo, who had traveled by different means and from the other direction," as the late biologist Stephen Jay Gould put it.[121]

Notes

I do not wish to minimize the task of locating some of the following sources. Many books are still in print and widely available. Others are not, or are found only in a few libraries, sometimes under closed reserve. Some libraries will allow you to make interlibrary loans to obtain books from distant libraries. A few researchers directed me to their Web sites for works in progress or in the midst of translation. For these I have, in most cases, listed the full Web address. However, as the experienced Internet traveler knows, these sites shift in and out of existence, and you may have luck one hour, frustration the next. One trick, if the full address doesn't work, is to type in just the basic Web site address; that is, the address only through the domain label "com," "net," "org," or "edu." Then try negotiating your way to the proper page by the seat of your pants. Good luck.

CHAPTER 1: A HISTORY OF SCIENCE

1. Bertrand Russell, *A History of Western Philosophy* (New York: Touchstone/Simon & Schuster, 1945), p. 131.
2. George Saliba, professor of Arabic and Islamic science at the department of Middle East and Asian languages and cultures, Columbia University, elaborates in a letter to the author of May 6, 2002, on the "point that Ptolemy called the 'center of the equalizer of motion,' later to be called the 'equant.' " Saliba writes: "The equant is in fact a sphere as imagined by Ptolemy, which he did not call equant. He simply called it 'equalizer of motion,' as you note. If one is cutting the Ptolemaic spheres for the planets across their equatorial planes, then the center of the equant sphere will indeed be a point. One can argue that the equant sphere can also be represented by a circle. What I do in such cases is to stress the absurdity of the situation by saying that Ptolemy was requiring a physical sphere to move in place around an axis that does pass through its center."
3. Lawrence W. Weiss (now retired), formerly of Pantaleoni Govens & Weiss, New York, N.Y.; Rod Berman, of Jeffer Mangelf Butler & Marmarc LLP.

4. On an early draft of the manuscript for this book, George Saliba noted: "I wrote my very first article on the meaning of the term 'algebra' way back in the sixties, and I would say here 'meaning compulsion,' as in compelling the unknown 'x' to assume a numerical value, and not 'bone setting,' which is mistaken but widely quoted in dictionaries and the like."
5. Multnomah County School Board, *Portland African-American Baseline Essays,* c. 1982, pp. S-52–S-53.
6. Ibid., pp. S-41, S-42.
7. David Park, *The Fire Within the Eye* (Princeton, N.J.: Princeton University Press, 1997), pp. 76–77.
8. Richard Powers, "Eyes Wide Open," *New York Times Magazine,* April 18, 1999, p. 83.
9. Robert Temple, *The Genius of China: 3,000 Years of Science, Discovery, and Invention* (London: Prion Books Limited, 1998), p. 81.
10. George Gheverghese Joseph, *The Crest of the Peacock: Non-European Roots of Mathematics* (New York: Penguin, 1992), pp. 8–12. Also see Jacob Bronowski, *The Ascent of Man* (London: BBC Books, 1973), p. 177.
11. George Saliba, "Islamic Precursors to Copernicus," lecture at Smith College, Northampton, Massachusetts, September 25, 2000.
12. Ibid.
13. Glen Bowersock, review of *Not Out of Africa* by Mary Lefkowitz, *The New York Times Book Review,* February 25, 1996.
14. Jim Holt, "Mistaken Identity Theory," *Lingua Franca* (Mar. 2000): 60.
15. Bronowski, *Ascent of Man,* pp. 158–164.
16. Joseph, *Crest of the Peacock,* pp. xi–xiii.
17. Ibid., p. 215.
18. Bronowski, *Ascent of Man,* p. 162.

CHAPTER 2: MATHEMATICS

1. Mathematician Robert Kaplan, author of *The Nothing That Is: A Natural History of Zero* (New York: Oxford University Press, 1999), explains, in a letter to the author dated July 19, 2000: "In most cases one infinite set isn't equal to another, because there are a lot of different 'sizes' of infinity. So if you take a ruler and cut up the 1 to 2 inch segment into 'infinitesimally small' segments, and you get as many slices as there are rational numbers (numbers of the form p/q, where p and q are integers and q isn't zero); and you do the same for the 2 to 12 inch segment, then you will indeed have as many slices in the first set as in the second. But if instead one of the segments were cut into as many slices as there are rationals but the other into as many as there are real numbers (rationals plus irrationals, like pi and the square root of 2), then this second set will have more slices than the first.

"The infinity of the rational numbers is the same size (called by Georg Cantor aleph null) as the size of the set of counting numbers; that of the reals (which he called C, for 'continuum') is larger; and we can go on getting larger and larger infinities (there are, for example, more functions on the reals than there are reals). Cantor was the first to recognize and prove this, and he paid a heavy price (madness and more).

"It's true that infinite subsets of the rationals, or of the counting numbers, will have exactly as many members as the set they came from—aleph null (so 12 or 16 or 20 times aleph null is aleph null, as the story about Taco Bell illustrates), and Galileo flirted with this idea before turning away. This is a fantastic insight, this seeing of one-to-one correspondences between two sets. We have to give Cantor his due as the first to recognize that there are different sizes of infinite sets, and how mind-bogglingly many sizes there are (by the way, strictly speaking I wouldn't talk about 'different infinities' but 'different sizes of infinity,' or 'sets of different sizes'—technically, 'different cardinalities')."

2. George Gheverghese Joseph, letter to author, March 16, 2001.
3. Tobias Dantzig, *Number: The Language of Science* (New York: Macmillan, 1967), p. 26.
4. Lancelot Hogben, *Mathematics for the Million* (London: Allen & Unwin, 1942), p. 245.
5. Joseph, letter to author, March 16, 2001.
6. Dantzig, *Number,* pp. 32–33.
7. Interview with Robert Kaplan, January 1, 2000.
8. George Gheverghese Joseph, "Mathematics," in Helaine Selin (ed.), *Encyclopaedia of the History of Science, Technology, and Medicine in Non-Western Cultures* (Dordrecht/Boston/London: Kluwer Academic Publishers, 1997), p. 604.
9. University of Chicago economist Robert Book, letter to author, December 17, 1997.
10. George Gheverghese Joseph, *The Crest of the Peacock: Non-European Roots of Mathematics* (London: Penguin Books, 1991), p. 60.
11. Ibid., p. 96.
12. Dantzig, *Number,* p. 77.
13. Joseph, *Crest of the Peacock,* p. 22.
14. Hogben, *Mathematics for the Million,* p. 44.
15. Joseph, letter to author, March 16, 2001.
16. Hogben, *Mathematics for the Million.*
17. Ibid.
18. Joseph, letter to author, March 16, 2001.
19. See ibid.
20. D. H. Fowler, *The Mathematics of Plato's Academy* (Oxford: Clarendon Press, 1987), pp. 283–284.

21. Joseph, *Crest of the Peacock*, p. 125.
22. Morris Kline, *Mathematics: A Cultural Approach* (New York: Addison-Wesley, 1962), p. 14.
23. Ibid., p. 30.
24. Ibid., p. 14.
25. Ibid.
26. Robert Kaplan, in a letter to the author, August 12, 2000, writes: "Ted Williams and Hank Aaron had incredible eye-hand coordination. The ancient Greeks had incredible intuition-proof coordination. We owe to them alone the concept of a proof in mathematics. Others could see what they saw, could intuit beautifully, and many saw what they didn't—but no others felt any need, or devised any way, of tying their insights down by proofs: i.e., by a network of logical connections to a small set of 'fundamental' insights. Without proof what you have is rumor, gossip, politics, I-said-he-said, mistaken and brilliant insights jumbled up together. It is this architecture that Morris Kline rightly admires, but too crudely exalts above the insights others had. It's crucial to see how important the issue of proof is in mathematics. There are too many cases in math—namely, an infinite number—ever to be able to verify any statement in it empirically. The claim, for example, that all numbers are less than 60,000 works for the first 59,999 positive integers, and the infinite number of negative integers, and the larger infinity of reals less than 60,000, but it turns out to be false after all. How could you ever test empirically whether or not pi was rational; or whether there was a last prime; or whether all Mersenne primes are in fact prime? How could you test empirically whether the three angle bisectors of any triangle all meet in one point? Careless drawing might make it look as if they did in one case and didn't in another, and the most careful drawing would, after all, be of only one particular triangle. What's always needed is a proof that covers all possible cases at once—and it was just this that the Greeks invented. It is the very fact that math's subject matter isn't just large but in fact always infinite that makes math without proof like baseball without its beloved umpires."

George Gheverghese Joseph, a mathematician at the University of Manchester (U.K.), responds: "The comment that follows confines itself to the proof tradition in Indian mathematics only. The situation is very complicated and Western math historians have not even started grappling with it. The sources of information on traditional proofs, rationales, derivations and demonstrations in Indian mathematics and astronomy are found in commentaries on the basic texts. Now, many commentaries restrict themselves to the explanation of the words of the texts and do not go further. But there are a certain number of commentaries which also explain rationales partly or fully. Then there are works which, mainly based on earlier siddhanta texts that introduce revisions, innovations, and methodologies, all aimed at arriving at better and more accurate results. Often these texts give the rationales as well. Finally, there are works wholly devoted to the elucidation of mathematical and astronomical rationale, and also short independent texts which take up for elucidation some topic or other. Sometimes marginalia and post-colophonic statements in manuscripts give valuable information. There are also large numbers of short tracts which demonstrate the rationale of minor points or specific topics. It has to be remembered here that in technical literature, as a rule, rationales, including innovations and inventions, generally form part of the intimate instruction from the teacher to the pupil, and are not always put on record in commentaries or in manuscripts.

"Consider just one example, the commentaries on the works of Aryabhata (c. A.D. 500). You may well have chosen the commentaries on the works of Brahmagupta or Bhaskara 2 (Bhaskaracharya) instead. The works of Bhaskara 1, the great Aryabhata commentator are absolutely essential in understanding the methods and rationale of Aryabhatiya. And similarly are some of the works in Kerala. From about 7th century onwards, if not earlier, Kerala became the centre of the Aryabhatan School of astronomy and mathematics. Let me mention a few texts that go into a detailed rationale and proofs on a whole host of topics, of which the work on infinite series relating to circular and trigonometric functions is the most notable, being one of the two strands that went into the creation of what we would describe as modern mathematics. Any list of such texts would include Nilakanthaís Aryabhatiyabhasya (c. A.D. 1500), Kriyakrmakari of Narayanan and Sankara Variyar, Yuktibhasa of Jysthadeva. If you study any of these texts you will come across uppapitis (or demonstrations) of a whole series of results. To ignore these works and state that there were no proofs in the Indian tradition would be going against all the facts.

"What would be more productive instead of categorizing different mathematical traditions as intuitive, rational, empirical, heuristic, rigorous, etc., would be to consider the philosophical underpinnings of the proof methods in different traditions. Why, for instance, was the method of indirect proof rejected by the Indian tradition except in the rare case of showing the non-existence of a mathematical object but never the existence of an object? Why, in the Indian tradition, was there no philosophical crisis that accompanied the discovery of irrational numbers? These are some of the issues that need to be addressed rather than categorizing different traditions by such labels as empirical, heuristic, inexact, intuitive, rational, rigorous, etc.

"To take another example, why do the Chinese procedures of solving equations of different orders incline more towards the constructivist approach rather than the analytical approach? Now, we know that constructivists claim that what we know and our patterns of meaning are a result of an active and purposeful construction by us interacting dialectically with our world. This is in line with the naturalistic philosophical tradition of Yin and Yang of China, which had little appreciation of ideal entities and their associated ontological realm of stasis found in the Greek tradition. Yet there is a paradox here. Why, for instance, in the Chinese method of demonstration is the scope for a dialectical type of argument so limited?

"Another area worth exploring in any comparative study of different methods of proof was what was perceived as the function of proof in different traditions. An Indian uppapati had a social, pedagogical, and philosophical function. The purpose of proof was to convince different audiences."

For further reading, Joseph suggests three of his articles: "Different Ways of Knowing: Contrasting Styles of Argument in Indian and Greek Mathematical Traditions," in P. Ernest (ed.), *Mathematics, Education and Philosophy: An International Perspective* (London: Falmer Press, 1994), pp. 194–204; "Different Ways of Knowing: Contrasting Styles of Argument in India and the West," in D. F. Robitalle et al. (eds.), *Selected Lectures from the Seventh International Conference on Mathematical Education* (Quebec: Les Presses de L'Universitè Laval, 1994), pp. 183–197; and "What Is a Square Root? A Study of Geometrical Representation in Different Mathematical Traditions," *Mathematics in Schools* 23 (May 1997): 4–9.

27. Interview with Ayele Bekerie, May 5, 1998.
28. W. W. Rouse Ball, *A Short Account of the History of Mathematics,* 4th ed. (New York: Dover, 1960 [1908]), p. 1.
29. Kline, *Mathematics,* pp. 9–10.
30. Joseph, *Crest of the Peacock,* pp. 227–228.
31. Ibid., p. 5.
32. Dantzig, *Number,* p. 4.
33. Telephone interview, June 10, 1998, with Joe Nickell, an editor at *Skeptical Inquirer* magazine and former magician-in-residence at the Houdini Magical Hall of Fame.
34. Karl Menninger, *Number Words and Number Symbols* (New York: Dover, 1969), p. 33.
35. Several interviews with Herbert Terrace, winter 1989.
36. George Gheverghese Joseph, "Mathematics," in *Encyclopaedia,* pp. 604–605.
37. Joseph, *Crest of the Peacock,* p. 27.
38. James Ritter, "Mathematics in Egypt," *Encyclopaedia,* pp. 629–630.
39. Joseph, *Crest of the Peacock,* pp. 66 et seq.
40. Kline, Mathematics, p. 13.
41. Leon Lederman (with Dick Teresi), *The God Particle* (Boston: Houghton Mifflin, 1993), p. 52.
42. Kaplan, *The Nothing That Is,* p. 7.
43. Ibid., p. 9.
44. Ibid., p. 7.

45. Seven hundred according to Kaplan, *The Nothing That Is,* p. 12; 300 according to Joseph, *Crest of the Peacock,* p. 98.
46. Joseph, *Crest of the Peacock,* p. 99.
47. Ibid., p. 102.
48. Based on a translation by Taha Baquir (1950) of a tablet found at Tell Harmal in 1949, as cited by Joseph, *Crest of the Peacock,* p. 109.
49. Ibid., p. 102.
50. Ibid., pp. 106–107.
51. John Allen Paulos, *Beyond Numeracy* (New York: Vintage Books, 1992), pp. 205–207.
52. Ibid., p. 179.
53. Joseph, *Crest of the Peacock,* pp. 103–105.
54. Mathematician Robert Kaplan says that the Babylonians were not really "comfortable with irrationals" because they didn't really "know" irrationals. That is, because the Babylonians did not construct proofs as the Greeks did, they did not really grasp such concepts. In a letter to the author dated August 23, 2000, Kaplan writes, "There is a very big, very important point here, which it is crucial not to stumble on. Anyone comfortable with irrationals is ignorant or naive. The big point is that we have (or at least I have) no evidence that the Babylonians or Egyptians or Indians, for that matter, had any idea that the square root of two, or pi, are irrational—or any idea, as far as I know, that there were such things as irrationals. You can get better and better approximations to irrationals, but unless you know, via a proof, that they are irrationals, you won't realize that you can never have anything but an approximation, since there is no 'closed form,' no nice repeating pattern, to the decimal expansion of an irrational. The Greeks are distinguished by having come up with a proof in Pythagorean times that the square root of two was irrational. They knew it, and this fact haunted their philosophy (Plato) and their math ever after. And notice: you can't show experimentally that a number is irrational: its decimal form might start repeating (making it rational) just after you stop getting one more decimal place. By the way, we didn't know pi was irrational (although many suspected it) until late in the nineteenth century. It is on this difference between understanding the existence and nature of the irrationals, and just going on comfortably getting one more decimal place of pi, that the whole issue rests."

    Mathematician George Gheverghese Joseph counters the notion that non-Western mathematicians used irrational numbers without grasping what they really were. Joseph, in a letter to the author dated May 18, 2001, writes, "Let me quote from the work of the Kerala (India) mathematician, Nilakantha: 'Why is only the approximate value (of circumference) given here? Let me explain. Because the real value cannot be obtained. If the diameter can be measured without a remainder, the circumference measured by the same unit (of measurement) will leave a remainder. Similarly, the unit which measures the circumference without a remainder will leave a remainder when used for measuring the diameter. Hence, the two measured by the same unit will never be without a remainder. Though we try very hard we can reduce the remainder to a small quantity but never achieve the state of 'remainderless-ness.' This is the problem."

If this does not show some understanding of the irrationality of pi, what does it show?"

55. Joseph, *Crest of the Peacock,* p. 113.
56. Ibid., pp. 115–117.
57. Ibid., pp.116–117.
58. Kaplan, *The Nothing That Is,* p. 17.
59. Dantzig, *Number.*
60. Ibid, p. 29.
61. Ibid.
62. Ibid, p. 29.
63. Fowler, *Mathematics of Plato's Academy.*
64. Ibid.
65. Robert Kaplan, letter to author, August 23, 2000.
66. Hogben, *Mathematics for the Million,* p. 63.
67. Ibid.
68. Kline, Mathematics, p. 103.
69. Dantzig, *Number,* p. 13.
70. Lederman, *The God Particle,* p. 70.
71. Peter Machamer, *The Cambridge Companion to Galileo* (Cambridge: Cambridge University Press, 1998), p. 63.
72. Ibid., pp. 64–65.
73. Fowler, *Mathematics of Plato's Academy,* p. 21.
74. Iamblichus is quoted by Mary Lefkowitz in *NOT Out of Africa* (New York: Basic Books, 1996), p. 76.
75. Also see Kathleen Freeman, *The Pre-Socratic Philosophers* (New York: Oxford University Press, 1953).
76. Joseph, *Crest of the Peacock,* pp. 221–222.
77. Ibid., p. 224.
78. Ibid., p. 5.
79. Ibid., pp. 226–228.
80. Ibid., p. 229.
81. Ibid., pp. 229–230.
82. Ibid., pp. 234–235.
83. Ibid., pp. 238–239.
84. Ibid., p. 249.
85. Ibid., p. 241.

86. Ibid., pp. 256–257.
87. Ibid., p. 218.
88. Kaplan, *The Nothing That Is,* p. 56.
89. Takao Hayashi, "Bakhshali Manuscript," in *Encyclopaedia,* p. 147.
90. Joseph, *Crest of the Peacock,* p. 257.
91. Ibid., p. 130.
92. Ibid., pp. 140–141.
93. George Gheverghese Joseph in a letter to the author, May 18, 2001.
94. Ulrich Libbrecht, "Mathematics in China," in *Encyclopedia,* pp. 626–27.
95. Joseph, *Crest of the Peacock,* p. 131.
96. Libbrecht, "Mathematics in China," in *Encyclopedia,* p. 626.
97. Ibid.
98. Kaplan, letter to the author, August 23, 2000.
99. Joseph, *Crest of the Peacock,* pp. 146–147.
100. Joseph, letter to author, May 18, 2001.
101. David Berlinski, *A Tour of the Calculus* (New York: Pantheon Books, 1995), p. 7.
102. David E. Mungello, *Leibniz and Confucianism: The Search for Accord* (Honolulu: University of Hawaii Press, 1977).
103. Joseph, *Crest of the Peacock,* p. 301.
104. Simon Singh, *The Code Book* (New York: Doubleday, 1999), pp. 14–22.
105. Ibid., p. 26.
106. Joseph, *Crest of the Peacock,* p. 303.
107. Singh, *The Code Book,* pp. 15–16; Joseph, *Crest of the Peacock,* p. 303.
108. Ibid., p. 305.
109. Ibid.
110. Jan P. Hogenduk, "Mathematics in Islam," in *Encyclopedia,* p. 638.
111. Joseph, *Crest of the Peacock,* p. 11.
112. Again, here is Saliba's comment to me: "I wrote my very first article on the meaning of the term 'algebra' way back in the sixties, and I would say here, 'meaning compulsion,' as in compelling the unknown 'x' to assume a numerical value and not 'bone setting,' which is mistaken but widely quoted in dictionaries and the like."
113. Joseph, *Crest of the Peacock,* p. 11.
114. Ibid.
115. Ibid., p. 305.
116. Ibid., p. 306.
117. Ibid.
118. Ibid., p. 319.
119. Ibid., p. 309.
120. Hogenduk, in *Encyclopedia,* p. 638.
121. Joseph, *Crest of the Peacock,* p. 309.

122. Joseph, letter to author, May 18, 2001.
123. *Encyclopaedia,* p. 638.
124. Joseph, *Crest of the Peacock,* pp. 318–319.
125. *Encyclopaedia,* p. 638.
126. Joseph, *Crest of the Peacock,* p. 345.
127. *Encyclopaedia,* p. 650.
128. Joseph, *Crest of the Peacock,* p. 49.
129. Michael P. Closs, "Mathematics of the Maya," in *Encyclopaedia,* pp. 648–649.
130. Ibid., p. 647.
131. Ibid., p. 648.
132. Interview with Marjorie Senechal, December 10, 1999.
133. Kaplan, *The Nothing That Is,* p. 96.
134. Ibid., p. 92.
135. Joseph, letter to author, May 18, 2001.
136. Closs, "Mathematics of the Maya," p. 649.
137. Ibid., p. 650; George Gheverghese Joseph, lecture at the American Association for the Advancement of Science, Boston, 1993.
138. Judith Herrin, *The Formation of Christendom* (Princeton, N.J.: Princeton University Press), p. 85.
139. L. E. Doggett, "Calendars," in P. Kenneth Seidelmann (ed.), *Explanatory Supplement to the Astronomical Almanac* (Herndon, Va.: University Science Books, 1992), sec. 1.4.
140. Interview with Barry Mazur, February 6, 1997.
141. Dantzig, *Number,* p. 35.
142. Ibid., p. 249.
143. Joseph, *Crest of the Peacock,* p. 15.
144. Dantzig, *Number,* pp. 31–32.
145. Ibid., p. 31.
146. Ibid., p. 32.
147. Ibid., p. 31.
148. Ruth Freitag, *The Battle of the Centuries* (Washington D.C.: Government Printing Office, 1995), p. 2.
149. Edwin R. Thiele, *The Mysterious Numbers of the Hebrew Kings: A Reconstruction of the Chronology of the Kingdoms of Israel and Judah* (Chicago: University of Chicago Press, 1951).
150. Kaplan, *The Nothing That Is,* p. 59.
151. Ibid., p. 11.
152. Ibid., p. 12.
153. Ibid., p. 17.
154. Ibid., p. 20.
155. Ibid., pp. 26–27.
156. Ibid., pp. 56–57.

157. Ibid., pp. 61, 62.
158. Interview with Robert Kaplan, Jan. 1, 2000.
159. Joseph, *Crest of the Peacock,* p. 315, and Kaplan, *The Nothing That Is.*
160. Hogben, *Mathematics for the Million,* p. 50.
161. Kaplan, *The Nothing That Is,* pp. 78–79.
162. Hogben, *Mathematics for the Million.*
163. George Joseph is writing a paper on the Indian zero. He shares a passage from that (at this time) unpublished work: "What is generally recognized as the greatest contribution of India to world mathematics is the place value system of written numbers (numerals) incorporating a zero, which is the ancestor of our world-wide system of number representation. There are historical records of only three other numbers systems which were based on the positional principle. Predating all other systems was the Babylonian system, which must have evolved around the third millennium B.C. A sexigesimal scale was employed, with a simple collection of the correct number of symbols, to write numbers less than 60. But it was imperfectly developed, being partly additive and partly place valued as within the base of 60 a decimal system was used. Also, the absence of a symbol for zero until the early Hellenistic period limited the usefulness of the system for computational and representational purpose.

"The Chinese rod numeral system was essentially a decimal base system. The numbers 1, 2, . . . 9 are represented by rods whose orientation and location determine the place value of the number represented and whose color shows whether the quantity was positive or negative. In terms of computation, the representation of zero by a blank space posed no problem, for, unlike the Babylonian system, the blank was itself a numeral. The third positional system was the Mayan, essentially a vigesimal (base 20) system incorporating a symbol for zero that was recognised as a numeral in its own right. The system had, however, a serious irregularity since its units were 1, 20, 18 × 20, 18 × 202, 18 × 203, . . . and so on. This anomaly reduces its efficiency in arithmetical calculation. For example, one of the most useful facilities of our number system is the ability to multiply a given number by 10 by adding a zero to it. An addition of a Mayan zero to the end of a number would not in general multiply the number by 20 because of the mixed base system employed.

"The discussion so far of the different positional system highlights two important points. First, a place-value system could and did exist without any symbol for zero. But the zero symbol as part of the numerical system never existed and could not have come into being without place value. Second, the relative strength of the Indian number system is inextricably tied up with the Indian concept of zero. It would, therefore, be useful to examine the origins and use of zero in Indian mathematics.

"The Sanskrit word for zero, sunya, means 'void or empty.' Its derivative, Sunyata, is the Buddhist doctrine of Emptiness, being the spiritual practice of emptying the mind of all impressions. This is a course of action prescribed in a wide range of creative endeavours. For example, the practice of sunyata is recommended in writing poetry, composing a piece of music, in producing a painting, or in any activity that comes out of the mind of the artist. An architect is told in the silpa sutras (traditional manuals of architecture) that designing a building involves the organization of empty space, for 'it is not the walls which make a building but the empty spaces created by the walls.' The whole process of creation is well described in the following quotation from a Tantric Buddhist text: 'First the realization of the void (sunya) / Second the seed in which all is concentrated / Third the physical manifestation / Fourth one should implant the syllable (Havraja Tantra).

"The mathematical correspondence was soon established. Just as emptiness of space is a necessary condition for the appearance of any object, the number zero being no number at all is the condition for the existence of all numbers. A discussion of the mathematics of the sunya involves three issues: (i) the concept of the sunya within a place-value system, (ii) the symbols used for sunya, and (iii) the mathematical operations with the sunya. Material from early texts are used as illustrations.

"The word sunya is derived from suna which is the past participle of svi, 'to grow.' In one of the early Vedas, Rgveda, occurs another meaning: the sense of 'lack or deficiency.' It is possible that the two different words were fused to give 'sunya' a single sense of 'absence of emptiness' with the potential for growth.

"It was soon recognized that the sunya denoted notational place (place holder) as well as the 'void' or absence of numerical value in a particular notational place. Consequently all numerical quantities, however great they may be, can be represented with just ten symbols. A twelfth century text (Manasollasa) states: 'Basically, there are only nine digits, starting from "one" and going to "nine." By the adding of zeros these are raised successively to tens, hundreds and beyond.'

"And in a commentary on Patanjali's Yogasutra there appears in the seventh century the following analogy: 'Just as the same sign is called a hundred in the "hundreds" place, ten in the "tens" place and one in the "units" place, so is one and the same woman referred to (differently) as mother, daughter or sister.'

"The earliest mention of a symbol for zero occurs in the Chandahsutra of Pingala (fl.3rd century B.C.). It discusses a method for calculating the number of arrangements of long and short syllables in a meter containing a certain number of syllables (i.e., the number of combinations of two items from a total of n items, repetitions being allowed). It began as a dot (bindu), found in inscriptions in India as well as in Cambodia and Sumatra, and then became a circle (chidra or randra meaning a hole). The association between zero and its symbol had become well-known around the early centuries of the Christian era, as the following quotation shows: 'The stars shone forth, like zero dots (sunya-bindu)—scattered in the sky as if on the blue rug, the Creator reckoned the total with a bit of the moon for chalk.' (Vasavadatta, ca. A.D. 400)

"Sanskrit texts on mathematics/astronomy from the time of Brahmagupta usually contain a section called 'sunya-ganita' or computations involving zero. While the discussion in the arithmetical texts (patiganita) is limited only to the addition, subtraction and multiplication with zero, the treatment in algebra texts (bijaganita) covers such questions as the effect of zero on the positive and negative signs, division with zero and more particularly the relation between zero and infinity (ananta).

"Take as an example, Brahmagupta's Brahmasupta-Siddhanta. In it he treats the zero as a separate entity from the positive (dhana) and negative (rhna) quantities, implying that sunya is neither positive nor negative but forms the boundary line between the two kinds, being the sum of two equal but opposite quantities. He states that a number, whether positive or negative, remains unchanged when zero is added to or subtracted from it. In multiplication with zero, the product is zero. A zero divided by zero or by some number becomes zero. Likewise the square and square root of zero is zero. But when a number is divided by zero, the answer is an undefinable quantity 'that which has that zero as the denominator.'

"The earliest inscription in India of a recognizable antecedent of our numeral system is found in an inscription from Gwalior dated 'Samvat 933' (A.D. 876). The spread of these numerals westwards is a fascinating story. The Arabs were the leading actors in this drama. Indian numerals probably arrived at Baghdad in 773 A.D. with the diplomatic mission from Sind to the court of Caliph al-Mansur. In about 820 al-Khwarizmi wrote his famous *Arithmetic,* the first Arab text to deal with the new numerals. The text contains a detailed exposition of both the representation of numbers and operations using Indian numerals. Al-Khwarizmi was at pains to point out the usefulness of a place-value system incorporating zero, particularly for writing large numbers. Texts on Indian reckoning continued to be written and by the end of the eleventh century, this method of representation and computation was widespread from the borders of Central Asia to the southern reaches of the Islamic world in North Africa and Egypt.

"In the transmission of Indian numerals to Europe, as with almost all knowledge from the Islamic world, Spain and (to a lesser extent) Sicily played the role of intermediaries, being the areas in Europe which had been under Muslim rule for many years. Documents from Spain and coins from Sicily show the spread and the slow evolution of the numerals, with a landmark being their appearance in an influential mathematical text of medieval Europe, Liber Abaci, written by Fibonacci (1170–1250), who learnt to work with Indian numerals during his extensive travels in North Africa, Egypt, Syria and Sicily. And the spread westwards continued slowly, displacing Roman numerals, and eventually, once the contest between the abacists (those in favour of the use of the abacus or some mechanical device for calculation) and the algorists (those who favoured the use of the new numerals) had been won by the latter, it was only a matter of time before the final triumph of the new numerals occurred with bankers, traders and merchants adopting the system for their daily calculations."

164. David Wells, *The Penguin Dictionary of Curious and Interesting Numbers,* rev. ed. (London: Penguin, 1997), pp. 199–200.
165. Kaplan, *The Nothing That Is,* p. 97.
166. Joseph, letter to author, May 18, 2001.
167. Telephone interview with Robert Kaplan, November 5, 1990.
168. *Encyclopaedia,* p. 647.
169. Interview with Barbara Fash, Harvard University, Peabody Museum, April 15, 1977.
170. Kaplan, *The Nothing That Is,* p. 89.
171. Dantzig, *Number,* p. 35.
172. Interview with Barry Mazur, Dec. 28, 1999.

CHAPTER 3: ASTRONOMY

1. Fred Adams and Greg Laughlin, *The Five Ages of the Universe* (New York: Free Press, 1999), p. 40.
2. Ibid., p. 84.
3. Ibid., p. xv. One billion trillion is 10^{22}. Anthony Aveni, professor of astronomy and anthropology at Colgate University, and the author of *Conversing with the Planets* and other works of archaeoastronomy, puts the figure closer to 1 collision per 10^{10} years per galaxy. Either way, the odds are long. (Aveni wrote this and other comments directly on the manuscript between May 25, 2001, and July 5, 2001; such comments will hereafter be called "Aveni commentary.")
4. Frank Drake and Dava Sobel, *Is Anyone Out There?* (New York: Delacorte Press, 1992), p. 180.
5. Aveni commentary, June 2001.
6. Aveni commentary, June 2001.
7. Interview with Aveni, Mar. 3, 2001.
8. Susan Milbrath, *Star Gods of the Maya* (Austin: University of Texas Press, 1999), p. 1.
9. Martha Macri, "Astronomy in Mesoamerica," in Helaine Selin (ed.), *Encyclopaedia of the History of Science, Technology, and Medicine in Non-Western Cul-*

tures (Dordrecht/Boston/London: Kluwer Academic Publishers, 1997), p. 134.

10. Anthony Aveni, *Stairways to the Stars: Skywatching in Three Great Ancient Cultures* (New York: Wiley, 1997), p. 95.
11. See the Web site of Michiel Berger: www.michielb.nl/Maya.
12. Milbrath, *Star Gods of the Maya,* p. 46.
13. Macri, "Astronomy in Mesoamerica," p. 134. Aveni dates recordings to A.D. 85 at La Mojarra—eclipses and Venus observations implied (Aveni commentary, June 2001).
14. Milbrath, *Star Gods of the Maya,* p. 196.
15. Aveni, *Stairways to the Stars,* p. 97.
16. Michael D. Coe, *The Maya* (New York: Thames and Hudson, 1999), p. 199.
17. Aveni, *Stairways to the Stars,* p. 98.
18. Coe, *The Maya,* pp. 199–200.
19. Ibid., p. 199.
20. Ibid., p. 193.
21. Ibid., p. 240.
22. Ibid., pp. 263–266, citing Aveni, *Stairways to the Stars,* p. 80.
23. For the complete account, go to the following Web address: http://www.ridgecrest.ca.us/~n6tst/maya.
24. Coe, *The Maya,* p. 217.
25. Aveni, *Ancient Astronomers,* ed. Jeremy A. Sabloff (Washington, D.C.: Smithsonian Books, 1993), p. 117.
26. Michael E. Smith, *The Aztecs* (Melrose, Mass.: Blackwell Publishers, 1996), p. 45.
27. Ibid., p. 260.
28. R. Tom Zuidema, "The Inca Calendar," in Anthony F. Aveni (ed.), *Native American Astronomy* (Austin: University of Texas Press, 1977), p. 220.
29. It was called Temple of the Sun by the Spanish, who obliterated much of it to build the currently standing Church of Santo Domingo.
30. Aveni, *Ancient Astronomers,* p. 138.
31. Zuidema, "The Inca Calendar," p. 221.
32. Aveni, *Stairways to the Stars,* p. 8.
33. Aveni, *Ancient Astronomers,* p. 141.
34. Zuidema, "The Inca Calendar," p. 221.
35. Ibid., p. 143.
36. Ibid.
37. Aveni commentary, June 2001.
38. Press release, University of Chicago, Sept. 24, 1998. Bauer and Dearborn's research on the Island of the Sun is a continuation of their joint research on Inca astronomy. They are the authors of *Astronomy and Empire in the Ancient Andes* (Austin: University of Texas Press, 1995).

39. E. C. Krupp, *Echoes of the Ancient Skies: The Astronomy of Lost Civilizations* (New York: Harper & Row, 1983), p. 50.
40. Ibid., p. 152.
41. Ibid., p. 154.
42. Dorothy Mayer, "An Examination of Miller's Hypothesis," in Aveni (ed.), *Native American Astronomy,* p. 180.
43. Aveni, *Ancient Astronomers,* p. 130. Aveni credits Florence Haweley Ellis for this suggestion.
44. Von Del Chamberlain, "Reflections on Rock Art and Astronomy," *The Quarterly Bulletin of the Center For Archaeoastronomy,* No. 14, December Solstice, 1994.
45. Paula Giese, "Lakota Star Knowledge," http://www.kstrom.net/isk/stars/startabs.html.
46. Louis Lord, "The Year 1000," *U.S. News and World Report,* Aug. 16, 1999. Also at www.usnews.com/usnews/issue/990816/cahokia.htm.
47. Aveni, *Ancient Astronomers,* p. 126.
48. Charles C. Mann, "1491," unpublished article (eventually published in abridged form in *The Atlantic Monthly,* March 2002).
49. John A. Eddy, "Medicine Wheels and Plains Indian Astronomy," in Aveni (ed.), *Native American Astronomy,* p. 148.
50. Aveni commentary, June 2001. Aveni cites David Vogt as a leading critic. See David Vogt, "Medicine Wheel Astronomy," in Clive L. N. Ruggles and Nicholas J. Saunders (eds.), *Astronomies and Cultures: Papers Derived from the Third "Oxford" International Symposium on Archaeoastronomy, St. Andrews, UK, September 1990* (Niwot, Colo.: University Press of Colorado, 1993), p. 163.
51. Krupp, *Echoes of the Ancient Skies,* p. 142.
52. Ibid.
53. Paula Giese, "Bighorn Medicine Wheel 3: How They Work," http://www.kstrom.net/isk/stars/starkno8.html.
54. John A. Eddy, "Astronomical Alignment of the Big Horn Medicine Wheel," *Science* 184 (June 1974): 1035–1043.
55. Krupp, *Echoes of the Ancient Skies,* p. 145.
56. Eddy, "Medicine Wheels and Plains Indian Astronomy," p. 168.
57. Aveni commentary, June 2001.
58. Waldo R. Wedel, "Native Astronomy and the Plains Caddoans," in Aveni (ed.), *Native American Astronomy,* pp. 133–134.
59. Aveni, *Ancient Astronomers,* p. 132.
60. Wedel, "Native Astronomy and the Plains Caddoans," pp. 134–136.
61. Aveni, *Ancient Astronomers,* p. 135.
62. Wayne Orchiston, "Astronomy of Polynesia," in Christopher Walker (ed.), *Astronomy Before the Telescope* (New York: St. Martin's Press, 1966).

63. A. Grimble, "Gilbertese Astronomy and Astronomical Observances," *Journal of the Polynesian Society* 40 (1931): 197–224.
64. Aveni, *Ancient Astronomy,* pp. 149–150.
65. Aveni, *Ancient Astronomers,* pp. 151–152.
66. Ibid., p. 153.
67. W. Coote, *Wanderings, South and East* (London: Sampson Low, 1882), available at Austronesian Navigation and Migration, AsiaPacificUniverse.com, http://www.geocities.com/Tokyo/Temple/9845/austro.htm.
68. Johann Reinhold Forster, *Observations Made During a Voyage Round the World (in the Resolution 1771–5)* (London, 1777), available at Austronesian Navigation and Migration, AsiaPacificUniverse.com, http://www.geocities.com/Tokyo/Temple/9845/austro.htm.
69. Ibid.
70. Aveni, *Ancient Astronomers,* p. 154.
71. Polynesian Voyaging Society, at http://leahi.kcc.hawaii.edu/org/pvs/navigate/latitude.html.
72. Sioni Ake Mokofisi, "The Polynesian Gift to Utah," courtesy of the Literature and Arts Guild of Polynesia, http://polynesia2000.bizhosting.com or http://www.kued.org/pbs/index.html.
73. N. M. Swerdlow (ed.), *Ancient Astronomy and Celestial Divination* (Cambridge, Mass.: MIT Press, 1999), p. 2.
74. John Malcolm Russell, "Robbing the Archaeological Cradle," *Natural History* (Feb. 2001): 53.
75. John Britton and Christopher Walker," Astronomy and Astrology in Mesopotamia," in Walker (ed.), *Astronomy Before the Telescope,* p. 42.
76. Ibid., pp. 45–46.
77. "Ancient and Lost Civilizations: Sumerian Calendars," http://www.crystalinks.com/sumercalendars.html.
78. Britton and Walker, "Astronomy and Astrology in Mesopotamia," p. 50.
79. Swerdlow, *Ancient Astronomy and Celestial Divination,* p. 14.
80. Aveni commentary, June 2000.
81. Britton and Walker, "Astronomy and Astrology in Mesopotamia," p. 42.
82. F. Rochberg, "Babylonian Horoscopy: The Texts and Their Relations," in Swerdlow, *Ancient Astronomy and Celestial Divination,* p. 40.
83. Ibid.
84. Britton and Walker, "Astronomy and Astrology in Mesopotamia," p. 56.
85. Otto Neugebauer, "The History of Ancient Astronomy: Problems and Methods," in *Astronomy and History: Selected Essays* (New York: Springer Verlag, 1983), p. 52.
86. Otto Neugebauer, "Exact Science in Antiquity," in *Astronomy and History: Selected Essays,* p. 28.

87. Britton and Walker, "Astronomy and Astrology in Mesopotamia," p. 49.
88. Aveni, *Ancient Astronomers,* p. 45.
89. Ibid.
90. Britton and Walker, "Astronomy and Astrology in Mesopotamia," p. 52.
91. Otto Neugebauer, "The History of Ancient Astronomy: Problems and Methods," in *Astronomy and History: Selected Essays,* p. 47.
92. Aveni, *Ancient Astronomers,* p. 46.
93. Neugebauer, "Problems and Methods," p. 47.
94. John P. Britton, "Lunar Anomaly in Babylonian Astronomy: Portrait of an Original Theory," in N. M. Swerdlow (ed.), *Ancient Astronomy and Celestial Divination* (Cambridge, Mass.: MIT Press, 2000), p. 187.
95. Ibid.
96. Ibid., p. 244.
97. N. M. Swerdlow, "The Derivation of the Parameters of Babylonian Planetary Theory," in Swerdlow, *Ancient Astronomy and Celestial Divination,* p. 293.
98. Britton and Walker, "Astronomy and Astrology in Mesopotamia," pp. 66–67.
99. Otto Neugebauer, "Mathematical Methods in Ancient Astronomy," in *Astronomy and History: Selected Essays,* p. 101.
100. Otto Neugebauer, "The Origins of the Egyptian Calendar," in *Astronomy and History: Selected Essays,* p. 196.
101. Ibid., pp. 196–197.
102. Aveni, *Ancient Astronomers,* p. 41.
103. Otto Neugebauer, "The Origins of the Egyptian Calendar," in *Astronomy and History: Selected Essays,* p. 197.

    Barbara C. Sproul, director of the religion program at Hunter College, CUNY, points out in a letter to the author dated May 24, 2002, that Otto Neugebauer, the greatest scholar of non-Western science, has not been adequately lionized. Thus she wrote the following poetic tribute:

    When writing a book without motto,
    Always go for a fellow called Otto
    To quote from a source
    Of Neugebauer's force
    Will make further words n'obbligato . . .

    For who could know more about stars
    Or Sumerian ideas of Mars?
    'Twas Otto the scholar—
    Yes, bet bottom dollar—
    He read all those writings in jars.

104. Ibid., pp. 200–201.
105. Otto Neugebauer, "The Egyptian 'Decans,' " *Astronomy and History: Selected Essays,* p. 205.
106. Ronald A. Wells, "Astronomy in Egypt," in Walker (ed.), *Astronomy Before the Telescope,* p. 33.
107. Ibid., p. 38.
108. Aveni, *Ancient Astronomers,* p. 42.
109. Wells, "Astronomy in Egypt," p. 38.
110. Neugebauer, "The Egyptian 'Decans,' " in *Astronomy and History: Selected Essays,* pp. 208–209.
111. Wells, "Astronomy in Egypt," p. 39.
112. Ibid., p. 35.
113. E. C. Krupp, *Echoes of the Ancient Skies* (New York: Harper & Row, 1983), p. 102.
114. Kate Spence, "Ancient Egyptian Chronology and the Astronomical Orientation of Pyramids," *Nature,* November 16, 2000, pp. 320–324.
115. Aveni commentary, June 2000.
116. Wells, "Astronomy in Egypt," p. 37.
117. Krupp, *Echoes of the Ancient Skies,* p. 105.
118. "And," says Aveni, "Pingree has photocopies of most of them in his office—huge piles!" (Aveni commentary, June 2000).
119. David Pingree at www.vigyanprasar.com/dream/sept99/article2.htm.
120. David Pingree at http://asnic.utexas.edu/asnic/cas/davidpingree.html.
121. David Pingree, "Astronomy in India," in Walker (ed.), *Astronomy Before the Telescope,* p. 142.
122. Sudheer Birodkar at http://www.crystalinks.com/indiastronomy.html.
123. http://www.indiaheritage.org/science/astro.htm.
124. Pingree, "Astronomy in India," pp. 123–244.
125. Ibid., p. 125.
126. Otto Neugebauer, in *Astronomy and History: Selected Essays,* p. 425.
127. Otto Neugebauer, in *Astronomy and History: Selected Essays,* p. 436.
128. Sudheer Birodkar, trans. Ebenezer Burgess, at http://india.coolatlanta.com/GreatPages/sudheer/.
129. Pingree, "Astronomy in India," p. 127.
130. Howard R. Turner, *Science in Medieval Islam* (Austin: University of Texas Press 1995), p. 61.
131. Pingree, "Astronomy in India," p. 128.
132. Ibid., p. 133.
133. J. J. O'Connor and E. F. Robertson at http://www-groups.dcs.st-and.ac.uk/~history/Mathematicians/Aryabhata_I.html.
134. J. J. O'Connor and E. F. Robertson at www.crystalinks.com/indiastronomy.html.

135. Pingree, "Astronomy in India," pp. 126–277.
136. http://education.eth.net/scientists/framepages/ccorner_scientist23.htm.
137. http://britannica.com/bcom/eb/article/0/0,5716,16380+1,00.html. At this writing, only the first few words of the article are available to nonsubscribers. You must subscribe to view complete articles.
138. O'Connor and Robertson at http://www-groups.dcs.st-and.ac.uk/~history/Mathematicians/Brahmagupta.html.
139. Pingree, "Astronomy in India," pp. 136–137.
140. Ibid., p. 139.
141. Ibid., p. 142.
142. David A. King, "Islamic Astronomy," in Walker (ed.), *Astronomy Before the Telescope,* p. 144.
143. Aveni, *Ancient Astronomers,* pp. 69–70.
144. Turner, *Science in Medieval Islam,* p. 59.
145. King, "Islamic Astronomy," p. 147.
146. Ibid., p. 148.
147. Ibid., p. 151.
148. O'Connor and Robertson at http://www-groups.dcs.st-and.ac.uk/~history/Mathematicians/Yunus.html. The authors, at the School of Mathematics and Statistics, University of St. Andrews, Scotland, report that Ibn Yunus predicted the date of his own death to be in seven days' time when he was in good health. He tidied up his business affairs, locked himself in his house, and recited the Qur'an until he died on the day he predicted.
149. Aveni, *Ancient Astronomers,* pp. 70–71.
150. King, "Islamic Astronomy," pp. 160–161.
151. Ibid., p. 161.
152. Ibid., p. 156.
153. Turner, *Science in Medieval Islam,* p. 63.
154. Ibid., p. 64.
155. Ibid., p. 63.
156. King, "Islamic Astronomy," p. 158.
157. O'Connor and Robertson at http://www-groups.dcs.st-and.ac.uk/~history/Mathematicians/Al-Biruni.html.
158. Ibid.
159. Aveni, *Ancient Astronomers,* p. 65.
160. Ibid.
161. Turner, *Science in Medieval Islam,* pp. 66–67.
162. Aveni, *Ancient Astronomers,* pp. 66–67.
163. Ibid., p. 67.
164. King, "Islamic Astronomy," pp. 165–166.
165. Aveni, *Ancient Astronomers,* p. 67.
166. Turner, *Science in Medieval Islam,* p. 67.

167. Ibid., p. 65.
168. Aydin Sayili, *The Observatory in Islam: And Its Place in the General History of the Observatory* (Ankara: Turk Tarih Kurumu Basimevi, 1960), p. 51.
169. Turner, *Science in Medieval Islam,* p. 65.
170. Sahoor at http://users.erols.com/zenithco/beg.html.
171. O'Connor and Robertson at http://www-groups.dcs.stand.ac.uk/~history/Mathematicians/Beg.html.
172. Sahoor at http://users.erols.com/zenithco/beg.html.
173. *Dictionary of Scientific Biography,* ed. Charles C. Gillispie (New York : Scribner, 1980); Leon Lederman (with Dick Teresi), *The God Particle* (New York: Houghton Mifflin, 1993), p. 81.
174. Aveni commentary, June 2001.
175. Sahoor at http://users.erols.com/gmqm/islmoil1.html#tashkent.
176. Turner, *Science in Medieval Islam,* p. 65.
177. Sayili, *The Observatory in Islam,* p. 392.
178. King, "Islamic Astronomy," p. 148.
179. Turner, *Science in Medieval Islam,* p. 68.
180. Ibid.
181. King, "Islamic Astronomy," p. 148.
182. Ibid., pp. 170–171.
183. Colin Ronan, "Astronomy in China, Korea and Japan," in Walker (ed.), *Astronomy Before the Telescope,* p. 245.
184. Ibid., pp. 245–246.
185. Aveni, *Ancient Astronomers,* p. 76.
186. Sun Xiaochun, "Stars in Chinese Science," in *Encyclopaedia,* p. 909.
187. Ronan, "Astronomy in China, Korea and Japan," p. 247.
188. Aveni, *Ancient Astronomers,* p. 77.
189. NASA, news release, June 12, 1989.
190. Aveni, *Ancient Astronomers,* p. 77.
191. http://www.wcslc.edu/pers_pages/e-k0027/bone.htm.
192. Nathan Sivin, "Science and Medicine in China's Past," http://www.sas.upenn.edu/~nsivin/index/html.
193. Robert Temple, *The Genius of China* (London: Prion Books, 1986), pp. 29–30.
194. Ibid., p. 30.
195. Ibid., pp. 33–34.
196. Ibid., pp. 247.
197. Sivin at http://www.sas.upenn.edu/~nsivin/index/html.
198. Ibid.
199. Aveni, *Ancient Astronomers,* p. 83.
200. Ronan, "Astronomy in China, Korea and Japan," p. 250.
201. Aveni, *Ancient Astronomers,* p. 79.

202. Ronan, "Astronomy in China, Korea and Japan," pp. 255–256.
203. Temple, *The Genius of China,* p. 30.
204. Ibid., pp. 31–33.
205. Ibid., p. 35.
206. Ronan, "Astronomy in China, Korea and Japan," p. 256.
207. Temple, *The Genius of China,* pp. 35–36.
208. Aveni commentary, June 2001.
209. Ronan, "Astronomy in China, Korea and Japan," pp. 257–259.
210. Ibid., p. 86.
211. Temple, *The Genius of China,* p. 36.
212. Ibid.
213. Ibid., pp. 37–38.
214. Ibid., p. 38.
215. Ronan, "Astronomy in China, Korea and Japan," p. 261.

CHAPTER 4: COSMOLOGY

1. Anthony L. Peratt, "Plasma Cosmology," *Sky and Telescope,* February 1992, p. 136.
2. Interview with Edward Harrison, Mar. 15, 1995.
3. Edward Harrison, *Masks of the Universe* (New York: Macmillan, 1985), p. 1.
4. Ibid., p. 2.
5. Leon Lederman (with Dick Teresi), *The God Particle* (Boston: Houghton Mifflin, 1993), p. 286.
6. Interview with Edward Harrison, May 7, 1999.
7. Harrison, *Masks of the Universe,* p. 11.
8. Gale E. Christianson, *Edwin Hubble: Mariner of the Nebulae* (New York: Farrar, Strauss & Giroux, 1995), p. 138.
9. Ibid., pp. 148–149.
10. Ibid., pp. 185–186.
11. Ibid., p. 151.
12. Ibid., pp. 155–159.
13. Lederman, *The God Particle,* p. 385.
14. Ibid., p. 386.
15. Ibid.
16. Ibid., p. 396.
17. Ibid., p. 399.
18. Interview with George Greenstein, Sept. 18, 1996.
19. George Greenstein, *The Symbiotic Universe* (New York: Morrow, 1988), pp. 255–258.
20. John Leslie, *Universes* (New York: Routledge, 1989), pp. 13–14. In a letter to the author dated April 17, 2002, Barbara Sproul responds to Leslie's

statement: "This is the worst sort of theology, presuming an anthropomorphized God (separate, finite, "with intentions toward the universe"), a kind of kindergarten theology the rest of the world got over long ago partly *because* they didn't make the split between science and religion we do. My point being that you can't tar all religion with the anthropomorphic daddy-God image when most religions have gone way beyond it. Our scientists, as sophisticated as they often are, don't have any idea what religion really is."

21. Conversation with physicist-turned-Anglican-priest John Polkinghorne, November 30, 2000, Chancellor's House, University of Massachusetts, Amherst.
22. Jim Holt, "War of the Worlds," *Lingua Franca,* December 2000–January 2001).
23. David Hume, *Dialogues Concerning Natural Religion* (1779), ed. J. V. Price (New York: Oxford University Press, 1976).
24. Rocky Kolb, *Blind Watchers of the Sky* (Reading, Mass.: Addison-Wesley, 1996), p. vii. Barbara Sproul, in a letter to the author, April 17, 2002, writes: "The western view grounded in science? Of course not. When the Centers for Disease Control or the National Institutes of Health decide to put money into AIDS research rather than the study of termite parasites, they are making religious decisions—value decisions—not scientific ones."
25. Christianson, *Hubble: Mariner of the Nebulae,* p. 351.
26. Ibid., p. 347.
27. Ewa Wasilewska, *Creation Stories of the Middle East* (London: Jessica Kingsley Publishers, 2000), p. 49.
28. E. O. James, *Creation and Cosmology* (Leiden: E. J. Brill, 1969), p. 23.
29. Barbara C. Sproul, *Primal Myths: Creating the World* (San Francisco: Harper & Row, 1979), p. 94.
30. Wayne Horowitz, *Mesopotamian Cosmic Geography* (Winona Lake, Ind.: Eisenbrauns, 1998), p. 112.
31. Ibid., pp. 112–114.
32. Ibid., pp. 114–116.
33. Anthony Aveni, *Conversing with the Planets* (New York: Kodansha International, 1992), p. 53.
34. Horowitz, *Mesopotamian Cosmic Geography,* pp. 114–116.
35. Ibid., pp. 118–119.
36. Sproul, *Primal Myths,* p. 91.
37. Ibid.
38. Aveni wrote this and other comments directly on the manuscript between May 25, 2001, and July 5, 2001; such comments will hereafter be called "Aveni commentary."
39. James, *Creation and Cosmology,* p. 21.

40. Aveni, *Conversing with the Planets,* pp. 152, 51, 53, and Wasilewska, *Creation Stories of the Middle East,* pp. 45, 46.
41. Wasilewska, *Creation Stories of the Middle East,* pp. 45–50.
42. Ibid., p. 50.
43. Ibid.
44. Carl Sagan, *Cosmos* (New York: Ballantine Books, 1980), p. 213.
45. *The Encyclopedia of Religion*, ed. Mircea Eliade (New York: Macmillan, 1987), 4:110.
46. Sagan, *Cosmos,* pp. 213, 214.
47. Aveni, *Conversing with the Planets,* p. 152.
48. Sagan, *Cosmos,* p. 213.
49. Albert Schweitzer, *Indian Thought and Its Development* (New York: Holt, 1936), p. 29.
50. D. M. Bose (ed.), *A Concise History of Science in India* (New Delhi: Indian National Science Academy, 1971), pp. 458, 459; S. N. Dasgupta, *Yoga Philosophy in Relation to Other Systems of Indian Thought* (Delhi: Motilal Banarsidass, 1930), p. 77.
51. Eliade (ed.), *Encyclopedia of Religion,* 6:348.
52. Greg Bailey, *The Mythology of Brahma* (New York: Oxford University Press, 1983), pp. 10, 90.
53. Mircea Eliade, *The Myth of the Eternal Return* (New York: Pantheon Books, 1949), p. 78.
54. Sproul, *Primal Myths,* p. 181.
55. James, *Creation and Cosmology,* p. 37.
56. Eliade, *The Myth of the Eternal Return,* p. 78.
57. James, *Creation and Cosmology,* p. 37.
58. Sproul, *Primal Myths,* pp. 186, 187.
59. Ibid.
60. Ibid., p. 188.
61. Ibid., pp. 188, 198.
62. Schweitzer, *Indian Thought and Its Development,* p. 30.
63. Ibid., p. 26.
64. Bailey, *The Mythology of Brahma,* pp. 3–7.
65. Sproul, *Primal Myths,* pp. 192–194.
66. Ibid., p. 336.
67. Ibid., pp. 330–358.
68. Alfred Gell, "Closure and Multiplication: An Essay on Polynesian Cosmology and Ritual," in Daniel De Coppet and André Iteanu (eds.), *Cosmos and Society in Oceania* (Oxford: de Coppet, Berg, 1995), p. 21.
69. Ibid., p. 23.
70. Ibid., quoting Teuira Henry, *Ancient Tahiti* (Honolulu: Bishop Museum Press, 1928).

71. Teuira Henry, *Ancient Tahiti,* as quoted in Sproul, *Primal Myths,* pp. 350, 351.
72. Michael Kioni Dudley, *Man, Gods, and Nature: A Hawaiian Nation* (Honolulu: Na Kane O Ka Malo Press, 1990), p. 10; David Malo, *Hawaiian Antiquities,* trans. Nathaniel Emerson (Honolulu: Bishop Museum, 1951), pp. 9, 12.
73. Dudley, *Man, Gods, and Nature,* pp. 10, 11, 28.
74. Sproul, *Primal Myths,* p. 331.
75. Ibid.
76. Dudley, *Man, Gods, and Nature,* p. 16.
77. Charles Long, *Alpha: The Myths of Creation* (New York: George Braziller, 1963), pp. 58, 59.
78. Dudley, *Man, Gods, and Nature,* p. 16.
79. Ibid., pp. 10, 15.
80. Sproul, *Primal Myths,* pp. 338, 339.
81. Ibid., pp. 353–357.
82. Ibid.
83. Gell, "Closure and Multiplication," p. 23.
84. Ibid.
85. Dennis Tedlock (trans.), *Popul Vuh: The Mayan Book of the Dawn of Life* (New York: Simon & Schuster, 1996), pp. 30, 63–64, 221.
86. Ibid., pp. 65, 66.
87. Ibid., pp. 31, 32.
88. Aveni, *Conversing with the Planets,* p. 50.
89. Ibid.
90. Tedlock (trans.), *Popul Vuh,* pp. 34, 36, 37.
91. Ibid., pp. 32, 67.
92. Ibid., p. 68.
93. Ibid., pp. 70–72.
94. Tedlock (trans.), *Popul Vuh,* pp. 145–148.
95. Interview of Linda Schele by Kathleen McAuliffe for *Omni,* February 1995, p. 103; Aveni, *Conversing with the Planets,* p. 152.
96. Barbara Tedlock, *Time and the Highland Maya* (Albuquerque: University of New Mexico Press, 1982), p. 181.
97. Interview with Schele, February 1995.
98. Ibid.
99. B. Tedlock, *Time and the Highland Maya,* p. 181.
100. D. Tedlock, *Popul Vuh,* pp. 16, 72.
101. Ibid., pp. 16, 236.
102. Ibid., p. 237.
103. Aveni, *Conversing with the Planets,* pp. 64–68.
104. Ibid., pp. 64–68.
105. Timothy Ferris, *Coming of Age in the Milky Way* (New York: Morrow, 1988), pp. 19–20.

106. Kolb, *Blind Watchers of the Sky,* p. 291.
107. Two telephone interviews with Edward Harrison; one in May 1999, another in August 1999.
108. Kolb, *Blind Watchers of the Sky,* p. 274.
109. Ibid., p. 272.
110. Ibid., p. 271.
111. Lederman, *The God Particle,* pp. 253, 277.
112. Aveni commentary, June 2001.
113. Actually, no one really measures how many optimum collisions occur. The 600 figure is how many events occurred that produced top quarks, which is what Fermilab was looking for. Top quarks are produced near maximum energy, but many of the 600 events were doubtless below this level.
114. Interviews with Henry Frisch, Nov. 29, Dec. 5, 2000.
115. To which Anthony Aveni says, "Amen." Aveni commentary, June 2001.
116. Helge Kragh, *Cosmology and Controversy* (Princeton, N.J.: Princeton University Press, 1996), p. vi.
117. Christianson, *Edwin Hubble: Mariner of the Nebulae,* pp. 340–341.
118. Eliot Marshall, "Science Beyond the Pale," *Science,* July 6, 1990, pp. 14–15.
119. George Smoot and Keay Davidson, *Wrinkles in Time* (New York: Morrow, 1993), p. 9.
120. Regarding the words "like seeing God," Barbara Sproul comments, in a letter to the author, April 17, 2002: "Another instance of baby theology."

CHAPTER 5: PHYSICS

1. David Park, emeritus professor of physics at Williams College, points out (in a letter to the author, October 18, 2001) that the chasm between theorists and experimenters is relatively new. "As a theorist in graduate school," he says, "I was surrounded and trained by theorists who knew all about the experiments that were relevant to our work. An experiment in particle physics is not performed without a lot of hard theoretical work on possible outcomes." The author responds that today, much of that theoretical preparation is conducted by the experimenters themselves. Roy Schwitters, an experimenter who was the director of the ill-fated superconducting super collider (SSC) in Texas, is one of many who bemoan the lack of theoretical guidance today. I once asked Alvin Tollestrup, an experimenter at Fermilab, what role theorists played at the lab. He pointed to a room with shelves of dusty unread journals, saying, "They fill up all these journals for us."
2. Leon Lederman (with Dick Teresi), *The God Particle* (Boston: Houghton Mifflin, 1993), p. 152.
3. Park, letter to author, October 18, 2001.
4. Ibid.

5. Park writes (letter to author, October 18, 2001): "I'm uncomfortable with the way you treat Aristotle as if he weren't very smart. If you read Aristotle carefully, his logical dissections of what other philosophers say as well as his modest attempts to lay out a natural philosophy of his own, you realize that this is about the smartest mind you have ever encountered. Writers about physics who make fun of him are saying, for all to see, nothing more than 'I haven't bothered to read him.' Aristotle's program was to find a path leading toward a science that one can be sure is correct. He often refers to 'the science we are seeking,' never to 'here it is.' It is to be constructed, à la Euclid, step by step, by logical deduction, from undoubted premises. Of course, the trouble was to find them. . . . Experiment leads one to make guesses, maybe good guesses, maybe get it exactly right, but how could you ever know? Modern scientists know that this is quite correct, that if you want truth you have to go to a theologian, not a scientist, that all we deal with is varying degrees of plausibility. We are used to it, it is part of our methodology, and it has taken us about 2,500 years to learn it. (If you can stand a fairly windy and Germanic way of saying all this, read the only modern philosopher that most modern scientists respect, Karl Popper.)

"If you want to know whether Aristotle was capable of observing what actually happens in nature, read his biological studies, which are a large fraction of his total works; they're awesome. He was perhaps the greatest biologist, certainly one of the best, who ever lived.

"As to his erroneous law of motion, get on your bicycle. If you want to go faster you push harder. If you stop pushing, it stops presently, and you fall off. The stuff about natural motions is less insistent in Aristotle than in the medievals. He had no theory of motion (which for him included every kind of change) other than simple contact. He was surrounded by motion; he had to organize his thoughts in some way. Enough.

"No, more. A simple example:

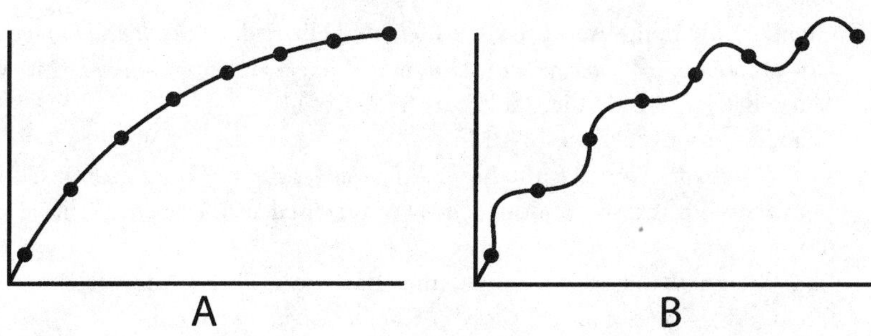

"Suppose someone takes some data and draws a smooth parabolic curve (a) through them and announces a law. Aristotle says this isn't good enough, nothing has been proved. How do you know the curve isn't (b)? Take all the data you want, your parabolic curve is plausible and might even be right, but you will never know it. It is logically impossible to prove a general law from specific experiments."

6. Steven Weinberg, *Dreams of a Final Theory* (New York: Pantheon Books, 1992), p. 7.
7. Lederman, *The God Particle,* p. 1.
8. Ibid., p. 33.
9. David Park, *The Fire Within the Eye: A Historical Essay on the Nature and Meaning of Light* (Princeton, N.J.: Princeton University Press, 1997), p. 49.
10. Ibid., pp. 36, 37, 49.
11. Park, letter to author, October 18, 2001.
12. Ibid.
13. Colin A. Ronan and Joseph Needham, *The Shorter Science and Civilization in China* (Cambridge: Cambridge University Press, 1981), 2:327.
14. Ibid.
15. Joseph Needham, *Science and Civilization in China,* vol. 5: *Chemistry and Chemical Technology* (Cambridge: Cambridge University Press, 1976), p. 150.
16. Ibid., p. 149.
17. Ronan and Needham, *The Shorter Science,* 5:348.
18. Ibid., p. 349.
19. Dai Nianzu, "Acoustics," in Institute of the History of Natural Sciences, Chinese Academy of Sciences (ed.), *Ancient China's Technology and Science* (Beijing: China Books & Periodicals, 1983), pp. 139, 140.
20. Ibid., p. 141.
21. Cheng-yih Chen, "The Generation of Chromatic Scales in the Chinese Bronze Set-Bells of the Fifth Century," in Cheng-yih Chen (ed.), *Science and Technology in Chinese Civilization* (Singapore: World Scientific, 1978), pp. 155, 157.
22. Ibid., p. 158.
23. Ibid., p. 160.
24. Cheng-yih Chen, "Acoustics," in Helaine Selin (ed.), *Encyclopedia of the History of Science, Technology, and Medicine in Non-Western Cultures* (Norwell, Mass.: Kluwer Academic Publishers, 1997), p. 11.
25. Nianzu, "Acoustics," pp. 144, 145.
26. A. C. Graham, *Later Mohist Logic, Ethics and Science* (Hong Kong: Chinese University Press, and London: School of Oriental and African Studies, 1978), p. 3.
27. Jing-Guang Wang, "Optics in China Based on Three Ancient Books," in Chen (ed.), *Science and Technology in Chinese Civilization,* p. 143.

28. Jin Qiupeng, "Optics," in *Ancient China's Technology and Science,* pp. 166, 167.
29. Ibid., p. 167.
30. A. C. Graham and Nathan Sivin, "A Systematic Approach to the Mohist Optics," in Shigeru Nakayam and Nathan Sivin (eds.), *Chinese Science: Explorations of an Ancient Tradition* (Cambridge, Mass.: MIT Press, 1973), p. 113.
31. Qiupeng, "Optics," p. 172.
32. Ibid., p. 174.
33. Wang, "Optics in China," p. 144.
34. Ibid., p. 145.
35. Ibid., p. 147.
36. Ibid., pp. 150–152. See also Qiupeng, "Optics," pp. 167–170.
37. Wang, "Optics in China," p. 152; Qiupeng, "Optics," pp. 167, 168, 170.
38. Wang, "Optics in China," p. 152.
39. Graham, *Later Mohist Logic, Ethics and Science,* p. 387.
40. Dai Nianzu, "Mechanics," in *Ancient China's Technology and Science,* pp. 124, 126.
41. Ibid., pp. 126, 127.
42. Ibid., p. 127.
43. Graham, *Later Mohist Logic, Ethics and Science,* p. 386.
44. Nianzu, "Mechanics," p. 129.
45. Graham, *Later Mohist Logic, Ethics and Science,* p. 385.
46. Ibid., p. 396.
47. Ibid.
48. Physicist Victor Weisskopf, press conference at Fermilab, March 14, 1979.
49. Nianzu, "Mechanics," p. 130.
50. Ibid., p. 137.
51. Ronan and Needham, *The Shorter Science,* 5:340, 341.
52. Robert M. Hazen and James Trefil, *Science Matters* (New York: Doubleday, 1991), p. 6.
53. Park, letter to author, October 18, 2001.
54. Zhang Yinzhi, "Mohist Views of Time and Space: A Brief Analysis," in J. T. Fraser, N. Lawrence, and F. C. Haber (eds.), *Time, Science, and Society in China and the West: The Study of Time* (Amherst: University of Massachusetts Press, 1986), pp. 207, 208.
55. Ibid.
56. D. M. Bose, S. N. Sen, and B. V. Subbarayappa (eds.), *A Concise History of Science in India* (New Delhi: Indian National Science Academy, 1971), p. 448.
57. Ibid., p. 453.
58. Debiprasad Chattopdhyaya, *History of Science and Technology in Ancient India* (Calcutta: Firma KLM PVT, 1991), p. 57.
59. Ibid.

60. Dick Teresi, review of Paul Halpern, *The Pursuit of Destiny, The Wall Street Journal,* January 10, 2001.
61. Park, letter to author, October 18, 2001.
62. Bose, Sen, and Subbarayappa (eds.), *A Concise History of Science in India,* pp. 455, 456.
63. Ibid., pp. 456, 457.
64. Ibid., pp. 458, 459; S. N. Dasgupta, *Yoga Philosophy in Relation to Other Systems of Indian Thought* (Delhi: Motilal Banarsidass, 1930), p. 77.
65. Dasgupta, *Yoga Philosophy,* pp. 24, 25.
66. Bose, Sen, and Subbarayappa (eds.), *A Concise History of Science in India,* pp. 452, 453.
67. Mrinal Kanti Gangopadhyay, "The Atomic Hypothesis," in Chattopdhyaya, *History of Science and Technology* p. 289.
68. Bose, Sen, and Subbarayappa (eds.), *A Concise History of Science in India,* pp. 448, 450, 451.
69. Ibid., pp. 463, 465.
70. John W. Hill and Doris K. Kolb, *Chemistry for Changing Times* (Upper Saddle River, N.J.: Prentice Hall, 1998), p. 37.
71. Bose, Sen, and Subbarayappa (eds.), *A Concise History of Science in India,* p. 466.
72. N. L. Jain, "Chemical Theories of the Jains," in Henry M. Leicester (ed.), *Chymia: Annual Studies in the History of Chemistry* (Philadelphia: University of Pennsylvania Press, 1966), 11:13, 14.
73. Ibid., pp. 13, 14.
74. Ibid., pp. 11–13.
75. Ibid., pp. 13, 14.
76. Ibid., pp. 14–16.
77. Hill and Kolb, *Chemistry for Changing Times,* p. 48.
78. Bose, Sen, and Subbarayappa (eds.), *A Concise History of Science in India,* pp. 466, 467.
79. Gangopadhyay, in Chattopdhyaya, *History of Science and Technology,* p. 283.
80. Lederman, *The God Particle,* p. 103.
81. Dick Teresi, "The Last Great Experiment of the Twentieth Century," *Omni,* January 1984): 46.
82. John Maxson Stillman, *The Story of Alchemy and Early Chemistry* (New York: Dover, 1960), pp. 105–111.
83. Jain, "Chemical Theories of the Jains," pp.14–16.
84. Bose, Sen, and Subbarayappa (eds.), *A Concise History of Science in India,* p. 467.
85. Jain, "Chemical Theories of the Jains," pp. 16–17.
86. Bose, Sen, and Subbarayappa (eds.), *A Concise History of Science in India,* pp. 468, 469.
87. Ibid., p. 469.
88. Dasgupta, *Yoga Philosophy,* pp.73, 74.

89. Stillman, *The Story of Alchemy,* pp. 105–111.
90. Prapphulla Chandra Ray, *A History of Hindu Chemistry from the Earliest Times to the Middle of the Sixteenth Century* A.D. (London: London: Williams and Norgate, 1902–09), 1:7–9.
91. Ibid., pp. 7, 8
92. Jacques Duchesne-Guillemin, *Symbols and Values in Zoroastrianism* (New York: Harper & Row, 1966), pp. 5–9.
93. Ibid., pp. 25, 26.
94. Ibid., p. 28.
95. Ibid., p. 17.
96. Park, *The Fire Within the Eye,* p. 13.
97. Park, letter to author, October 18, 2001.
98. Park, *The Fire Within the Eye,* pp. 23–25.
99. Arthur Zajonc, *Catching the Light* (New York: Bantam Books, 1993), pp. 41, 42.
100. Ibid.
101. Duchesne-Guillemin, *Symbols and Values,* p. 72.
102. Zajonc, *Catching the Light,* pp. 42, 43.
103. Jean Kellens, *Essays on Zarathustra and Zoroastrianism,* trans. Prods Oktor Skjærvø (Costa Mesa, Calif.: Mazda Publishers, 2000), pp. 48, 49.
104. Zajonc, *Catching the Light,* pp. 42, 43.
105. Duchesne-Guillemin, *Symbols and Values,* p. 65.
106. Park, *The Fire Within the Eye,* p. 24.
107. Zajonc, *Catching the Light,* p. 48.
108. Ibid., pp. 48, 49.
109. Ibid., pp. 53, 54.
110. Park, letter to author, October 18, 2001.
111. Park, *The Fire Within the Eye,* p. 73.
112. Park, letter to author, October 18, 2001.
113. Alfred L. Ivry, "Al-Kindi's *On First Philosophy* and Aristotle's *Metaphysics,*" in George F. Hourani (ed.), *Essays on Islamic Philosophy and Science* (Albany: State University Press of New York, 1975), p. 15.
114. Ibid., p. 18.
115. Park, letter to author, October 18, 2001.
116. Park, *The Fire Within the Eye,* pp. 73–75.
117. Saleh Beshara Omar, *Ibn al-Haytham's Optics: A Study of the Origins of Experimental Science* (Chicago: Bibliotheca Islamica, 1977), pp. 20–37.
118. Park, *The Fire Within the Eye,* pp. 73–75.
119. Ibid., pp. 76, 77.
120. John D. Cutnell and Kenneth W. Johnson, *Physics* (New York: Wiley, 1995), pp. 817, 818.
121. Park, *The Fire Within the Eye,* pp. 76, 77.

122. Ibid., p. 77.
123. Seyyed Hossein Nasr, *Science and Civilization in Islam* (Cambridge, Mass.: Harvard University Press, 1968), pp. 130, 131.
124. Omar, *Ibn al-Haytham's Optics,* pp. 67, 69.
125. Nasr, *Science and Civilization,* p. 129.
126. Ibid.
127. Park, *The Fire Within the Eye,* pp. 77, 78.
128. Ibid., p. 80.
129. Ibid., p. 84.
130. Ibid., pp. 80–82.
131. Concept contributed by the mathematician Robert Kaplan (during an interview by the author, January 1, 2000).
132. C. A. Qadir, *Philosophy and Science in the Islamic World* (Croom Helm: London, 1988), pp. 141–143.
133. Nasr, *Science and Civilization,* p. 130.
134. Qadir, *Philosophy and Science,* pp. 42, 43.
135. Alnoor Dhanani, "Atomism in Islamic Thought," in Selin (ed.), *Encyclopedia of the History of Science, Technology, and Medicine in Non-Western Cultures,* pp. 139, 140.
136. Timothy McGrew, "Physics in the Islamic World," in Selin (ed.), *Encyclopedia of the History of Science, Technology, and Medicine in Non-Western Cultures,* p. 820.
137. Shlomo Pines, *Studies in Arabic Versions of Greek Texts and in Medieval Science* (Jerusalem: Magnes Press, Hebrew University, 1986), pp. 192, 193.
138. Max Jammer, *Concepts of Space: The History of Theories of Space in Physics* (New York: Dover, 1954), pp. 63, 65.
139. Qadir, *Philosophy and Science,* p. 67.
140. Jammer, *Concepts of Space,* p. 63.
141. Ibid., pp. 63, 64.
142. Ibid., p. 64.
143. Qadir, *Philosophy and Science,* p. 67.
144. Seyyed Hossein Nasr, *An Introduction to Islamic Cosmological Doctrines* (Cambridge, Mass.: Harvard Univ. Press, 1964), p. 64.
145. Jammer, *Concepts of Space,* pp. 91, 92.
146. Ibid., pp. 64, 65.
147. Pines, *Studies in Arabic Versions of Greek Texts and in Medieval Science,* pp. 356, 357.
148. Marshall Clagett, *The Science of Mechanics in the Middle Ages* (Madison: University of Wisconsin Press, 1959), pp. 511–513.
149. Pines, *Studies in Arabic Versions of Greek Texts and in Medieval Science,* pp. 356, 357.

CHAPTER 6: GEOLOGY

1. Kenneth F. Weaver, "The Search for Our Ancestors," *National Geographic* 168 (Nov. 1985): 616.
2. Gordon Childe, "The Prehistory of Science: Archaeological Documents," in Guy S. Metraux and François Crouzet (eds.), *The Evolution of Science: Readings from the History of Mankind* (New York: New American Library, Mentor Books, 1963), pp. 39, 40.
3. Ibid.
4. Weaver, "Search for Our Ancestors," p. 616.
5. Childe, "Prehistory of Science," pp. 39, 40.
6. Ibid., p. 72.
7. V. V. Tikhomirov, "The Development of the Geological Sciences in the USSR from Ancient Times to the Middle of the Nineteenth Century," in Cecil J. Schneer (ed.), *Toward a History of Geology: Proceedings of the New Hampshire Inter-Disciplinary Conference on the History of Geology, Sept. 7–12, 1967* (Cambridge, Mass.: MIT Press, 1969), pp. 357–359.
8. Childe, "Prehistory of Science," pp. 66, 67.
9. Debiprasad Chattopadhyaya, *History of Science and Technology in Ancient India: The Beginnings* (Calcutta: Firm Klm, 1986), pp. 316, 317.
10. Tikhomirov, "Development of the Geological Sciences in the USSR," pp. 357–359.
11. R. Campbell Thompson, *A Dictionary of Assyrian Chemistry and Geology* (Oxford: Clarendon Press, 1936), pp. xix, xx.
12. Samuel Noah Kramer, *The Sumerians: Their History, Culture, and Character* (Chicago: University of Chicago Press, 1963), p. 103.
13. H. J. J. Winter, *Eastern Science: An Outline of Its Scope and Contribution* (Westport, Conn.: Greenwood Press, 1952, reprint ed., 1985), pp. 5, 6.
14. Kramer, *The Sumerians,* p. 90.
15. A. Leo Oppenheim, *Ancient Mesopotamia: Portrait of a Dead Civilization* (Chicago: University of Chicago Press, 1977), p. 247.
16. R. Campbell Thompson, *Dictionary of Assyrian Chemistry and Geology,* pp. xix, xx.
17. Ibid., pp. 200–205.
18. Ibid., pp. xxxvi, xxxvii.
19. Ibid., p. xxi.
20. Oppenheim, *Ancient Mesopotamia,* pp. 41, 42, 84.
21. S. Terry Childs, "Metallurgy in Africa," in Helaine Selin (ed.), *Encyclopaedia of the History of Science, Technology, and Medicine in Non-Western Cultures* (Norwell, Mass.: Kluwer Academic Publishers, 1997), p. 721.
22. Winter, *Eastern Science,* p. 14.
23. Childe, "Prehistory of Science," p. 721.

24. Marshall Clagett, *Ancient Egyptian Science: A Source Book* (Philadelphia: American Philosophical Society, 1989), pp. 49, 50.
25. Ibid., pp. 68, 69.
26. Ibid., p. 109.
27. Ibid., pp. 109–113.
28. Ibid., pp. 237–241.
29. Ibid. pp. xi, xii.
30. Ibid., pp. 247, 248; p. 28.
31. Chattopadhyaya, *History of Science and Technology in Ancient India,* pp. 70–72.
32. Winter, *Eastern Science,* p. 16.
33. D. M. Bose, S. N. Sen, and B. V. Subbarayappa (eds.), *A Concise History of Science in India* (New Delhi: Indian National Science Academy, 1971), p. 570.
34. Chattopadhyaya, *History of Science and Technology in Ancient India,* pp. 295, 296.
35. R. C. Majumdar, "Scientific Spirit in Ancient India," in Metraux and Crouzet (eds.), *Evolution of Science,* p. 85.
36. Chattopadhyaya, *History of Science and Technology in Ancient India,* p. 330.
37. Bose, Sen, and Subbarayappa (eds.), *A Concise History of Science in India,* p. 280.
38. Ibid., p. 290.
39. Majumdar, "Scientific Spirit in Ancient India," p. 85.
40. Bose, Sen, and Subbarayappa (eds.), *A Concise History of Science in India,* pp. 299, 300.
41. S. Warren Carey, *Theories of the Earth and Universe: A History of Dogma in the Earth Sciences* (Stanford, Calif.: Stanford University Press, 1988), p. 338.
42. Susan J. Thompson, *A Chronology of Geological Thinking from Antiquity to 1899* (Metuchen, N.J.: Scarecrow Press, 1988), p. 1.
43. Bose, Sen, and Subbarayappa (eds.), *A Concise History of Science in India,* p. 450.
44. K. S. Murty, "History of Geoscience Information in India," in Anthony P. Harvey and Judith A. Diment (eds.), *Geoscience Information: A State-of-the-Art Review: Proceedings of the First International Conference of Geological Information, London, 10–12 April, 1978* (Heathfield, England: Broad Oak Press, 1979), p. 51.
45. Chattopadhyaya, *History of Science and Technology in Ancient India,* p. 278.
46. Ibid., pp. 295, 296.
47. Carey, *Theories of the Earth and Universe,* p. 14.
48. Thompson, *Chronology of Geological Thinking,* p. 1.
49. Carey, *Theories of the Earth and Universe,* pp. 16, 17.
50. Winter, *Eastern Science,* pp. 37–40.
51. Ashok K. Dutt, "Geography in India," in Selin (ed.), *Encyclopaedia of the History of Science, Technology, and Medicine in Non-Western Cultures,* p. 353.
52. Lin Wenzhao, "Magnetism and the Compass," in Institute of the History of Natural Sciences (ed.), *Ancient China's Technology and Science* (Beijing: China Books & Periodicals, 1983), pp. 160, 161.

53. Tong B. Tang, *Science and Technology in China* (London: Longman, 1984), Introduction and chapter 7; Joseph Needham, *Science and Civilization in China* (Cambridge: Cambridge University Press, 1976), 2:295.
54. Ibid., p. 323; Yang Wenheng, "Rocks, Mineralogy and Mining," in Institute of the History of Natural Sciences (ed.), *Ancient China's Technology and Science,* p. 262.
55. Tang, *Science and Technology,* Introduction and chap. 6.
56. Wenheng, "Rocks, Mineralogy and Mining," p. 259.
57. Needham, *Science and Civilization in China,* 2:306, 307.
58. Ibid., pp. 307, 308.
59. Ibid., p. 299.
60. Ibid., p 296.
61. Carey, *Theories of the Earth and Universe,* p. 35.
62. Needham, *Science and Civilization in China,* 2:297.
63. Carey, *Theories of the Earth and Universe,* p. 35.
64. Frank Dawson Adams, *The Birth and Development of the Geological Sciences* (Baltimore: Williams & Wilkins, 1938), p. 12.
65. Needham, *Science and Civilization in China,* 2:292.
66. Tang, *Science and Technology,* Introduction and chap. 5.
67. Tang Xiren, "Earthquake Forecasting, Precautions Against Earthquakes and Anti-Seismic Measures," in Institute of the History of Natural Sciences (ed.), *Ancient China's Technology and Science,* p. 273.
68. Needham, *Science and Civilization in China,* 2:301.
69. Tang, *Science and Technology,* Introduction and chap. 5.
70. Needham, *Science and Civilization in China,* 2:301.
71. Ibid., pp. 301, 302.
72. Ibid., pp. 302ff.
73. Edward J. Tarbuck and Frederick K. Lutgens, *Earth: An Introduction to Physical Geology,* 5th ed. (Upper Saddle River, N.J.: Prentice Hall, 1996), p. 381.
74. Needham, *Science and Civilization in China,* 3:3.
75. Wen-yuan Qian, *The Great Inertia: Scientific Stagnation in Traditional China* (London: Croom Helm, 1985), pp. 78, 79.
76. Lin Wenzhao, "Magnetism and the Compass," in Institute of the History of Natural Sciences (ed.), *Ancient China's Technology and Science,* p. 153.
77. Kiyosi Yabuuti, "Sciences in China from the Fourth to the End of the Twelfth Century," in Metraux and François Crouzet (eds.), *The Evolution of Science,* pp. 126, 127.
78. Carey, *Theories of the Earth and Universe,* p. 27.
79. Ibid.
80. Needham, *Science and Civilization in China,* 3:11.
81. Ibid., pp. 9, 10; Wenzhao, "Magnetism and the Compass," p. 154.

82. Yabuuti, "Sciences in China," pp. 126, 127.
83. Needham, *Science and Civilization in China,* 3:14, 15.
84. Carey, *Theories of the Earth and Universe,* p. 15.
85. Needham, *Science and Civilization in China,* 2:238.
86. Ibid., pp. 238–241.
87. Tang, *Science and Technology,* Introduction and chap. 5.
88. Needham, *Science and Civilization in China,* 2:245.
89. Ibid., pp. 261, 262.
90. Ibid., pp. 224–225.
91. Ibid.
92. Ibid., p. 231.
93. J. M. Millas-Vallicrosa, "Translations of Oriental Scientific Works," in Metraux and François Crouzet (eds.), *The Evolution of Science,* p. 128.
94. Ibid., p. 130.
95. Abdul Latif Samian, "Al-Biruni," in Selin (ed.), *Encyclopaedia of the History of Science, Technology, and Medicine in Non-Western Cultures,* p. 157.
96. Winter, *Eastern Science,* p. 71; Seyyed Hossein Nasr, *Science and Civilization in Islam* (Cambridge, Mass.: Harvard University Press, 1968), p. 231.
97. Medhi Aminrazavi, "Ibn Sina," in Selin (ed.), *Encyclopaedia of the History of Science, Technology, and Medicine in Non-Western Cultures,* p. 434.
98. Adams, *Birth and Development of the Geological Sciences,* p. 19. See also Winter, *Eastern Science,* p. 72.
99. Tikhomirov, "Development of the Geological Sciences in the USSR," pp. 361, 362.
100. Ibid., p. 360.
101. R. J. Forbes, *Studies in Early Petroleum History* (Leiden, Netherlands: E. J. Brill, 1958), pp. 154, 155.
102. Nasr, *Science and Civilization in Islam,* pp. 152, 153; Winter, *Eastern Science,* p. 69.
103. W. C. Krumbein and L. L. Sloss, *Stratigraphy and Sedimentation* (San Francisco: Freeman, 1951).
104. Winter, *Eastern Science,* p. 71.
105. Nasr, *Science and Civilization in Islam,* p. 114.
106. Ibid., pp. 115, 116.
107. Tikhomirov, "Development of the Geological Sciences in the USSR," pp. 360, 361.
108. Adams, *Birth and Development of the Geological Sciences,* pp. 333–335, quoting from translation.
109. Tarbuck and Lutgens, *Earth,* p. 116.
110. E. J. Homeyard and D. C. Mandeville (eds. and trans.), *Avicenne de Congelatione et Conglutinatione Lapidum: Being Sections of the Kitab al-Shifa* (Paris: *Librairie orientaliste Paul Geuthner),* pp. 20, 22.

111. S. Thompson, *Chronology of Geological Thinking,* p. 15; also in Winter, *Eastern Science,* p. 69.
112. Tarbuck and Lutgens, *Earth,* p. 266.
113. Ibid., p. 142.
114. Tikhomirov, "Development of the Geological Sciences in the USSR," p. 360.
115. Adams, *Birth and Development of the Geological Sciences,* p. 254.
116. S. Thompson, *Chronology of Geological Thinking,* p. 16.
117. Adams, *Birth and Development of the Geological Sciences,* pp. 333–335, quoting from translation.
118. Arthur N. Strahler and Alan H. Strahler, *Modern Physical Geography* (New York: Wiley, 1978), p. 291.
119. Quoted in Frank Dawson Adams, *The Birth and Development of the Geological Sciences* (Baltimore: William & Wilkins, 1938), pp. 333–335, quoting from translation.
120. Tarbuck and Lutgens, *Earth,* p. 132.
121. Ibid., p. 146.
122. Tikhomirov, "Development of the Geological Sciences in the USSR," pp. 359, 360.
123. Ibid., p. 361.
124. Nasr, *Science and Civilization in Islam,* pp. 99, 101.
125. Ibid., p. 98.
126. Ibid., p. 106.
127. Strahler and Strahler, *Modern Physical Geography,* pp. 4, 5; see also Carey, *Theories of the Earth and Universe,* p. 15.
128. Nasr, *Science and Civilization in Islam,* pp. 107, 108.
129. K. V. Sarma, "Varahamihira," in Selin (ed.), *Encyclopaedia of the History of Science, Technology, and Medicine in Non-Western Cultures,* pp. 999, 1000.
130. S. Thompson, *Chronology of Geological Thinking,* pp. 13, 14.
131. Ibid., pp. 13, 14.
132. K. S. Murty, "History of Geoscience Information in India," p. 52.
133. Needham, *Science and Civilization in China,* 2:306, 307.
134. Sheila Seaman, associate professor of geology at the University of Massachusetts at Amherst, in a letter to the author, fall 2001.
135. Tong B. Tang, *Science and Technology in China,* Introduction and chap. 6.
136. Wenzhao, "Magnetism and the Compass," p. 263.
137. Seaman, letter to author, fall 2001.
138. Needham, *Science and Civilization in China,* 2:291.
139. Ibid., p. 299.
140. Carey, *Theories of the Earth and Universe,* p. 35; Winter, *Eastern Science,* p. 13; Needham, *Science and Civilization in China,* 2:299; Seaman, letter to author, fall 2001.

141. Needham, *Science and Civilization in China,* 2:299.
142. Ibid., pp. 292–293.
143. Seaman, letter to author, fall 2001; Needham, *Science and Civilization in China,* 2:290–292.
144. Ibid., p. 267, pp. 226–227.
145. Tang, *Science and Technology*, chap. 5 and Introduction; Needham, *Science and Civilization in China,* 2:236.
146. Ibid., pp. 233–235.
147. Ibid., p. 236.
148. Dava Sobel, *Galileo's Daughter* (New York: Walker, 1999), pp. 74–76.
149. Tarbuck and Lutgens, *Earth,* p. 123.
150. G. Milne, "Normal Erosion as a Factor in Soil Profile Development," [letter] *Nature,* Sept. 26, 1936, pp. 548–549.
151. Paul Richards, "Agriculture in Africa," in Selin (ed.), *Encyclopaedia of the History of Science, Technology, and Medicine in Non-Western Cultures,* p. 15.
152. Gary A. Wright, *People of the High Country: Jackson Hole Before the Settlers* (New York: Peter Lang, 1984), pp.12–18.
153. Ibid., p. 46.
154. Barry Holstun Lopez, *Giving Birth to Thunder, Sleeping with His Daughter: Coyote Builds North America* (New York: Avon Books, 1977), pp. 9, 10.
155. Joseph Weixelman, "The Power to Evoke Wonder: Native Americans and the Geysers of Yellowstone National Park," Merrill G. Burlingame Special Collections, Montana State University, Bozeman, July 19, 1992, p. 21.
156. Seaman, letter to author, fall 2001.
157. Leslie B. Davis, Stephen A. Aaberg, James G. Schmitt, and Ann M. Johnson, *The Obsidian Cliff Plateau Prehistoric Lithic Source, Yellowstone National Park, Wyoming,* Selections from the Division of Cultural Resources No. 6, Rocky Mountain Region, National Park Service, Denver, Colorado, 1995.
158. Richard Erodes and Alfonso Ortiz (eds.), *American Indian Trickster Tales* (New York: Viking, 1998), pp. 23–24.
159. William R. Gray, "The Northwest," in Robert L. Breeden (ed.), *America's Majestic Canyons* (Washington, D.C.: National Geographic Society, 1979), p. 26.
160. Weixelman, "The Power to Evoke Wonder," p. 37.
161. Ibid., pp. 51, 52.
162. Roman Pina Chan, *The Olmec: Mother Culture of Mesoamerica,* ed. Laura Laurencich Minelli (New York: Rizzoli, 1989), p. 70.
163. Ibid., p. 46.
164. Joyce Marcus and Kent V. Flannery, *Zapotec Civilization: How Urban Society Evolved in Mexico's Oaxaca Valley* (New York: Thames and Hudson, 1996), pp. 96, 97, 101–103, 109, 110.

165. Warwick Bray, John L. Sorenson, and James R. Moriarty III, *Metallurgy in Ancient Mexico* (Greeley, Colo.: University of Northern Colorado, Museum of Anthropology, 1982), p. 17.
166. Dorothy Hosler, *The Sounds and Colors of Power* (Cambridge, Mass.: MIT Press, 1994), pp. 13–16.
167. Bray, Sorenson, and Moriarty, *Metallurgy in Ancient Mexico,* p. 2.
168. Ibid., p. 5.
169. Dennis Tedlock (trans.), *Popul Vuh: The Mayan Book of the Dawn of Life and the Glories of Gods and Kings* (New York: Simon & Schuster, 1996), p. 224.
170. D. G. A. Whitten, with J. R. V. Brooks, *The Penguin Dictionary of Geology* (New York: Penguin Books, 1972).
171. Jacques Soustell, *Daily Life of the Aztecs: On the Eve of the Spanish Conquest,* trans. Patrick O'Brian (Stanford, Calif.: Stanford University Press, 1961), p. xv.
172. Ibid., p. xix.
173. Bernardino de Sahagun, *Florentine Codex: General History of the Things of New Spain,* Book 11: *Earthly Things,* trans. Charles Dibble and Arthur Anderson (Santa Fe, N.M.: School of American Research/University of Utah, 1963), no. 14, pt. XII, p. 229.
174. Ibid., p. 222.
175. Barbara J. Williams, "Pictorial Representation of Soils in the Valley of Mexico: Evidence from the Codex Vergara," in William V. Davidson and James J. Parsons (eds.), *Geoscience and Man,* vol. 21: *Historical Geography of Latin American* (Baton Rouge: Louisiana State University Press, 1980), p. 60.
176. de Sahagun, *Florentine Codex* no. 14, pt. XII, p. 247.
177. David Freidel, Linda Schele, and Joy Parker, *Maya Cosmos: Three Thousand Years on the Shaman's Path* (New York: Morrow, 1993), p. 132–135.
178. Ibid., pp. 135, 139.
179. J. Eric Thompson, *Maya History and Religion* (Norman, Okla.: Univ. of Oklahoma Press, 1970), pp. 183, 184, 245.
180. Marcus and Flannery, *Zapotec Civilization,* p. 95.
181. Tedlock, *Popul Vuh,* pp. 240, 241.
182. Ibid., p. 77.
183. Ibid., p. 241.
184. John B. Carlson, "Lodestone Compass: Chinese or Olmec Primacy?" *Science,* Sept. 5, 1975, p. 753.
185. Ibid., p. 758.
186. Vincent H. Malmstrom, "Knowledge of Magnetism in Pre-Columbian Mesoamerica," *Nature,* Feb. 5, 1976, pp. 390–391.
187. Tedlock, *Popul Vuh,* p. 220.
188. Bray, Sorenson, and Moriarty, *Metallurgy in Ancient Mexico,* p. 6.

189. Ibid., pp. 7, 8.
190. Salvador Palomino, "Three Times, Three Spaces in Cosmos Quechua," in Inter Press Service (comp.), *Story Earth: Native Voices on the Environment* (San Francisco: Mercury House, 1993), p. 59.
191. Francisco de Avila (comp.), *The Huarochiri Manuscript: A Testament of Ancient and Colonial Andean Religion,* trans. Frank Salomon and George L. Urioste (Austin: University of Texas Press, 1991), p. 15.
192. William Sullivan, *The Secret of the Incas: Myth, Astronomy, and the War Against Time* (New York: Crown, 1996), pp. 23, 303.
193. de Avila, *The Huarochiri Manuscript,* p. 15.
194. Gary Urton, *At the Crossroads of the Earth and the Sky: An Andean Cosmology* (Austin: University of Texas Press, 1981), p. 64.
195. Ibid., pp. 88, 89.
196. de Avila, *The Huarochiri Manuscript,* p. 15.
197. Ibid., pp. 140, 141 (sec. 422–430), and annotations.
198. Ibid., pp. 141, 142 (sec. 431–436), and annotations.
199. Sullivan, *Secret of the Incas,* pp. 233, 234.
200. Ibid.
201. J. Eric Thompson, *Maya History and Religion,* pp. 262–263.
202. Tarbuck and Lutgens, *Earth,* pp. 383, 384.
203. Sullivan, *Secret of the Incas,* p. 303.
204. Loren McIntyre, "Lost Empire of the Incas," *National Geographic* 144 (Dec. 1973): 757.
205. Sonia P. Juvik and James O. Juvik (eds.), *Atlas of Hawaii* (Honolulu: University of Hawaii Press, 1998), pp. 161–162.
206. David Malo, *Hawaiian Antiquities (Moolelo Hawaii),* trans. Nathaniel B. Emerson (Honolulu: Bernice P. Bishop Museum, 1951), p. 2.
207. Ibid., pp. 19, 20.
208. Ibid., pp. 2, 3.
209. Michael Kioni Dudley, *A Hawaiian Nation: Man, Gods, and Nature* (Honolulu: Na Kane O Ka Malo Press, 1990), pp. 10, 11.
210. Malo, *Hawaiian Antiquities,* p. 132.
211. Seaman, letter to author, fall 2001.
212. Ibid.
213. Dudley, *A Hawaiian Nation,* p. 11.
214. Malo, *Hawaiian Antiquities,* p. 2.
215. Nathaniel B. Emerson, *Pele and Hiiaka: A Myth from Hawaii* (Honolulu: 'Al Pohaku Press, 1915), pp. 213, 214.
216. Ibid., pp. 215–224.
217. Seaman, letter to author, fall 2001.
218. Dudley, *A Hawaiian Nation,* pp. 20–25.
219. Ibid., pp. 15–19.

220. Nancy Hudson-Rodd, "Geographical Knowledge," in Selin (ed.), *Encyclopaedia of the History of Science, Technology, and Medicine in Non-Western Cultures,* p. 349.
221. Stanley Breeden, "The First Australians," and Joseph Judge, "Child of Gondwana," *National Geographic* 173 (Feb 1988): 176, 177, 270, 274, 280.
222. David Suzuki and Peter Knudtson, *Wisdom of the Elders: Sacred Native Stories of Nature* (New York: Bantam Books, 1992), p. 173.
223. Damien Arabagali, as told to Herbert Paulzen, "They Trampled on Our Taboos," in Inter Press Service (comp.), *Story Earth,* pp. 80, 81.
224. Tarbuck and Lutgens, *Earth,* p. 540.

CHAPTER 7: CHEMISTRY

1. Leon Lederman (with Dick Teresi), *The God Particle* (Boston: Houghton Mifflin, 1993), pp. 108–110.
2. Wang Kuike, "Alchemy in Ancient China," in Institute of the History of Natural Sciences, Chinese Academy of Sciences (ed.), *Ancient China's Technology and Science* (Beijing: China Books & Periodicals, 1983), pp. 214, 215.
3. Ralph E. Lapp, *Matter,* LIFE Science Library (New York: Time Inc., 1963), p. 33; John W. Hill and Doris K. Kolb, *Chemistry for Changing Times* (Upper Saddle River, N.J.: Prentice Hall, 1998), p. 34.
4. H. J. J. Winter, *Eastern Science: An Outline of its Scope and Contribution* (Westport, Conn.: Greenwood Press, 1952, reprint ed., 1985), p. 24.
5. Ibid.
6. Burton Feldman, *The Nobel Prize* (New York: Arcade Publishing, 2000), pp. 207, 235.
7. *Dictionary of Scientific Biography;* and Lederman, *The God Particle,* pp. 182–184.
8. Feldman, *The Nobel Prize,* p. 134.
9. Ibid., p. 135.
10. Emilio Segre, *Enrico Fermi, Physicist* (Chicago: University of Chicago Press, 1970), pp. 98–99.
11. Feldman, *The Nobel Prize,* p. 162.
12. Eduard Farber, *The Evolution of Chemistry* (New York: Ronald Press, 1952), p. 15.
13. A. Leo Oppenheim, *Ancient Mesopotamia: Portrait of a Dead Civilization* (Chicago: University of Chicago Press, 1977), p. 321.
14. John Read, *Prelude to Chemistry* (New York: Macmillan, 1937), p. xxi.
15. Ibid., p. 2 (quoting from M. A. Atwood).
16. Ibid., pp. 6, 9.
17. Ibid., p. 4.
18. Arthur Greenberg, *A Chemical History Tour* (New York: Wiley-Interscience, 2000), p. 43.

19. Edward Bruce Bynum, *The African Unconscious: Roots of Ancient Mysticism and Modern Psychology* (New York: Teachers College Press, 1999), pp. 37, 38.
20. John Maxson Stillman, *The Story of Alchemy and Early Chemistry* (New York: Dover, 1960), p. 137.
21. Ibid., p 135.
22. Farber, *The Evolution of Chemistry,* p. 16.
23. Ibid., pp. 16, 17.
24. Read, *Prelude to Chemistry,* pp. 14–16; Farber, *The Evolution of Chemistry,* pp. 33–35.
25. Ibid.
26. Will Durant, *The Story of Civilization,* vol. 1: *Our Oriental Heritage* (New York: Simon & Schuster, 1953), p. 150.
27. R. Campbell Thompson, *A Dictionary of Assyrian Chemistry and Geology* (Oxford: Clarendon Press, 1936), p. xxx.
28. R. J. Forbes, *Studies in Ancient Technology* (Leiden: E. J. Brill, 1965), 3:181.
29. Cathy Cobb and Harold Goldwhite, *Creations of Fire: Chemistry's Lively History from Alchemy to the Atomic Age* (New York: Plenum Press, 1995), p. 15.
30. Ibid.
31. David Malo, *Hawaiian Antiquities,* trans. Nathaniel B. Emerson (Honolulu: Bernice B. Bishop Museum, 1951), p. 97.
32. Marshall Clagett, *Ancient Egyptian Science: A Source Book* (Philadelphia: American Philosophical Society, 1989), 1:229–234.
33. Stillman, *The Story of Alchemy,* pp. 6–8.
34. Cobb and Goldwhite, *Creations of Fire,* p. 7.
35. Stillman, *The Story of Alchemy,* pp. 78–79.
36. Ibid., p. 81.
37. Ibid., p. 82.
38. Ibid., pp. 87–98.
39. Quoted in Forbes, *Studies in Ancient Technology,* 3:2.
40. Ibid., pp. 6, 7–10, 11.
41. Ibid., p. 15.
42. Ibid., pp. 6, 17
43. Hill and Kolb, *Chemistry for Changing Times,* p. 496.
44. Forbes, *Studies in Ancient Technology,* 3:183.
45. Stillman, *The Story of Alchemy,* p. 95.
46. Cobb and Goldwhite, *Creations of Fire,* p. 13.
47. Stillman, *The Story of Alchemy,* pp. 94–97.
48. Farber, *The Evolution of Chemistry,* p. 19.
49. Giovanni Curatola, *The Simon and Schuster Book of Oriental Carpets* (New York: Simon and Schuster, 1981), p. 17.
50. Forbes, *Studies in Ancient Technology,* 4:110.

51. Stillman, *The Story of Alchemy,* p. 86.
52. John S. Mbiti, *An Introduction to African Religion* (London: Heinemann, 1975), p. 17.
53. J. Olumide Lucas, *The Religion of the Yorubas* (Brooklyn, N.Y.: Athelia Henrietta Press, 1996), pp. 235, 349–354.
54. Rose Egbinládé Sakey-Milligan, priestess of the Ifá-Òrìsà religion, personal communication to the author, Nov. 12, 2000.
55. Afolabi A. Epega and Philip John Neimark, *The Sacred Ifa Oracle* (Brooklyn, N.Y.: Athelia Henrietta Press, 1995), p. xvi.
56. M. J. Field, *Religion and Medicine of the Ga People* (New York: Oxford University Press, 1961), p. 111.
57. Gloria Thomas-Emeagwali, "Textile Technology in Nigeria in the Nineteenth and Early Twentieth Centuries," in Gloria Thomas-Emeagwali (ed.), *African Systems of Science, Technology, and Art: The Nigerian Experience* (London: Arnak House, 1993), pp. 25, 26, 133.
58. Ibid., p. 26.
59. Okediji Moyo, "The Mythic Mechanics: Art and Technology in Western Nigeria," in Thomas-Emeagwali, *African Systems,* p. 110.
60. Thomas-Emeagwali, "Textile Technology in Nigeria," p. 26.
61. Curatola, *The Simon and Schuster Book of Oriental Carpets,* p.18.
62. Richard Okagbue, "The Scientific Basis of Traditional Food Processing in Nigerian Communities," in Thomas-Emeagwali, *African Systems,* p. 66.
63. Ibid., pp. 69–72.
64. John Mann, *Murder, Magic, and Medicine* (New York: Oxford University Press, 1992), p. 48.
65. Peter Holmes, *The Energetics of Western Herbs: Integrating Western and Oriental Herbal Medicine Traditions* (Boulder, Colo.: Artemis, 1989), 1:21–24. Today, Western medicine is increasingly incorporating some of Eastern medicine's holistic approaches, especially Chinese acupuncture; Western medicine has always been extremely effective in curing acute, identifiable problems but less successful with complex, vague, or chronic complaints.
66. Field, *Religion and Medicine of the Ga People,* p. 131.
67. Ibid., pp. 114, 115.
68. Ibid., p. 121.
69. Ibid., p. 125.
70. Malidoma Patrice Some, *Of Water and the Spirit* (New York: Penguin, 1994), p. 263.
71. Hill and Kolb, *Chemistry for Changing Times,* p. 227.
72. Jeremy Narby, *The Cosmic Serpent: DNA and the Origins of Knowledge* (New York: Tarcher/Putnam, 1998), pp. 114, 195.
73. Quoted in ibid., p. 195, from W. I. B. Beveridge, "The Art of Scientific Investigation" (1950).

74. Mann, *Murder, Magic, and Medicine,* pp. 58, 62.
75. Ibid., pp. 31–35.
76. Hill and Kolb, *Chemistry for Changing Times,* p. 329.
77. Campbell Thompson, *A Dictionary of Assyrian Chemistry and Geology,* p. xvi.
78. Forbes, *Studies in Ancient Technology,* 1:131, 132.
79. Lucas, *The Religion of the Yorubas,* p. 239.
80. Campbell Thompson, *A Dictionary of Assyrian Chemistry and Geology,* pp. xi, xii.
81. Forbes, *Studies in Ancient Technology,* 1:128.
82. Lederman, *The God Particle,* p. 108.
83. Campbell Thompson, *A Dictionary of Assyrian Chemistry and Geology,* pp. 140, 141.
84. Forbes, *Studies in Ancient Technology,* 3:219.
85. Campbell Thompson, *A Dictionary of Assyrian Chemistry and Geology,* pp. 102, xxxxii, xxxiii.
86. R. J. Forbes, *A Short History of the Art of Distillation* (Leiden: E. J. Brill, 1970), pp. 1–16.
87. Forbes, *Studies in Ancient Technology,* 5:120.
88. Ibid., pp. 120, 131, 132.
89. Hill and Kolb, *Chemistry for Changing Times,* p. 282.
90. A. Leo Oppenheim, *Ancient Mesopotamia: Portrait of a Dead Civilization* (Chicago: Univ. of Chicago Press, 1977), p. 231.
91. Campbell Thompson, *A Dictionary of Assyrian Chemistry and Geology,* pp. xxiii.
92. Farber, *The Evolution of Chemistry,* p. 21.
93. Forbes, *Studies in Ancient Technology,* 5:135, 136.
94. Ibid., pp. 136, 138.
95. H. Moore, "Reproductions of an Ancient Babylonian Glaze," *Iraq* 10 (1948), quoted in Forbes, *Studies in Ancient Technology,* 5:136.
96. Campbell Thompson, *A Dictionary of Assyrian Chemistry and Geology,* pp. xxiii–xxvi.
97. Forbes, *Studies in Ancient Technology,* 5:141.
98. Ibid., pp. 138–141, 143,144.
99. Ahmad Y. al-Hassan and Donald R. Hill, *Islamic Technology: An Illustrated History* (Cambridge: Cambridge University Press, 1987).
100. Martin Levey, "Chemical Notions of the Early Ninth-Century," in Henry M. Leicester (ed.), *Chymia: Annual Studies in the History of Chemistry,* vol. 11, ed. (Philadelphia: Univ. of Pennsylvania Press, 1966), pp. 29, 33.
101. Gareth Roberts, *The Mirror of Alchemy: Alchemical Ideas and Images in Manuscripts and Books from Antiquity to the Seventeenth Century,* ed. Henry M. Leicester (Toronto: University of Toronto Press, 1994), p. 26.
102. Al-Hassan and Hill, *Islamic Technology,* p. 133; Hamed Abdel-reheem Ead (ed.), "Alchemy in Islamic Times," www.levity.com/alchemy/islam01.html.

103. Joseph Needham, *Science in Traditional China: A Comparative Perspective* (Cambridge, Mass.: Harvard University Press, 1981), p. 68.
104. S. Nomanul Haq, "Jabir ibn Hayyan," in Helaine Selin (ed.), *Encyclopedia of the History of Science, Technology, and Medicine in Non-Western Cultures* (Norwell, Mass.: Kluwer Academic Publishers, 1997), p. 459.
105. Roberts, Mirror of Alchemy, pp. 45, 47–50.
106. Winter, *Eastern Science,* p. 64.
107. Roberts, Mirror of Alchemy, p. 51.
108. Seyyed Hossein Nasr, *An Introduction to Islamic Cosmological Doctrines* (Cambridge, Mass.: Harvard University Press, 1964), p. 247.
109. Frank Dawson Adams, *The Birth and Development of the Geological Sciences* (Baltimore: Williams & Wilkins, 1938), p. 19. See also Winter, *Eastern Science,* p. 72.
110. Winter, *Eastern Science,* p. 83.
111. Ralph E. Lapp, *Matter,* LIFE Science Library series (New York: Time Inc., 1963), p. 34.
112. National Institutes of Health, "Islamic Culture and the Medical Arts: Pharmaceutics and Alchemy," www.nlm.nih.gov/exhibition/islamic_medical/islamic_11.html.
113. Martin Levey and Noury Al-Khaledy, "Chemistry in the Medical Formulary of al-Samarqani," in Henry M. Leicester (ed.), *Chymia,* p. 41.
114. Hill and Kolb, *Chemistry for Changing Times,* pp. 64–69; Lapp, *Matter,* pp. 126–149.
115. Lapp, *Matter,* p. 37.
116. Winter, *Eastern Science,* p. 64.
117. Greenberg, *Chemical History Tour,* p. 46.
118. Seyyed Hossein Nasr, *Science and Civilization in Islam* (Cambridge, Mass.: Harvard Univ. Press, 1968), p. 269.
119. Ibid., pp. 273–277.
120. P. C. Ray, "History of Hindu Chemistry," quoted in R. C. Majumdar, "Scientific Spirit in Ancient India," in Guy S. Metraux and François Crouzet (eds.), *The Evolution of Science* (New York: New American Library, 1963), p. 83.
121. Durant, *The Story of Civilization,* p. 529.
122. Read, *Prelude to Chemistry,* p. 19.
123. Majumdar, "Scientific Spirit in Ancient India," p. 86.
124. Stillman, *The Story of Alchemy and Early Chemistry,* pp. 105–111.
125. Lederman, *The God Particle,* pp. 128, 374.
126. Feldman, *The Nobel Prize,* p. 367.
127. Winter, *Eastern Science,* p. 28.
128. Wang Kuike, "Alchemy in Ancient China," Institute of the History of Natural Sciences (ed.), *Ancient China's Technology and Science* (Beijing: Foreign Language Press, 1983), p. 214.
129. Read, *Prelude to Chemistry,* p. 6.

130. Kuike, "Alchemy in Ancient China," p. 213.
131. Joseph Needham, *Science & Civilization in China,* vol. 5, pt. 3: "Chemistry and Chemical Technology" (Cambridge: Cambridge Univ. Press, 1976), p. 145.
132. Read, *Prelude to Chemistry,* p. 6.
133. Needham, *Science and Civilization in China,* 5:144.
134. Ibid., pp. 149, 150.
135. Winter, *Eastern Science,* p. 25.
136. Joseph Needham, *Science in Traditional China* (Cambridge, Mass.: Harvard University Press, 1981), p. 29.
137. Kuike, "Alchemy in Ancient China," pp. 220–222.
138. Ibid. pp. 221, 222.
139. Mann, *Murder, Magic, and Medicine,* pp. 111–112.
140. Hill and Kolb, *Chemistry for Changing Times,* p. 242.
141. Richard Evans Schultes, "Amazonian Ethnobotany and the Search for New Drugs," in Ciba Foundation, *Ethnobotany and the Search for New Drugs* (New York: Wiley, 1994), pp. 107, 108.
142. Richard Evans Schultes and Robert F. Raffauf, *The Healing Forest: Medicinal and Toxic Plants of the Northwest Amazonia* (Portland, Ore.: Dioscorides Press, 1990), p. 35.
143. Schultes, "Amazonian Ethnobotany," pp. 108, 109.
144. Mann, *Murder, Magic, and Medicine,* pp. 60–64.
145. Ibid.
146. Dennis J. McKenna, L. E. Luna, and G. N. Towers, "Biodynamic Constituents in Ayahuasca Admixture Plants: An Uninvestigated Folk Pharmacopeia," in Richard Evans Schultes and Siri von Reis (eds.), *Ethnobotany: Evolution of a Discipline* (Portland, Ore.: Dioscorides Press, 1995), p. 351.
147. Schultes, "Amazonian Ethnobotany," pp. 108, 109.
148. Quoted in Narby, *The Cosmic Serpent,* pp. 10, 11.
149. Richard Evans Schultes and Robert F. Raffauf, *Vine of the Soul: Medicine Men, Their Plants and Rituals in the Colombian Amazonia* (Oracle, Ariz.: Synergistic Press, 1992).
150. Schultes, "Amazonian Ethnobotany," pp. 108, 109.
151. Mann, *Murder, Magic, and Medicine,* p. 19.
152. Schultes and Raffauf, *The Healing Forest,* pp. 264–270, 302–310.
153. Mann, *Murder, Magic, and Medicine,* p. 39.
154. Schultes and Raffauf, *The Healing Forest,* pp. 264–270.
155. Mann, *Murder, Magic, and Medicine.* p. 21.
156. Narby, *The Cosmic Serpent,* p. 40.
157. Ibid., p. 171.
158. Schultes and Raffauf, *The Healing Forest,* pp. 303–309.
159. Schultes, "Amazonian Ethnobotany," pp. 109, 110, 112.

160. Michael J. Balick, "Ethnobotany, Drug Development and Biodiversity Conservation—Exploring the Linkages," in Ciba Foundation, *Ethnobotany and the Search for New Drugs,* Symposium 185 (Chichester: John Wiley, 1994), p. 5.
161. M. M. Iwu, quoted in ibid., p. 20.
162. Walter H. Lewis and Memory P. Elvin-Lewis, "Basic, Quantitative and Experimental Research Phases of Future Ethnobotany with Reference to the Medicinal Plants of South America," in Ciba Foundation, *Ethnobotany and the Search for New Drugs,* pp. 65–67.
163. Narby, *The Cosmic Serpent,* pp. 28, 29.
164. Ibid., pp. 26, 27.
165. F. Trupp (1981), quoted in Richard Evans Schultes & Robert F. Raffauf, *Vine of the Soul: Medicine Men, Their Plants and Rituals in the Colombian Amazonia* (Oracle, Ariz.: Synergetic Press, 1992), p. 22.
166. T. McKenna, quoted in Schultes and Raffauf, *Vine of the Soul,* p. 58.
167. Schultes and Raffauf, "Vine of the Soul," p. 58.
168. Linda Schele Freidel and Joy Parker, *Maya Cosmos: Three Thousand Years on the Shaman's Path* (New York: Morrow, 1993), pp. 234, 244.
169. Ibid., pp. 210, 211.
170. Ibid., pp. 248, 249, 455 note 31.
171. Dennis Tedlock (trans.), *Popul Vuh: The Mayan Book of the Dawn of Life* (New York: Simon & Schuster, 1996), pp. 34, 340.
172. Bernardino de Sahagún, *Florentine Codex: General History of the Things of New Spain,* Book 11: *Earthly Things,* trans. Charles Dibble and Arthur Anderson (Santa Fe, N.M.: School of American Research/University of Utah, 1963), no. 14, pt. XII, pp. 239, 240.
173. Forbes, *Studies in Ancient Technology,* 4:104.
174. de Sahagún, *Earthly Things,* p. 243.
175. Elizabeth Andros Foster (ed. and trans.), *Motolinía's History of the Indians of New Spain* (Berkeley, Calif.: Cortes Society, 1950), p. 220.
176. Ibid., pp. 2, 69.
177. Ibid., pp. 218, 219; de Sahagún, *Florentine Codex,* Book 10: *The People,* trans. Charles Dibble and Arthur Anderson, pt. 11, p. 90.
178. *American Heritage Dictionary* (Boston: Houghton Mifflin, 1970); de Sahagún, *The People,* p. 89.
179. Anthony P. Andrews, *Mayan Salt Production and Trade* (Tucson: Univ. of Arizona Press, 1983), p. 11.
180. Forbes, *Studies in Ancient Technology,* 5:19.
181. David L. Feigman, *Legal Alchemy* (New York: Freeman, 2000), p. lx.

CHAPTER 8: TECHNOLOGY

1. Multnomah County School Board, *Portland African-American Baseline Essays* (Portland, Ore.: Multnomah County School Board, c. 1982).
2. Robert Patton, "Ooparts," *Omni* (Sept. 1982): 54.
3. Charles C. Mann, "1491," *Atlantic Monthly* (Mar. 2002).
4. Alfred Crosby, *Ecological Imperialism: The Biological Expansion of Europe, 900–1900* (Cambridge: Cambridge University Press, 1986), p. 22.
5. Ibid., p. 17.
6. Peter James and Nick Thorpe, *Ancient Inventions* (New York: Ballantine Books, 1994), p. 355.
7. Ibid., p. 206.
8. Ibid., p. 355.
9. Arnold Pacey, "Technology," in *Encyclopaedia of the History of Science, Technology, and Medicine in Non-Western Cultures,* ed. Helaine Selin (Dordrecht/Boston/London: Kluwer Academic Publishers, 1997), p. 937.
10. Howard R. Turner, *Science in Medieval Islam: An Illustrated Introduction* (Austin: University of Texas Press, 1995), p. 165.
11. Pacey, in *Encyclopaedia,* p. 937.
12. Alfred Crosby, professor of history at the University of Texas, in a letter to the author, December 29, 2000.
13. James and Thorpe, *Ancient Inventions,* p. 384.
14. Arnold Pacey, *Technology in World Civilization: A Thousand Year History,* Cambridge, Mass.: MIT Press, 1990), p. 8.
15. Donald R. Hill, "Technology in the Islamic World," in *Encyclopaedia,* p. 948.
16. Ibid., p. 949.
17. Pacey, *Technology in World Civilization,* pp. 10–11.
18. James and Thorpe, *Ancient Inventions,* p. 393.
19. Ahmad Y. Al-Hassan and Donald R. Hill, *Islamic Technology: An Illustrated History* (Cambridge: Cambridge University Press, 1987), p. 264.
20. Pacey, *Technology in World Civilization,* p. 34.
21. James and Thorpe, *Ancient Inventions,* p. 139.
22. al-Hassan and Hill, *Islamic Technology,* p. 45.
23. Ibid., pp. 59, 90, 61.
24. Turner, *Science in Medieval Islam,* p. 188.
25. James and Thorpe, *Ancient Inventions,* p. 140.
26. Pacey, *Technology in World Civilization,* p. 58.
27. Jack Weatherford, *Indian Givers* (New York: Fawcett Columbine, 1988), p. 204.
28. Michael D. Coe, *The Maya* (London: Thames and Hudson, 1999), p. 114.
29. Ibid., p. 6.
30. Weatherford, *Indian Givers,* p. 221.
31. Coe, *The Maya,* p. 30.

31. Ibid., p. 29.
32. Mann, "1491."
33. Coe, *The Maya,* p. 118.
34. Linda Schele and Mary Ellen Miller, *The Blood of Kings: Dynasty and Ritual in Maya Art* (London: Sotheby's, 1986), pp. 241–243.
35. Dorothy Hosler et al., "Prehistorical Polymers: Rubber Processing in Ancient Mesoamerica," *Science,* June 18, 1999, pp. 1988–1989.
36. Ibid.
37. Ibid.
38. Michael E. Smith, *The Aztecs* (Oxford: Blackwell, 1998), pp. 86–89.
39. Coe, *The Maya,* p. 103.
40. Ibid., p. 73.
41. Ibid.
42. Ibid., p. 114.
43. Ibid., p. 78.
44. Ibid.
45. Smith, *The Aztecs,* p. 231.
46. Ibid., p. 228.
47. Ibid., pp. 34, 36.
48. Ibid., p. 31.
49. Ibid., p. 199.
50. Ibid., pp. 76, 69.
51. Thomas F. Lynch, "The Identification of Inca Posts and Roads from Catarpe to Rio Frio," in Michael A. Malpass (ed.), *Provincial Inca: Archaeological and Ethnohistorical Assessment of the Impact of the Inca State* (Iowa City: University of Iowa Press, 1993), p. 123.
52. Sue Grosbol, ". . . And He Said in the Time of the Ynga, They Paid Tribute and Served the Ynga," in Malpass (ed.), *Provincial Inca,* p. 50.
53. Susan A. Niles, "The Provinces in the Heartland: Stylistic Variation and Architectural Innovation near Inca Cuzco," in Malpass (ed.), *Provincial Inca,* p. 153.
54. Katharina J. Schreiber, "The Inca Occupation of the Province of Andamarca Lucanas, Peru," in Malpass (ed.), *Provincial Inca,* p. 87.
55. Alfred Crosby in a letter to author, December 29, 2000.
56. Jared Diamond, *Guns, Germs, and Steel* (New York: Norton, 1997), p. 377.
57. Martin Bernal, *Black Athena,* vol. 2: *The Archaeological and Documentary Evidence* (New Brunswick, N.J.: Rutgers University Press, 1996).
58. Stanley Burstein (ed.), *Ancient African Civilizations* (Princeton, N.J.: Marcus Wiener Publishers, 1998), pp. 3–4.
59. Ibid., p. 4.
60. Ibid., p. 13.
61. Basil Davidson, *The Lost Cities of Africa* (Boston: Little, Brown, 1987), p. 65.
62. Burstein, *Ancient African Civilizations,* p. 31.

63. Davidson, *Lost Cities of Africa,* pp. 46, 47, 36.
64. Ibid., p. 67.
65. Burstein, *Ancient African Civilizations,* p. 14.
66. Davidson, *Lost Cities of Africa,* pp. 81–87.
67. Ibid., p. 143, 154, 165.
68. Norimitsu Onishi, "A Wall, a Moat, Behold! A Lost Yoruba Kingdom," *The New York Times,* September 20, 1999, p. A4.
69. Alfred Crosby, in a letter to the author, December 29, 2000.
70. Onishi, "A Wall, a Moat," p. A4.
71. James and Thorpe, *Ancient Inventions,* p. 445.
72. Ibid., pp. 442–445, 455.
73. Ibid., p. 362.
74. Jonathan Mark Kenoyer, *Ancient Cities of the Indus Valley Civilization* (Oxford: Oxford University Press, 1998).
75. Pacey, *Technology in World Civilization,* p. 29.
76. William Broad, "Fascination is Forever: The Arts and Sciences of Diamonds," *New York Times,* October 31, 1997, p. E31.
77. Pacey, *Technology in World Civilization,* pp. 80–81.
78. Al-Hassan and Hill, *Islamic Technology,* p. 248.
79. O. P. Jaggi, *History of Science and Technology in India,* vol. 7: *Science and Technology in Medieval India* (Delhi: Atma Ram, 1969), p. 150.
80. Ibid., pp. 150–154.
81. Ibid., p. 167.
82. Pacey, *Technology in World Civilization,* pp. vii, 117.
83. Ibid., pp. 166, 120.
84. Joseph Needham, *Science in Traditional China* (Cambridge, Mass.: Harvard University Press, 1981), pp. 27–28. Needham, the fount of all information in the West on science and technology in China, published a series of huge volumes on the subject, *Science and Civilisation in China,* to which the word *monumental* was applied. Literary and social critic George Steiner called these ground-breaking works the only successor of Proust, as an attempt to recreate in memory a vanished world. One on-line pundit responds that he likes Needham better than Proust, not least because he has no desire to kick the narrator in the pants.
85. Ibid., p. 55.
86. Ibid.
87. Derek J. de Solla Price, *Science Since Babylon* (New Haven, Conn.: Yale University Press, 1961), p. 124.
88. Needham, *Science in Traditional China,* p. 28.
89. Robert Temple, *The Genius of China: 3,000 Years of Science, Discovery, and Invention* (London: Prion Books Limited, 1998), p. 242.
90. Ibid., p. 224; Pacey, *Technology in World Civilization,* p. 47.

91. Needham, *Science in Traditional China,* p. 40.
92. Pacey, *Technology in World Civilization,* p. 47.
93. Ibid., p. 45.
94. Temple, *The Genius of China,* pp. 229–235.
95. Pacey, *Technology in World Civilization,* p. 45.
96. Needham, *Science in Traditional China,* p. 43.
97. Hans Breuer, *Columbus Was Chinese: Discoveries and Inventions of the Far East,* trans. Salvator Attanasio (New York: Herder and Herder, 1972), p. 128.
98. Ibid.
99. Alfred Crosby, in a letter to author, December 29, 2000.
100. Breuer, *Columbus Was Chinese,* p. 144.
101. Fred L. Wilson, Rochester Institute of Technology, http://www.rit.edu/~flwstv/china.html.
102. Diamond, *Guns, Germs, and Steel,* p. 111.
103. Qiu Lianghui, "A Preliminary Study of the Characteristics of Metallurgical Technology in Ancient China," in Fan Daidian and Robert S. Cohen (eds.), *Chinese Studies in the History and Philosophy of Science and Technology* (Norwell, Mass.: Kluwer Academic Publishers, 1996), p. 238.
104. Temple, *The Genius of China,* p. 42.
105. Ibid.
106. Ibid., p. 16.
107. Ibid., p. 49.
108. Ibid., pp. 49–50.
109. Ibid., p. 68.
110. Ibid., pp. 58–61.
111. Ibid., p. 72.
112. Breuer, *Columbus Was Chinese,* p. 135.
113. Diamond, *Guns, Germs, and Steel,* pp. 231–235.
114. Ibid., pp. 231–235.
115. Temple, *The Genius of China,* p. 81.
116. Ibid., p. 84.
117. Philip Morrison, "A Great Explorer," *New York Review of Books,* Dec. 12, 1974.
118. Gail Collins, "Pre-2K Thanksgiving," *New York Times,* November 23, 1999, p. A27.
119. Alfred Crosby, *Ecological Imperialism: The Biological Expansion of Europe, 900–1900* (Cambridge: Cambridge University Press, 1986), p. 106.
120. Linda Shaffer, "China, Technology & Change," *World History Bulletin,* fall/winter 1986/87, http//acc6itf.brooklyn.cuny.edu/~phalfall/texts/shaffer.html
121. Stephen Jay Gould, "A Cerion for Christopher," *Natural History* 105 (Oct. 1996): 22.

Selected Bibliography

Ancient and medieval non-Western science is memorialized in many fine sources, but negotiating through them is dicey. Even in outstanding books and articles, one finds exaggerations or otherwise dubious claims. (The same can be said for works on Western science.) I had nine scholars to help guide me through the material. The reader will not have this advantage. My advice: be skeptical, use common sense, and compare with other sources.

I have put a star (★) next to those sources that I found especially valuable. You'll note that books written by members of my board of advisers all have asterisks. While this may seem self-serving or biased, I in fact sought out these consultants primarily on the basis of their published work. It is only natural that their books are among the most useful.

There are only two books in the field that attempt to be comprehensive. Both are the work of Helaine Selin, science librarian at Hampshire College in Amherst, Massachusetts. One is a bibliography:

★ Selin, Helaine. *Science Across Cultures: An Annotated Bibliography of Books on Non-Western Science, Technology, and Medicine.* New York and London: Garland Publishing, Inc., 1992. [Selin lists 836 books on the topic, and while many are devoted to health and other soft or new-age sciences, there is much of value here. Selin's crisp summaries transform this volume from a bibliography to a readable book per se. Book is somewhat pricey at $72, and is out of print, but used copies may be available.]

The other is a one-volume encyclopedia:

Selin, Helaine (ed.). *Encyclopaedia of the History of Science, Technology, and Medicine in Non-Western Cultures.* Dordrecht/Boston/London: Kluwer Academic Publishers, 1997. [This 1,118-page, oversized volume is a collection of 600 articles rather than a comprehensive encyclopedia. Quality and coverage are uneven. One university librarian noted its "inconsistency." Some contributors, such as George Gheverghese Joseph and Takao Hayashi, are first-rate, while another contributor claims that Indian alchemists literally turned mercury into gold using fruit, herbs, mud, and a charcoal fire. Still, there is nothing else like it. Expensive, at a list price of $572, but cheaper used copies are available, and a few libraries carry the book.]

CHAPTER 1: A HISTORY OF SCIENCE

Bernal, Martin. *Black Athena: The Afroasiatic Roots of Classical Civilization.* Vols. I and II. New Brunswick, N.J.: Rutgers University Press, 1991.

Bernal, Martin. *Black Athena Writes Back.* Durham and London: Duke University Press, 2001.

Boorstin, Daniel J. *The Discoverers: A History of Man's Search to Know His World and Himself.* New York: Random House, 1983.

* Bowersock, Glen. "Rescuing the Greeks." *The New York Times Book Review,* February 25, 1996. [This is a review of Mary Lefkowitz's *Not Out of Africa* (see entry this chapter). In his review, Bowersock nicely summarizes how history has been revised to favor ancient Greece.]

Bronowski, J. *The Ascent of Man.* Boston: Little, Brown and Company, 1973.

Gross, Paul R., and Levitt, Norman. *Higher Superstition: The Academic Left and its Quarrels with Science.* Baltimore: The Johns Hopkins University Press, 1994. [The authors believe people of color and women are ruining academia and science.]

Kuhn, Thomas S. *The Structure of Scientific Revolutions.* Chicago: University of Chicago Press, 1962.

* Lefkowitz, Mary. *NOT Out of Africa.* New York: Basic Books, 1996. [This book is noteworthy in the way it portrays the anger of some scholars toward non-Western history.]

* Powers, Richard. "Eyes Wide Open," *The New York Times Magazine,* April 18, 1999. [An article, but worthwhile.]

Russell, Bertrand. *A History of Western Philosophy.* New York: Touchstone/Simon & Schuster, 1945.

* Sagan, Carl. *Cosmos.* New York: Random House, 1980. [Of all the science popularizers, Sagan is least dismissive of non-Western culture, and most frank about gaps in ancient Greek science and math.]

Stengel, Marc K. "The Diffusionists Have Landed." *The Atlantic Monthly,* January 2000. [Wacky tales of how King Arthur and other Europeans came to America.]

CHAPTER 2: MATHEMATICS

Ball, W. W. Rouse. *A Short Account of the History of Mathematics.* New York: Dover, 1960; London & New York: Macmillan, 1893.

Bell, E. T. *Men of Mathematics.* New York: Simon & Schuster, 1937. [All about Western scientists only, from Zeno to Poincaré, but worth reading for context.]

★ Dantzig, Tobias. *Number: The Language of Science.* New York: The Free Press, 1930. [Albert Einstein wrote: "This is beyond doubt the most interesting book on the evolution of mathematics which has ever fallen into my hands."]

Fowler, D. H. *The Mathematics of Plato's Academy.* Oxford: Clarendon Press, 1987.

★ Joseph, George Gheverghese. *The Crest of the Peacock: Non-European Roots of Mathematics.* London: Penguin Books, 1991. [The most comprehensive book on non-European math. Joseph also explores the revisionist history of mathematics.]

★ Kaplan, Robert. *The Nothing That Is: A Natural History of Zero.* New York: Oxford University Press, 2000. [A detailed, delightful, and thoughtful "biography" of an essential number.]

Kline, Morris. *Mathematics: A Cultural Approach.* New York: Addison-Wesley, 1962. [Worth reading for the author's opinion of non-Western mathematics.]

Machmer, Peter (ed.). *The Cambridge Companion to Galileo.* Cambridge, England: Cambridge University Press, 1998. [How Galileo used ancient Greek geometry, not the algebraic notations given in modern textbooks.]

Marty, Martin E., and Jerald C. Brauer. *The Unrelieved Paradox: Studies in the Theology of Franz Bibfeldt.* Grand Rapids, Mich.: William B. Eerdmans Publishing Company, 1994. [An entertaining view of the Year Zero controversy.]

★ Menninger, Karl. *Number Words and Number Symbols: A Cultural History of Numbers.* Cambridge, Mass.: MIT Press, 1969. [An outstanding primer on the basics of numbers. Menninger provides, on page 360, additional details on the "Russian peasant" method of multiplication that I describe in this chapter.]

★ Sobel, Dava. *Galileo's Daughter.* New York: Walker & Company, 1999. [A new look at the West's first real physicist and the relationship of science and religion.]

Wells, David. *The Penguin Dictionary of Curious and Interesting Numbers.* London: Penguin Books, 1986.

CHAPTER 3: ASTRONOMY

★ Aveni, Anthony F. *Ancient Astronomers.* Montreal and Washington: St. Remy Press and Smithsonian Books, 1993.

★ Aveni, Anthony F. *Between the Lines: The Mystery of the Giant Ground Drawings of Ancient Nasca, Peru.* Austin: University of Texas Press, 2000.

★ Aveni, Anthony. *Stairways to the Stars: Skywatching in Three Great Ancient Cultures.* New York: John Wiley & Sons, Inc., 1997. [The Maya, Incas, and Great Britain.]

★ Coe, Michael D. *The Maya.* New York: Thames and Hudson, 1999. [A good general book about the Maya.]

Drake, Frank, and Dava Sobel. *Is Anyone Out There?* New York: Delacorte Press,

1992. [A book about SETI, the Search for Extraterrestrial Intelligence, that is valuable for facts about space.]

Krupp, E. C. *Echoes of the Ancient Skies.* New York: Harper & Row, 1983.

★ Neugebauer, Otto. *Astronomy and History: Selected Essays.* New York: Springer-Verlag, 1983. [Neugebauer is the pioneer in this field.]

★ Neugebauer, Otto. *The Exact Sciences in Antiquity.* New York: Cover Publications, Inc., 1969. [Valuable.]

Sullivan, William. *The Secret of the Incas: Myth, Astronomy, and the War Against Time.* New York: Crown Publishers, Inc., 1996.

★ Swerdlow, N. M. (ed.). *Ancient Astronomy and Celestial Divination.* Cambridge, Mass.: MIT Press, 1999. [Swerdlow is a major researcher in the field of ancient astronomy.]

Temple, Robert. *The Genius of China.* London: Prion Books, Limited, 1986.

Turner, Howard R. *Science in Medieval Islam.* Austin: University of Texas Press, 1995.

Walker, Christopher (ed.). *Astronomy Before the Telescope.* New York: St. Martin's Press, 1996.

CHAPTER 4: COSMOLOGY

★ Aveni, Anthony. *Conversing with the Planets: How Science and Myth Invented the Cosmos.* New York: Kodansha International, 1992.

Greenstein, George. *The Symbiotic Universe.* New York: William Morrow & Co. Inc., 1988.

Guth, Alan H. *The Inflationary Universe.* Reading, Mass.: Addison Wesley, 1997. [A lone physicist saves the big bang universe.]

★ Harrison, Edward. *Masks of the Universe.* New York: Macmillan Publishing Company, 1985. [One of the few books on cosmology admitting that today's cosmology must, like all others, fade eventually.]

Harrison, Edward. *Cosmology: The Science of the Universe.* 2nd ed. Cambridge, England: Cambridge University Press, 2000. [More technical than the above, but still readable.]

Kolb, Rocky. *Blind Watchers of the Sky.* Reading, Massachusetts: Addison-Wesley Publishing Company, 1996. [A readable account of the Western view of cosmology by a leading scientist.]

Kragh, Helge. *Cosmology and Controversy: The Historical Development of Two Theories of the Universe.* Princeton, N.J.: Princeton University Press, 1996. [How the big bang theory, which posits an expanding universe with a beginning in time, triumphed over the steady state theory, which posits a stationary universe of infinite age.]

Leslie, John. *Universes.* London and New York: Routledge, 1989. [An examination of cosmologies and the logic behind them.]

Lightman, Alan, and Roberta Brawer. *Origins: The Lives and Worlds of Modern Cosmologists.* Cambridge, Mass.: Harvard University Press, 1990. [Interviews with contemporary cosmologists. Notably missing is any sense of history or mention of past cosmologies, ancient or otherwise.]

Long, Charles. *Alpha: The Myths of Creation.* New York: George Braziller, 1963.

* Sproul, Barbara C. *Primal Myths: Creation Myths Around the World.* San Francisco: HarperSanFrancisco, 1979. [About as complete as it gets: myths of the Bushmen, Hottentot, Egyptians, Sumerians, Muslims, Jains, Buddhists, Mongolians, Assiniboine, Jivaro, Maori, many others.]

Tedlock, Barbara. *Time and the Highland Maya.* Albuquerque: University of New Mexico Press, 1982.

Turner, Howard R. *Science in Medieval Islam.* Austin: University of Texas Press, 1995. [Excellent black-and-white drawings and photographs.]

CHAPTER 5: PHYSICS

* Aveni, Anthony F. *Empires of Time: Calendars, Clocks and Cultures.* New York: Basic Books, 1989.

*Bose, D. M., S. N. Sen, & B.V. Subbarayappa (eds.). *A Concise History of Science in India.* New Delhi: Indian National Science Academy, 1971. [A comprehensive sourcebook for the history of all Indian sciences with contributions from Indian scientists.]

Charya, Sri Umasvami. *The Sacred Books of the Jainas: Tattvarthadhigama Sutra.* Vol. II, ed. J. L. Jaini. Arrah, India: Kumar Devendra Prasad, 1920.

Chattopdhyaya, Debiprasad. *History of Science and Technology in Ancient India.* Calcutta: Firma KLM PVT, 1991. [See especially the chapter "The Atomic Hypothesis."]

Clagett, Marshall. *Ancient Egyptian Science; A Source Book.* Philadelphia: American Philosophical Society, 1989.

* Cole, K. C. *The Hole in the Universe: How Scientists Peered Over the Edge of Emptiness and Found Everything.* New York: Harcourt, Inc., 2000. [An elegant and readable essay about the importance of the void in science.]

Dasgupta, S. N. *Yoga Philosophy in Relation to Other Systems of Indian Thought.* Delhi: Motilal Banarsidass, 1930.

Duchesne-Guillemin, Jacques. *Symbols and Values in Zoroastrianism.* New York: Harper & Row, 1966.

* Graham, A. C. *Later Mohist Logic, Ethics and Science.* Hong Kong: Chinese University Press; London: School of Oriental & African Studies, University of London, 1978. [Technical discussion of the Mohist school and its contributions to Chinese sciences.]

Hart, George. *A Dictionary of Egyptian Gods and Goddesses.* London: Routledge & Kegan, 1986.

★ Hill, John W., and Doris K. Kolb. *Chemistry for Changing Times.* Upper Saddle River, N.J.: Prentice Hall, 1998. [A chemistry text, but relevant to physics also.]

Hunt, Frederick Vinton. *Origins in Acoustics: The Science of Sound from Antiquity to the Age of Newton.* New Haven: Yale University Press, 1978. [Interesting, if tangential.]

★ Jain, N. L. "Chemical Theories of the Jains" in *Chymia: Annual Studies in the History of Chemistry.* Vol. 11, ed. Henry M. Leicester. Philadelphia: University of Pennsylvania Press, 1966. [A discussion of ancient Jainist theories of the atom and the nature of matter.]

Jammer, Max. *Concepts of Space: The History of Theories of Space in Physics.* New York: Dover Publications, 1954. [A classic; but note that the author of the Introduction, Albert Einstein, disagrees in part with the content of the book.]

Kroeber, Theodora. *Ishi in Two Worlds: a Biography of the Last Wild Indian in North America.* Berkeley: University of California Press, 1965.

Léon-Portilla, Miguel. *Time and Reality in the Thought of the Maya.* Norman: University of Oklahoma Press, 1988.

★ Nasr, Seyyed Hossein. *An Introduction to Islamic Cosmological Doctrines.* Cambridge, Mass.: Harvard University Press, 1964. [Islamic medieval philosophy and science are woven together in this account covering atomism, cosmology, and the nature of matter.]

Nasr, Seyyed Hossein. *Science and Civilization in Islam.* Cambridge, Mass.: Harvard University Press, 1968.

Needham, Joseph. *Science & Civilization in China.* Vol. 5: *Chemistry and Chemical Technology.* Cambridge, England: Cambridge University Press, 1976. [Needham is valuable, but I find that his enthusiasm sometimes clouds his objectivity.]

Omar, Saleh Beshara. *Ibn al-Haytham's Optics: A Study of the Origins of Experimental Science.* Chicago: Bibliotheca Islamica, 1977.

★ Park, David. *The Fire Within the Eye: A Historical Essay on the Nature and Meaning of Light.* Princeton: Princeton University Press, 1997. [A synthesis of early Western theories of light, covering Islamic as well as Greek contributions. Park follows light into the twentieth century.]

Qadir, C. A. *Philosophy and Science in the Islamic World.* London: Croom Helm, 1988.

★ Ronan, Colin A., and Joseph Needham. *The Shorter Science and Civilization in China.* Vol. 2. Cambridge, England: Cambridge University Press, 1981. [A clear, general introduction to Chinese science.]

Stillman, John Maxson. *The Story of Alchemy and Early Chemistry.* New York: Dover Publications, 1960.

Stone, R. M. "The Shape of Time in African Music," *Time, Science, & Society in China and the West: The Study of Time,* ed. J. T. Fraser, N. Lawrence, and F. C.

Haber. Amherst: University of Massachusetts Press, International Society for the Study of Time, 1986.

Zajonc, Arthur. *Catching the Light.* New York: Bantam, 1993.

CHAPTER 6: GEOLOGY

Adams, Frank Dawson. *The Birth and Development of the Geological Sciences.* Baltimore: The Williams & Wilkins Co., 1938.

★ Bose, D. M., S. N. Sen, and B. V. Subbarayappa (eds.). *A Concise History of Science in India.* New Delhi: Indian National Science Academy, 1971. [A comprehensive sourcebook for the history of all Indian sciences, with contributions by Indian scientists.]

Carey, S. Warren. *Theories of the Earth and Universe: A History of Dogma in the Earth Sciences.* Stanford, Ca.: Stanford University Press, 1988.

Chattopadhyaya, Debiprasad. *History of Science and Technology in Ancient India: The Beginnings.* Calcutta: Firma KLM, 1986.

Chattopadhyaya, Debiprasad. *History of Science and Technology in Ancient India: Formation of the Theoretical Fundamentals of Natural Science.* Calcutta: Firma KLM, 1991.

Clagett, Marshall. *Ancient Egyptian Science: A Source Book.* Philadelphia: American Philosophical Society, 1989.

de Avila, Francisco (compiler, ca. 1598). *The Huarochiri Manuscript: A Testament of Ancient and Colonial Andean Religion, trans.* Frank Salomon and George L. Urioste. Austin: University of Texas Press, 1991.

★ de Sahagun, Bernardino. *Florentine Codex: General History of the Things of New Spain,* Book 11: *Earthly Things, trans.* Charles Dibble and Arthur Anderson. Santa Fe: School of American Research/ University of Utah, 1963. [This is a primary source for Aztec knowledge of natural phenomena compiled by a Spanish priest at the time of Cortez.]

★ Dudley, Michael Kioni. *A Hawaiian Nation: Man, Gods, and Nature.* Honolulu: Na Kane O Ka Malo Press, 1990. [The beliefs and practices of ancient Hawaii are interpreted as a combination of scientific observation and spiritual belief, as seen by a modern Hawaiian.]

Forbes, R. J. *Studies in Early Petroleum History.* Leiden: E. J. Brill, 1958. [Forbes is a leading historian of science.]

★ Freidel, David, Linda Schele, and Joy Parker. *Maya Cosmos: Three Thousand Years on the Shaman's Path.* New York: William Morrow, 1993. [This book weaves Mayan spiritual beliefs with scientific and technological achievements. Authors are pioneers in Mayan research.]

Geikie, Sir Archibald. *The Founders of Geology.* London: Macmillan, 1905.

Kramer, Samuel Noah. *The Sumerians: Their History, Culture, and Character.* Chicago: University of Chicago Press, 1963.

Lopez, Barry Holstun. *Giving Birth to Thunder, Sleeping with His Daughter: Coyote Builds North America.* New York: Avon Books, 1977.

* Malo, David. *Hawaiian Antiquities (Moolelo Hawaii),* trans. Nathaniel B. Emerson. Honolulu: Bernice P. Bishop Museum, 1951. [The firsthand account of a Hawaiian raised before white contact at the time of Cook's invasion. European and ancient beliefs are compared.]

* Nasr, Seyyed Hossein. *Science and Civilization in Islam.* Cambridge, Mass.: Harvard University Press, 1968. [An account of Islamic understanding of geography and geology, among other sciences, by a noted Islamic philosopher and scholar.]

* Needham, Joseph. *The Shorter Science and Civilization in China: An Abridgment of Joseph Needham's Original Text,* abridged by Colin A. Ronan. Cambridge, England: Cambridge University Press, 1981. Vols. 2 and 3. [Needham is perhaps too apt to see proof of scientific understanding in everything, but this is one of the few texts on China that attributes any geologic concepts to premodern China.]

Popul Vuh: The Mayan Book of the Dawn of Life and the Glories of Gods and Kings, trans. Dennis Tedlock. New York: Simon & Schuster, 1996.

Sullivan, William. *The Secret of the Incas: Myth, Astronomy, and the War Against Time.* New York: Crown Publishers, 1996.

* Tarbuck, Edward J., and Frederick K. Lutgens. *Earth: An Introduction to Physical Geology.* 5th ed. Upper Saddle River, N.J.: Prentice Hall, 1996. [A good basic introduction to the science of geology.]

* Thompson, J. Eric. *Maya History and Religion.* Norman: University of Oklahoma Press, 1970. [A noted science scholar covers a wide range of Mayan belief and practices.]

Thompson, R. Campbell. *A Dictionary of Assyrian Chemistry and Geology.* Oxford: Clarendon Press, 1936.

* Thompson, Susan J. *A Chronology of Geological Thinking from Antiquity to 1899.* Metuchen, N.J.: Scarecrow Press, 1988. [A basic reference work chronicling geological discoveries of Old World civilizations.]

Urton, Gary. *At the Crossroads of the Earth and the Sky: An Andean Cosmology.* Austin: University of Texas Press, 1981.

* Weixelman, Joseph. *The Power to Evoke Wonder: Native Americans and the Geysers of Yellowstone National Park.* Bozeman: Merrill G. Burlingame Special Collections, Montana State University, 1992. [Historical accounts of Native American attitudes toward natural phenomena, mixed with interviews of contemporary Native Americans.]

CHAPTER 7: CHEMISTRY

Clagett, M. *Ancient Egyptian Science: A Source Book.* Vol. 1. Philadelphia: American Philosophical Society, 1989.

★Cobb, Cathy, and Harold Goldwhite. *Creations of Fire: Chemistry's Lively History from Alchemy to the Atomic Age.* New York and London: Plenum Press, 1995. [Good, clear introduction to the history of chemistry covering Egypt, Mesopotamia, and Greece.]

★ de Sahagun, Bernardino. *Florentine Codex: General History of the Things of New Spain.* Book 11: *Earthly Things,* trans. Charles Dibble and Arthur Anderson. Santa Fe: School of American Research/University of Utah, 1963.

Farber, Eduard. *The Evolution of Chemistry.* New York: Ronald Press Co., 1952.

★ Forbes, R. J. *Studies in Ancient Technology.* Vols. I, III, IV, V. Leiden: E. J. Brill, 1965. [Detailed technical discussion of Egyptian, Mesopotamian, and Islamic contributions to modern practical chemistry.]

★ Hill, John W., and Doris K. Kolb. *Chemistry for Changing Times.* Upper Saddle River, N.J.: Prentice Hall, 1998. [Good basic introduction to modern Western chemistry.]

★ Mann, John. *Murder, Magic, and Medicine.* Oxford: Oxford University Press, 1992. [The history of modern medicines, with detailed accounts of contributions from African and South American cultures.]

★ Needham, Joseph. *Science & Civilization in China.* Vol. 5: *Chemistry and Chemical Technology.* Cambridge, England: Cambridge University Press, 1976. [One of several immense volumes covering early Chinese alchemic and chemical processes.]

★ Schultes, Richard Evans, and Robert F. Raffauf. *The Healing Forest: Medicinal and Toxic Plants of the Northwest Amazonia.* Historical, Ethno- and Economic Botany series, vol. 2. Portland, Ore.: Dioscorides Press, 1990. [Schultes is a preeminent ethnobotanist. This is a comprehensive reference work of medicinal plants used by Amazonian Indians.]

★ Sivin, Nathan. *Chinese Alchemy: Preliminary Studies.* Cambridge, Mass.: Harvard University Press, 1968. [A noted expert in Chinese alchemy, Sivin attempts to understand alchemic tests in relation to demonstrable chemical reactions.]

Stillman, John Maxson. *The Story of Alchemy and Early Chemistry.* New York: Dover Publishing, 1960. [Includes actual recipes from ancient Egypt.]

Thomas-Emeagwali, Gloria. "Textile Technology in Nigeria in the 19th and Early 20th Centuries," in *African Systems of Science, Technology, and Art: The Nigerian Experience,* ed. Gloria Thomas-Emeagwali. London: Karnak House, 1993. [Sourcebook on traditional African food and textile processing.]

CHAPTER 8: TECHNOLOGY

al-Hassan, Ahmad Y., and Donald R. Hill. *Islamic Technology: An Illustrated History.* Cambridge, England: Cambridge University Press, 1986. [Lovely pictures.]

★ Coe, Michael D. *The Maya.* London: Thames and Hudson, 1999. [A good general text on the Maya.]

Selected Bibliography

★ Crosby, Alfred W. *Ecological Imperialism: The Biological Expansion of Europe, 900–1900.* Cambridge, England: Cambridge University Press, 1986. [This historian ties together Old World beginnings in Mesopotamia with Europe and the New World.]

Davidson, Basil. *The Lost Cities of Africa.* Boston: Little, Brown and Company, 1987.

★ Diamond, Jared. *Guns, Germs, and Steel: The Fates of Human Societies.* New York: W. W. Norton & Company, 1997. [The author writes: "Some readers may feel that I am going to the opposite extreme from conventional histories, by devoting too little space to western Eurasia . . ."]

★ James, Peter, and Nick Thorpe. *Ancient Inventions.* New York: Ballantine Books, 1994. [Not as thoroughly researched or authoritative as one might like, but highly entertaining with copious pleasant illustrations.]

★ Needham, Joseph. *Science in Traditional China.* Cambridge, Mass.: Harvard University Press, 1981.

Pace, Arnold. *Technology in World Civilization: A Thousand Year History.* Cambridge, Mass.: The MIT Press, 1990.

Price, Derek J. de Solla. *Science Since Babylon.* New Haven, Conn.: Yale Univ. Press, 1975.

Smith, Michael E. *The Aztecs.* Oxford: Blackwell Publishers, 1998.

Temple, Robert. *The Genius of China: 3,000 Years of Science, Discovery, and Invention.* London: Prion Books Limited, 1998.

Turner, Howard R. *Science in Medieval Islam.* Austin: University of Texas Press, 1995.

Weatherford, Jack. *Indian Givers.* New York: Fawcett Columbine, 1988.

Acknowledgments

Tim Onosko first suggested a "multicultural history of science," and handed me the idea. Judith Hooper, Janet MacFadyen, and Kathleen Stein contributed extensive research, ideas, criticism, fact-checking, and editing. While she was an editor at *Omni,* Stein assigned to me, at my request, the negative article on this topic that was the counter-impetus for my research.

Thanks to Alice Mayhew and Anja Schmidt and to Lynn Nesbit and Eric Simonoff. Bonnie Thompson copyedited a difficult manuscript with grace and rigor, and Loretta Denner guided the book through production.

William Bridegam, librarian for Amherst College, and his staff not only volunteered the use of the library, but tracked down needed materials for me at distant libraries and archives. Also valuable was the W. E. B. Du Bois Library at the University of Massachusetts at Amherst with its affiliated libraries. They house a remarkable collection of books on non-Western math and science.

Index

Index

Index

Index

Index

Index